The Comprehensive Guide to Surviving a Nuclear War

Andrew Parry

Published by Andrew Parry, 2024.

THE COMPREHENSIVE GUIDE TO SURVIVING A NUCLEAR WAR

First edition. September 14, 2024.

ISBN: 979-8227818683

Written by Andrew Parry.

Table of Contents

The Current Threat: Understanding the Nuclear Conflict Landscape

The world today stands on the brink of a nuclear conflict in ways that are both reminiscent of the Cold War and strikingly new. While the threat of mutually assured destruction once loomed over the heads of superpowers, today's nuclear landscape is more fragmented, with multiple nations now holding nuclear capabilities. Understanding the nuclear conflict landscape is key to preparing for the worst-case scenario—a full-scale nuclear exchange that could plunge the planet into a devastating nuclear winter.

First, it's important to acknowledge that nuclear weapons are no longer solely in the hands of the few superpowers. Nations like North Korea and Pakistan have developed their own nuclear arsenals, and the concern isn't limited to official government entities. Non-state actors and terrorist organizations also pose a threat, given the possibility of acquiring nuclear materials through illegal markets or even cyberterrorism. These factors combine to make the nuclear conflict landscape far more unpredictable than in the past. It's this unpredictability that heightens the need for preparation.

The breakdown of global diplomacy has further exacerbated these concerns. Traditional alliances have weakened, and global power dynamics have shifted. International treaties, such as the Treaty on the Non-Proliferation of Nuclear Weapons (NPT), which were designed to curb the spread of nuclear arms, are facing unprecedented challenges. Countries that were once vocal in their support of disarmament now find themselves increasing their military budgets and modernizing their nuclear arsenals. With the weakening of these agreements, the global community is moving closer to a potential arms race rather than stepping away from it. The rise of nationalism and isolationist policies in many regions also serves to strain relationships between nuclear-armed nations, further increasing the likelihood of conflict.

Another key element to consider is the rapid advancement of missile technology. Nuclear weapons delivery systems have evolved beyond traditional bombers and missiles. Now, nations possess highly sophisticated intercontinental ballistic missiles (ICBMs) and submarine-launched ballistic missiles (SLBMs) that can strike targets across the globe within minutes. The development of hypersonic missiles, capable of flying faster than the speed of sound, reduces reaction times for defense systems and increases the potential for a catastrophic mistake. This short window for decision-making means that a misunderstanding or miscalculation could quickly escalate into a nuclear confrontation.

One of the most concerning factors in the current nuclear landscape is the role of cyber warfare. As nations increasingly rely on computer networks to manage their weapons systems, the risk of a cyberattack that could trigger an unintended nuclear launch grows. The possibility of hacking into these systems and creating false alarms or manipulating information is no longer theoretical; it's a reality that military planners must consider. The interconnectedness of global technology and the reliance on data to make split-second decisions means that an actor with malicious intent could exploit weaknesses and push the world toward a nuclear conflict.

With this in mind, it's not just the political tensions or the technological advancements that must be understood—it's the human element as well. Decision-makers in high-stress environments can make irrational decisions.

History has shown us close calls where the judgment of a few individuals prevented nuclear war. In 1983, Soviet officer Stanislav Petrov famously chose not to launch a retaliatory strike after detecting a false warning of incoming U.S. missiles, averting what could have been a nuclear disaster. The question remains: will future leaders exhibit the same restraint in the heat of the moment?

In addition to nation-state actors, non-state threats must also be acknowledged. The potential for terrorist organizations to acquire nuclear weapons or radioactive materials presents a unique and alarming threat. While these groups may not have the technological capacity to build fully operational nuclear weapons, the spread of dirty bombs—devices that use conventional explosives to disperse radioactive material—poses a significant danger. Even a relatively low-tech attack of this nature could cause mass panic, economic collapse, and long-term environmental damage, depending on the scale of the contamination.

Furthermore, many regions of the world are experiencing increased tension over access to vital resources, which could drive future conflicts. Climate change is intensifying the competition for water, food, and energy, and as resources become scarcer, the possibility of conflict rises. In some cases, nuclear powers may find themselves in direct competition with each other over these dwindling resources, pushing already fragile relationships to the breaking point. The combination of environmental stress and nuclear capabilities creates a scenario where small skirmishes could rapidly escalate into a full-blown nuclear war.

In preparing for the possibility of nuclear conflict, it is essential to monitor these geopolitical and technological shifts closely. Preppers must stay informed of current events, understand the emerging risks, and anticipate potential hotspots where conflict could ignite. This includes keeping a close eye on regions where nuclear weapons are being developed or expanded, such as the Korean Peninsula, South Asia, and the Middle East.

Equally important is understanding the alliances and enmities that exist between nuclear-armed nations. For example, the longstanding tensions between India and Pakistan, both of which possess nuclear weapons, have created a powder keg situation in South Asia. Similarly, tensions between the U.S. and Russia, despite periods of diplomacy, have never truly dissipated and could reignite at any time, especially with the recent modernizations in both nations' nuclear forces.

In addition to global politics, individuals must also recognize the importance of local-level preparation. Cities with strategic military or economic importance are likely to be prime targets in a nuclear conflict. Understanding your own region's vulnerabilities—whether it's near a military base, a major city, or a nuclear facility—can help guide your survival strategy.

As we navigate this uncertain nuclear landscape, being proactive about preparation is not just a necessity, it's a responsibility. Knowledge, planning, and awareness of the global situation are your first lines of defense in ensuring your family's survival in the event of a nuclear conflict. While we hope that diplomacy prevails and disaster is averted, the possibility of a nuclear winter cannot be ignored. The threat is real, and understanding the landscape is the first step toward ensuring your safety.

A Brief History of Nuclear Testing and Weapon Development

Nuclear weapons have shaped the geopolitical landscape of the world since their inception, altering the course of wars and international relations. To understand the gravity of nuclear conflict today, it's important to examine how these weapons were developed, tested, and deployed throughout history. The story of nuclear weapon development is both a testament to human ingenuity and a sobering reminder of our capacity for destruction.

The dawn of the nuclear age began during World War II, when the fear of Nazi Germany developing an atomic bomb pushed the United States into action. This led to the top-secret Manhattan Project, which was a massive undertaking involving some of the world's greatest scientific minds. Directed by physicist J. Robert Oppenheimer, the Manhattan Project aimed to harness the power of atomic energy for military purposes. The research culminated in the first successful nuclear test on July 16, 1945, at a site in New Mexico, code-named "Trinity." This test marked the world's first nuclear explosion and opened the door to a new era of warfare.

Just weeks after the Trinity test, the United States dropped two atomic bombs on the Japanese cities of Hiroshima and Nagasaki in August 1945. The bombings were intended to bring a swift end to World War II, but the immediate and long-term consequences were devastating. Hiroshima was hit on August 6 with a uranium bomb nicknamed "Little Boy," and Nagasaki followed three days later with a plutonium bomb called "Fat Man." The death toll from the initial blasts and subsequent radiation exposure reached hundreds of thousands. These events remain the only times nuclear weapons have been used in warfare, and they left a lasting scar on human history.

In the aftermath of World War II, the world became acutely aware of the destructive potential of nuclear weapons. The United States held a temporary monopoly on nuclear arms, but this would not last long. In 1949, the Soviet Union successfully tested its first nuclear device, known as "RDS-1" or "Joe-1," igniting an arms race between the two superpowers. This rivalry between the U.S. and the Soviet Union, later named the Cold War, would dominate international politics for decades, with nuclear weapons at the center of the conflict.

Nuclear testing became a way for nations to demonstrate their military power and refine their weaponry. The United States conducted hundreds of tests, both atmospheric and underground, at sites such as the Nevada Test Site and the Pacific Proving Grounds. Notably, the "Castle Bravo" test in 1954 was the largest nuclear explosion ever detonated by the United States. The bomb, detonated in the Bikini Atoll in the Pacific Ocean, was much more powerful than anticipated, spreading radioactive fallout over a wide area and contaminating nearby islands and even distant locations. The incident led to international outrage and spurred early movements against nuclear testing.

The Soviet Union, in turn, embarked on its own extensive testing program. In 1961, they conducted the largest man-made explosion in history—the Tsar Bomba. This hydrogen bomb had a yield of 50 megatons, roughly 3,300 times more powerful than the bomb dropped on Hiroshima.

While the Tsar Bomba was not a practical weapon due to its size and destructive power, it was a clear signal of the Soviet Union's nuclear capabilities.

Throughout the 1950s and 1960s, nuclear weapon technology evolved rapidly, with both the U.S. and Soviet Union developing more sophisticated and powerful hydrogen bombs, or thermonuclear weapons. Unlike the atomic bombs of Hiroshima and Nagasaki, which relied on nuclear fission, thermonuclear bombs used fusion—the same process that powers the sun. These bombs had exponentially greater destructive power, pushing the world into an even more dangerous arms race.

Other nations soon joined the nuclear club. The United Kingdom tested its first atomic bomb in 1952, followed by France in 1960 and China in 1964. This spread of nuclear technology heightened global tensions, as more countries now had the capability to wage nuclear war. At the same time, public opposition to nuclear testing began to grow, particularly as the effects of radiation from atmospheric tests became more widely known. Fallout from tests carried out in remote locations often drifted across borders, contaminating land, water, and air, affecting human populations far from the test sites.

In response to growing public concern, several international agreements were established to limit nuclear testing and proliferation. The Partial Test Ban Treaty (PTBT) was signed in 1963 by the U.S., Soviet Union, and the United Kingdom, prohibiting nuclear tests in the atmosphere, outer space, and underwater. This marked an important first step toward controlling the spread of nuclear weapons and reducing radioactive fallout, but it did not end nuclear testing altogether. Underground tests continued, and nations outside the treaty, such as France and China, persisted with atmospheric testing for years.

The Nuclear Non-Proliferation Treaty (NPT) of 1968 was another milestone in efforts to prevent the spread of nuclear weapons. The NPT sought to prevent new nations from developing nuclear arms while promoting disarmament and encouraging the peaceful use of nuclear energy. Though it was signed by many nations, some nuclear-armed states, such as India, Pakistan, and Israel, refused to join, and efforts to fully implement disarmament have been met with mixed success. Despite these treaties, the world continued to live under the shadow of nuclear weapons throughout the Cold War.

The collapse of the Soviet Union in 1991 marked the end of the Cold War, but it did not eliminate the nuclear threat. Former Soviet republics, such as Ukraine and Kazakhstan, inherited nuclear weapons, although many of these were eventually dismantled or returned to Russia under international agreements. Still, nuclear weapons remained a significant part of global military strategy.

In the years that followed, countries like India and Pakistan conducted their own nuclear tests in the late 1990s, signaling their entry into the nuclear arms race. North Korea, after withdrawing from the NPT, conducted its first nuclear test in 2006, raising concerns about the proliferation of nuclear weapons to unstable regimes. Iran's nuclear program has also been the subject of intense international scrutiny, with many fearing that it could lead to further proliferation in the Middle East.

Nuclear weapon development has not been solely about increasing the yield or power of bombs. Advances in missile technology, such as intercontinental ballistic missiles (ICBMs), submarine-launched ballistic missiles (SLBMs), and hypersonic missiles, have made nuclear weapons more versatile and harder to defend against. Today, nuclear-armed states continue to modernize their arsenals, developing smaller, more tactical nuclear weapons that are theoretically more "usable" in a limited conflict. This blurs the line between conventional and nuclear warfare, making the prospect of nuclear war more likely, not less.

In the modern era, while full-scale atmospheric testing has been largely curtailed, nuclear weapons remain a persistent threat. As new nations develop nuclear capabilities, and old powers modernize their arsenals, the world remains at risk of a nuclear confrontation that could lead to catastrophic consequences. Understanding the history of nuclear testing and development is essential in preparing for the possibility of surviving a nuclear winter. The lessons learned from past tests, and the ongoing political and technological developments, inform us of the immense risks involved and the need for preparedness.

Lessons from Hiroshima and Nagasaki: The Aftermath of Nuclear Bombs

The bombings of Hiroshima and Nagasaki in August 1945 remain some of the most harrowing examples of the destruction that nuclear weapons can unleash. These events marked the first and only times that nuclear weapons have been used in warfare, and the devastation they caused offers critical lessons about the immediate and long-term consequences of such attacks. Beyond the staggering loss of life, the bombings of these two Japanese cities revealed the environmental, medical, and psychological effects that nuclear weapons can inflict on both individuals and societies.

The city of Hiroshima was the first to experience the full force of an atomic bomb on August 6, 1945. The bomb, nicknamed "Little Boy," was a uranium-based device that detonated approximately 600 meters above the city, releasing a blast equivalent to 15 kilotons of TNT. The bomb's immediate impact was catastrophic. An estimated 80,000 people were killed instantly as a result of the intense heat, pressure, and radiation from the explosion. Buildings, homes, and infrastructure within a one-mile radius of ground zero were obliterated, leaving the city in ruins.

Three days later, on August 9, Nagasaki was struck by a second atomic bomb, this one named "Fat Man," which used plutonium as its core. Although the topography of Nagasaki limited the spread of destruction, the bomb still resulted in an estimated 40,000 immediate deaths. As in Hiroshima, the blast caused widespread destruction, fires, and radiation poisoning that devastated the city.

While the immediate death toll in both cities was staggering, the longer-term effects of the bombings were perhaps even more disturbing. The survivors of the bombings, known as *hibakusha* in Japan, faced a host of medical issues due to radiation exposure. In the weeks and months that followed, tens of thousands more people died from injuries, burns, and acute radiation sickness, which was caused by the intense gamma radiation released by the bombs. Symptoms of radiation sickness included nausea, vomiting, diarrhoea, hair loss, and haemorrhaging.

Many of those who initially survived the blast succumbed to these symptoms, which ravaged their bodies as internal organs were damaged by radiation.

In addition to the immediate medical aftermath, long-term health consequences became apparent in the years and decades that followed. Survivors who had been exposed to high levels of radiation developed various forms of cancer, particularly leukemia and thyroid cancer. The incidence of solid cancers, including breast, lung, and stomach cancer, also increased significantly among survivors. Even those who had been relatively far from the blast site were not immune to these effects, as radioactive fallout contaminated the environment, spreading particles that were inhaled or ingested by the population.

The effects of the bombings were not limited to the people present during the explosions. Genetic damage caused by radiation also had an impact on future generations. Studies of the children of survivors showed a higher incidence of birth defects, miscarriages, and other genetic disorders, raising concerns about the long-term consequences of nuclear radiation on human populations. Although the exact mechanisms of radiation-induced genetic damage are still not fully understood, the bombings of Hiroshima and Nagasaki provided one of the first real-world cases of the dangers posed by ionizing radiation to future generations.

The environmental consequences of the bombings were also profound. The intense heat generated by the nuclear explosions ignited fires that swept through both cities, burning everything in their path. The resulting firestorms created their own weather patterns, with intense winds and rising temperatures exacerbating the destruction. In the immediate aftermath, the landscape of both Hiroshima and Nagasaki was transformed into a barren wasteland, with much of the vegetation and wildlife wiped out by the blast, heat, and radiation.

Radioactive fallout from the bombs further contaminated the soil and water, rendering large areas uninhabitable for months or even years. In both cities, the surviving population faced severe shortages of clean water, food, and medical supplies. Fallout also spread beyond the immediate blast zone, contaminating areas miles away from ground zero. In the years that followed, the environment began to recover, but the contamination of the soil and water had long-lasting effects, particularly in areas where radioactive particles had settled.

One of the most important lessons from Hiroshima and Nagasaki is the sheer scale of human suffering that nuclear weapons can inflict. Beyond the immediate physical injuries, the psychological toll on the survivors was immense. Many survivors faced post-traumatic stress disorder (PTSD), depression, and other mental health issues as a result of their experiences. The *hibakusha* were also often stigmatized within Japanese society, as there was widespread fear that they carried radioactive contamination or that their exposure to radiation would lead to long-term illness or genetic defects in their offspring. This stigma added an additional layer of trauma to an already devastated population.

The bombings also had profound implications for how the world views nuclear weapons. The sheer destructive power of the atomic bombs and the horrific effects they had on civilian populations became a central focus of post-war international discourse. Many of the early movements for nuclear disarmament were spurred by the recognition of the suffering caused by these weapons. The bombings served as a reminder that nuclear weapons, even if used in a military context, would inevitably lead to mass civilian casualties and environmental devastation.

In the years following the bombings, extensive research was conducted on the survivors in an effort to understand the long-term effects of radiation. These studies formed the basis for much of our current understanding of radiation's impact on human health, including the development of safety protocols for radiation exposure. However, the lessons from Hiroshima and Nagasaki extend far beyond the medical and scientific fields. They are also a powerful reminder of the moral and ethical questions surrounding the use of nuclear weapons.

Despite the devastation caused by these bombings, the world continued to develop and test even more powerful nuclear weapons in the decades that followed. The lessons learned from Hiroshima and Nagasaki seemed to fade as nations entered into an arms race, with both the United States and the Soviet Union stockpiling thousands of nuclear warheads. Yet, for the survivors of these bombings, the memory of that day remains a stark warning of the horrors that nuclear war could unleash.

For preppers preparing for the possibility of a nuclear conflict, the lessons of Hiroshima and Nagasaki are invaluable. They demonstrate the importance of taking radiation exposure seriously and the need for long-term planning when it comes to medical care, food, water, and shelter. The bombings also highlight the importance of mental and psychological preparedness, as the trauma of surviving a nuclear event can have lasting effects. Understanding the magnitude of suffering experienced by those who survived Hiroshima and Nagasaki underscores the importance of being as prepared as possible for the worst-case scenario.

Ultimately, Hiroshima and Nagasaki serve as a solemn reminder of the destructive power of nuclear weapons. They also remind us of the resilience of humanity in the face of unimaginable horror. While the survivors endured immense suffering, their stories have contributed to our understanding of the dangers posed by nuclear war and the need for preparedness in an uncertain world.

Environmental Devastation from Nuclear Warfare

The environmental devastation resulting from nuclear warfare is both immediate and long-lasting, with the potential to alter ecosystems, weather patterns, and human habitation across the globe. When a nuclear weapon is detonated, the explosion itself causes catastrophic damage, but the environmental impacts extend far beyond the initial blast. The destruction of infrastructure, the release of radioactive particles, and the long-term effects on climate and biodiversity all contribute to an environmental disaster that can persist for decades, or even centuries. Understanding these consequences is crucial for preppers and survivalists planning for a nuclear event, as the environment they will have to navigate post-conflict will be drastically altered.

The most immediate and visible form of environmental devastation comes from the blast zone itself. When a nuclear weapon is detonated, temperatures at the core of the explosion can reach millions of degrees, vaporizing everything in the immediate vicinity. In Hiroshima and Nagasaki, the cities were reduced to rubble in seconds, with buildings, homes, vegetation, and animals incinerated by the intense heat. The blast creates a fireball that expands outward, igniting fires that spread rapidly through flammable materials like wood, paper, and even metal. These firestorms can consume entire cities, as was seen in Hiroshima, where the fires continued to burn for days after the explosion. In a modern nuclear conflict, the impact would likely be even greater due to the higher yield of contemporary nuclear weapons and the density of urban areas.

These fires not only destroy local ecosystems and habitats but also release massive amounts of smoke, soot, and ash into the atmosphere. This is where the environmental devastation of nuclear warfare starts to affect regions far beyond the initial blast zone. The soot and ash particles, once lifted into the upper atmosphere, can block sunlight from reaching the Earth's surface, leading to a phenomenon known as "nuclear winter." This drastic reduction in sunlight can cause temperatures to plummet, leading to a collapse in agricultural production and potentially triggering a global famine. Crops would fail due to the lack of sunlight and cold temperatures, and ecosystems dependent on photosynthesis would begin to break down. Nuclear winter could last for months or even years, creating a scenario in which surviving the initial blast would be only the beginning of the challenge.

Nuclear explosions also produce a dangerous fallout of radioactive particles that can be carried by wind currents over vast distances. Fallout refers to the residual radioactive material propelled into the atmosphere following the explosion. As these radioactive particles settle back to Earth, they contaminate soil, water, and the air itself. Fallout can travel hundreds or even thousands of miles from the site of the explosion, spreading contamination far beyond the initial target. In the aftermath of Hiroshima and Nagasaki, radioactive fallout was found in areas much further than the blast radius, leading to long-term environmental and health consequences. Modern nuclear weapons, which are far more powerful, would create an even larger fallout zone.

One of the most severe consequences of fallout is its impact on agriculture. Radioactive isotopes, such as cesium-137 and strontium-90, can become embedded in the soil, contaminating crops and water sources for years to come. Once radioactive material enters the food chain, it can bioaccumulate in plants and animals, ultimately making its way into human bodies. This process of bioaccumulation means that even small amounts of radioactive contamination can become concentrated over time, posing a severe threat to both humans and wildlife. In a post-nuclear environment, finding safe, uncontaminated food and water would become one of the most critical challenges for survivors.

Water sources, including rivers, lakes, and groundwater, would be particularly vulnerable to contamination. Fallout can enter these systems through rain, surface runoff, or direct deposition from the atmosphere. Contaminated water can spread radiation over a wider area, affecting not just local ecosystems but also human populations that rely on these sources for drinking water, irrigation, and hygiene. Once water is contaminated with radioactive particles, it becomes extremely difficult to purify, and using such water for drinking or agricultural purposes could result in long-term radiation exposure.

The impact of nuclear warfare on biodiversity is another profound consequence. In the immediate aftermath of a nuclear blast, animal populations in the affected areas would be decimated, either by the explosion itself or by the resulting fires and radiation. Many species would be unable to survive the drastic changes in temperature, loss of habitat, and scarcity of food. The loss of plant life due to firestorms and radioactive contamination would further destabilize ecosystems, leading to a collapse in the food chain. Insects, birds, and small mammals would likely experience mass die-offs in areas affected by fallout, as they are particularly sensitive to changes in their environment.

Some species, however, may be more resilient to radiation. For example, studies conducted around the Chernobyl Exclusion Zone have shown that certain species of plants, animals, and even fungi have adapted to the radioactive environment, surviving and even thriving in areas where humans can no longer live. While this offers a glimpse of hope for nature's ability to adapt, it also underscores the stark reality that the post-nuclear environment would be radically different from what existed before the conflict.

One often overlooked but equally devastating effect of nuclear warfare on the environment is the long-term contamination of the atmosphere and oceans. In addition to the fallout from individual detonations, nuclear war could release large amounts of carbon dioxide, sulfur dioxide, and other greenhouse gases into the atmosphere, exacerbating global climate change. The smoke and soot from firestorms could cause a temporary cooling effect (nuclear winter), but once those particles settled, the greenhouse gases could contribute to long-term warming, further destabilizing the climate. The oceans, which absorb much of the Earth's heat and carbon, could experience changes in temperature and acidity, disrupting marine ecosystems and threatening species that depend on stable ocean conditions.

Moreover, nuclear detonations in or near the ocean would have direct and catastrophic effects on marine life. The blast itself would kill marine organisms within a large radius, and the resulting fallout would contaminate the water.

Radioactive isotopes can be absorbed by plankton and other small marine organisms, entering the food chain and affecting larger species, including fish that humans consume. In the aftermath of a nuclear conflict, the contamination of both freshwater and saltwater ecosystems would likely lead to the collapse of fisheries and the death of marine life, severely impacting food supplies.

For those preparing to survive a nuclear event, understanding the environmental devastation is critical. The post-nuclear landscape would be a hostile and unpredictable environment, where the challenges of finding safe food, water, and shelter would be compounded by the long-lasting effects of radiation and climate change. Survivors would need to be equipped with the knowledge and tools to detect and avoid radiation, purify contaminated resources, and navigate a world where ecosystems and weather patterns have been fundamentally altered.

In conclusion, the environmental devastation caused by nuclear warfare is far-reaching and multifaceted. From the immediate destruction of cities and ecosystems to the long-term contamination of soil, water, and air, nuclear war would leave a legacy of environmental damage that would affect human survival for years, if not decades, to come. The threat of nuclear winter, combined with the spread of radioactive fallout and the collapse of ecosystems, paints a grim picture of the post-conflict world. For preppers, the key to survival lies not only in protecting oneself from the initial blast but also in understanding and preparing for the environmental challenges that follow.

The Horrors of Death: Medical Consequences of a Nuclear Attack

The medical consequences of a nuclear attack are among the most terrifying aspects of nuclear warfare. In the immediate aftermath of a nuclear detonation, the human body is subjected to a range of lethal and debilitating effects. These range from thermal burns and blast injuries to radiation sickness and long-term genetic damage. The sheer scale of suffering caused by a nuclear explosion is difficult to fully comprehend, as it affects not only those within the blast zone but also those exposed to radioactive fallout over a wide area. For preppers and those preparing for the possibility of surviving a nuclear conflict, understanding the medical horrors that follow a nuclear attack is essential to planning for survival and recovery.

When a nuclear bomb is detonated, the first wave of devastation comes from the intense heat generated by the explosion. Temperatures near the epicenter can reach millions of degrees, instantly vaporizing anything in the blast zone, including human bodies. Those located further from the explosion will experience severe thermal burns as a result of the intense heat and radiation. In Hiroshima and Nagasaki, survivors who were miles away from the epicenter suffered third-degree burns over large portions of their bodies. The heat from the blast was so intense that it caused clothing and skin to fuse together in some cases. These burn injuries were not only excruciatingly painful but also often fatal due to infections and the lack of medical treatment in the aftermath.

Thermal burns are just one aspect of the immediate medical consequences. The blast wave generated by a nuclear explosion produces a shockwave of air that moves faster than the speed of sound, causing massive destruction.

Buildings are leveled, debris is thrown great distances, and anyone caught in the path of the shockwave faces severe trauma. The blast can cause internal injuries, such as ruptured organs and hemorrhaging, as well as broken bones and lacerations from flying debris. In a post-nuclear environment, where medical resources are likely to be scarce or non-existent, even treatable injuries can quickly become life-threatening.

However, the most insidious and long-lasting medical consequence of a nuclear attack comes from the exposure to ionizing radiation. Radiation is released in vast quantities during a nuclear explosion, and it penetrates the body's tissues, damaging cells and causing widespread biological harm. Radiation exposure can occur in several forms, including direct exposure to the initial blast, inhalation or ingestion of radioactive particles from fallout, and contact with contaminated surfaces or objects. The extent of the damage depends on the dose of radiation received, the duration of exposure, and the part of the body affected.

One of the most immediate effects of radiation exposure is acute radiation syndrome (ARS), also known as radiation sickness. ARS occurs when a person is exposed to a high dose of radiation over a short period of time. Symptoms of ARS can begin within hours or days of exposure and often resemble the symptoms of other illnesses, making it difficult to diagnose without specific medical knowledge or equipment. In the early stages, victims may experience nausea, vomiting, diarrhea, and fatigue. These initial symptoms may subside temporarily, leading some to believe they are recovering. However, this is often followed by a second, more severe phase of illness, characterized by fever, dizziness, confusion, hair loss, and severe infections due to the destruction of the immune system. In extreme cases, ARS can lead to death within days or weeks of exposure.

One of the reasons radiation is so dangerous is that it damages the DNA within cells, leading to cell death or mutations. In the short term, this can result in the failure of vital organs such as the bone marrow, gastrointestinal system, and cardiovascular system. Bone marrow failure is particularly deadly because it leads to a collapse of the body's immune defenses, leaving survivors vulnerable to infections that would normally be treatable but become fatal in a post-nuclear environment. Gastrointestinal damage leads to the breakdown of the intestines, causing internal bleeding and fluid loss, which can be fatal without immediate medical intervention.

For those who survive the initial radiation sickness, the long-term health consequences of radiation exposure are equally grim. Survivors of Hiroshima and Nagasaki developed a range of cancers, including leukemia, thyroid cancer, breast cancer, and lung cancer, at significantly higher rates than the general population. Radiation-induced cancers often take years or even decades to develop, meaning that even those who appeared to have escaped the worst effects of the blast may face a lifetime of health problems. These cancers are caused by the damage to DNA that radiation inflicts, leading to uncontrolled cell growth and tumor formation. The risk of cancer increases with the dose of radiation received, and there is no known threshold below which radiation is completely safe.

In addition to cancer, radiation exposure can cause long-term genetic damage, which may be passed on to future generations. Studies of the children of survivors from Hiroshima and Nagasaki revealed an increased incidence of birth defects, genetic mutations, and other hereditary health issues. The full extent of radiation's impact on human genetics is still not completely understood, but it is clear that exposure to nuclear fallout can have consequences that span multiple generations.

The medical horrors of a nuclear attack do not stop at physical injuries and radiation sickness. The psychological toll on survivors is immense. The trauma of witnessing the destruction of entire cities, the loss of loved ones, and the fear of radiation exposure can lead to long-lasting mental health issues such as post-traumatic stress disorder (PTSD), depression, and anxiety. In Hiroshima and Nagasaki, survivors, known as *hibakusha*, often struggled with these psychological effects for the rest of their lives. For those planning to survive a nuclear conflict, mental resilience and preparation for psychological trauma are just as important as physical survival skills.

Medical treatment in the aftermath of a nuclear attack presents its own set of challenges. Hospitals and medical infrastructure in the blast zone are likely to be destroyed or overwhelmed by the sheer number of casualties. Even in areas not directly hit by the blast, the influx of injured and irradiated survivors would strain resources to the breaking point. Medical personnel may be in short supply, and the treatments for radiation sickness, such as bone marrow transplants and blood transfusions, may not be available. In addition, the risk of infection in a post-nuclear environment is extremely high, as contaminated water, lack of sanitation, and a breakdown in public health systems contribute to the spread of disease.

For preppers, understanding the medical consequences of a nuclear attack is essential to preparing for long-term survival. Stockpiling medical supplies such as antibiotics, burn ointments, wound dressings, and potassium iodide (which helps block the absorption of radioactive iodine) can increase your chances of surviving the aftermath. However, it is equally important to recognize that no amount of preparation can fully shield you from the horrors of a nuclear event. The best strategy is to avoid exposure to radiation in the first place by staying in a well-constructed fallout shelter until radiation levels have subsided. Knowing the symptoms of radiation sickness and having a plan to address injuries from burns or the blast can also be lifesaving.

In conclusion, the medical consequences of a nuclear attack are among the most devastating aspects of nuclear warfare. From the immediate horrors of burns and blast injuries to the long-term effects of radiation sickness and cancer, the human toll is immense. Understanding these dangers is key to preparing for the worst, but the reality is that even the most prepared survivors will face unimaginable challenges in the aftermath of a nuclear event. The path to survival lies in knowledge, preparation, and a clear understanding of the medical risks posed by nuclear weapons.

Short-Term Effects of Nuclear Fallout on the Human Body

The short-term effects of nuclear fallout on the human body are devastating and can lead to immediate injury, illness, or death, depending on the level of exposure. Fallout consists of radioactive particles that are released into the atmosphere after a nuclear explosion and then settle back to the ground, contaminating everything they touch. These particles, carried by wind and other weather patterns, can travel long distances, making the area impacted by fallout much larger than the initial blast zone. For anyone caught in a fallout zone without adequate shelter or protection, the consequences are severe.

One of the most immediate short-term effects of nuclear fallout on the human body is radiation exposure, which can result in a condition known as acute radiation syndrome (ARS), or radiation sickness. This condition occurs when a person is exposed to a high dose of ionizing radiation in a short period, typically within hours or days after the fallout. The severity of ARS depends on the dose of radiation absorbed, which is measured in units called sieverts (Sv). Even low levels of radiation can cause sickness, but high doses can lead to rapid death.

Symptoms of ARS generally occur in three stages. The first stage, known as the prodromal phase, begins within minutes to hours after exposure and lasts for several days. During this time, the individual may experience nausea, vomiting, diarrhea, fatigue, and headache. These symptoms occur because radiation damages the cells lining the gastrointestinal tract, which are highly sensitive to radiation. The more severe the exposure, the faster and more intense these symptoms will appear.

After the initial phase, there is often a latent period in which symptoms subside, leading the affected individual to believe that they are recovering. However, this phase is deceptive. For those exposed to higher levels of radiation, this calm period is often followed by a much more serious stage of illness, as the damage to internal organs and the immune system becomes more pronounced. This second phase, known as the manifest illness phase, is marked by symptoms such as fever, dizziness, confusion, hair loss, infections, internal bleeding, and severe fatigue. The body's ability to produce blood cells is compromised because radiation destroys bone marrow, leading to a weakened immune system, making infections more likely and harder to treat.

If the radiation dose is extremely high, death can occur within days or weeks from organ failure, particularly in the gastrointestinal system or cardiovascular system. For survivors of the initial exposure, medical treatment—if available—can help manage symptoms, but even with treatment, recovery is a long and uncertain process, as the body struggles to rebuild damaged cells and tissues.

Another immediate short-term effect of nuclear fallout is radiation burns, which occur when radioactive particles land on the skin or when individuals are exposed to high levels of radiation near ground zero. These burns can range from mild redness and irritation to severe blistering and tissue damage, similar to thermal burns. Radiation burns are deceptive in that they may not appear immediately after exposure; it may take hours or even days for the full extent of the damage to become apparent. In severe cases, the skin can peel away, leaving raw tissue that is highly susceptible to infection.

Eyes are also highly sensitive to radiation, and exposure to fallout can result in a condition known as radiation-induced cataracts. Even low doses of radiation can damage the lens of the eye, causing clouding and vision impairment over time. While cataracts are a long-term effect, acute exposure can also cause immediate damage to the eyes, including swelling and inflammation of the cornea, leading to temporary or permanent blindness.

Breathing in radioactive particles, which can easily happen during fallout, leads to internal radiation exposure, affecting the lungs and respiratory system. Radioactive dust and debris can settle in the lungs, causing damage to the tissue. In the short term, this can result in symptoms such as coughing, shortness of breath, and chest pain. More dangerous, however, is the risk of internal radiation poisoning, as these particles continue to release radiation inside the body. This internal exposure increases the risk of developing radiation sickness and, in the long term, cancer.

Contamination of food and water sources by fallout presents an additional danger. Radioactive particles that settle on crops, soil, and water can make food and water unsafe for consumption. Ingesting contaminated food or water leads to internal exposure, where radiation continues to affect the body from within. In the short term, this can cause symptoms similar to those of ARS, including nausea, vomiting, and diarrhoea. Over time, consuming contaminated food and water increases the risk of developing cancers of the digestive system and other organs.

One of the most concerning short-term effects of nuclear fallout is the psychological toll it can take on survivors. The terror of a nuclear event, combined with the fear of radiation exposure, can lead to shock, anxiety, and panic. Survivors may experience post-traumatic stress disorder (PTSD), extreme fear of contamination, and depression, especially as they watch friends, family, or neighbors succumb to radiation sickness. This psychological impact is compounded by the widespread destruction, the collapse of infrastructure, and the lack of immediate medical assistance, which can make survivors feel helpless and overwhelmed.

In a post-fallout environment, the body's exposure to radioactive particles depends largely on the proximity to the explosion, the duration of exposure, and the availability of shelter. Fallout shelters can provide protection from external radiation, but only if they are well-constructed and sealed against the infiltration of radioactive particles. For those unable to reach adequate shelter, exposure can continue for hours or even days, leading to cumulative radiation doses that increase the severity of radiation sickness and other health effects.

One of the challenges of dealing with fallout is the difficulty in detecting and measuring radiation. Fallout is often invisible and odorless, so those affected may not realize they are being exposed until symptoms of radiation sickness begin to appear. Even those outside the immediate blast zone may still face significant exposure, as fallout can spread over large areas depending on wind and weather patterns.

The best strategy for surviving the short-term effects of nuclear fallout is immediate action to minimize exposure. This means getting to a fallout shelter as quickly as possible after the detonation and staying indoors for as long as necessary until radiation levels decrease.

Sealing windows and doors, using air filtration systems, and avoiding contact with contaminated surfaces are critical steps to reduce the risk of radiation poisoning. Those exposed to fallout should also remove contaminated clothing and wash their skin thoroughly to reduce the risk of burns and further exposure.

In summary, the short-term effects of nuclear fallout on the human body are devastating and multifaceted. From radiation sickness and burns to internal poisoning from contaminated food and water, the body is subjected to a barrage of life-threatening conditions. Understanding these effects is crucial for survival, as knowing how to minimize exposure and recognize symptoms early can be the difference between life and death. The key to surviving nuclear fallout is preparation, quick action, and access to proper shelter and supplies. Without these, the human body is highly vulnerable to the immediate dangers posed by radioactive fallout.

Long-Term Effects of Radiation Exposure on Health

The long-term effects of radiation exposure on health are often more insidious than the immediate impacts, as they unfold over months, years, and even decades after the initial exposure. Survivors of nuclear attacks, radiation accidents, or prolonged exposure to radioactive fallout face a heightened risk of developing a range of serious health issues, including cancers, genetic mutations, cardiovascular diseases, and chronic illnesses. Understanding these long-term effects is critical for those preparing to survive in a post-nuclear environment, as radiation exposure can continue to affect individuals and future generations well after the initial fallout has dissipated.

One of the most well-known long-term health consequences of radiation exposure is the increased risk of cancer. Ionizing radiation, such as that released during a nuclear explosion or from radioactive fallout, damages the DNA within cells. This damage can result in mutations that cause cells to grow uncontrollably, leading to cancer. The risk of developing cancer depends on several factors, including the dose of radiation received, the duration of exposure, and the age and health of the individual at the time of exposure. However, even low levels of radiation exposure can increase the risk of cancer over time.

The types of cancer most commonly associated with radiation exposure include leukemia (cancer of the blood-forming tissues), thyroid cancer, breast cancer, lung cancer, and cancers of the digestive system. Leukemia is particularly common among survivors of acute radiation exposure, such as those who lived through the bombings of Hiroshima and Nagasaki. Studies of the *hibakusha*—the survivors of these bombings—showed a significant increase in leukemia cases within just a few years of the attacks. Solid tumors, such as breast and lung cancers, tend to develop more slowly and may not appear for 10 to 20 years or longer after exposure. Thyroid cancer is another major concern, especially among those who were exposed to radioactive iodine, a common byproduct of nuclear fallout. Children and adolescents are especially vulnerable to developing thyroid cancer from exposure to radioactive iodine because their thyroid glands are still growing.

In addition to cancer, radiation exposure can lead to a range of cardiovascular and respiratory diseases. Long-term radiation exposure damages the blood vessels, heart, and lungs, increasing the risk of heart disease, stroke, and pulmonary conditions. For example, survivors of the Hiroshima and Nagasaki bombings have shown higher rates of heart attacks and strokes later in life, likely due to the long-term effects of radiation on the cardiovascular system. Radiation can cause inflammation and scarring in blood vessels, which contributes to the buildup of plaque and increases the risk of cardiovascular events.

The immune system is also significantly impacted by long-term radiation exposure. The bone marrow, which produces the body's blood cells, is highly sensitive to radiation. In the short term, this can lead to acute radiation sickness and a weakened immune response. Over time, chronic exposure can result in a compromised immune system, leaving survivors more susceptible to infections and diseases. In a post-nuclear environment, where medical care and hygiene may be limited, a weakened immune system could lead to a host of secondary illnesses that would otherwise be treatable under normal conditions.

Radiation exposure also has profound effects on human genetics, raising concerns about the potential for genetic mutations and birth defects in future generations. Radiation can cause changes to the DNA in reproductive cells, which may be passed on to offspring. Studies of the children of *hibakusha* have shown an increased risk of birth defects, developmental issues, and genetic abnormalities. Though the extent of radiation's impact on human genetics is still being studied, there is clear evidence that exposure can lead to mutations that affect not only the individual but also their descendants. In addition to birth defects, there may be an increased risk of genetic diseases in future generations, though the exact mechanisms and likelihood of this are still not fully understood.

Chronic radiation exposure can also lead to long-term damage to organs and tissues. For instance, radiation-induced cataracts are a common result of exposure to high levels of radiation. The lens of the eye is particularly sensitive to radiation, and damage can cause clouding of the lens, leading to vision impairment or blindness. Unlike many other radiation-induced health issues, cataracts can develop at relatively low doses of radiation, and the risk increases with prolonged exposure.

The skin, which is often directly exposed to radioactive particles during fallout, can suffer long-term damage as well. Chronic radiation dermatitis is a condition where the skin becomes thickened, discoloured, or ulcerated due to prolonged exposure. This is especially true for individuals who were exposed to fallout or worked in contaminated areas without proper protection. Radiation burns and skin damage that occur immediately after exposure can leave survivors with long-lasting scars and a higher risk of skin cancer.

For many survivors of radiation exposure, psychological health is another area of concern. The trauma of surviving a nuclear event, combined with the ongoing fear of radiation-related illness, can lead to long-term psychological distress. Studies of survivors from Chernobyl and the atomic bombings in Japan have shown higher rates of anxiety, depression, and post-traumatic stress disorder (PTSD).

The psychological burden is further compounded by the social stigma that radiation survivors often face, as people fear contamination from those who have been exposed. This social isolation can exacerbate mental health issues, making recovery even more difficult.

In a post-nuclear world, survivors may also face what is known as "radiophobia," or an intense fear of radiation. This fear can lead to irrational behaviors, such as avoiding areas that may not actually be contaminated or neglecting necessary medical treatments out of fear of further radiation exposure. Radiophobia is a real psychological condition that can hinder individuals from making rational decisions about their health and survival.

The long-term effects of radiation exposure are not limited to individuals; they also affect the environment and the food chain, which can indirectly impact human health. As radioactive particles settle into the soil, they are absorbed by plants and animals, contaminating the food supply. Consuming contaminated food over a long period of time leads to internal radiation exposure, which can further increase the risk of cancer and other health issues. In areas affected by nuclear fallout, such as Chernobyl, wildlife has been observed to suffer from genetic mutations, reproductive issues, and shorter lifespans. While some species have adapted to the radiation, the long-term ecological effects are still being studied, and the risks to human health from consuming contaminated food remain a concern.

One of the most alarming aspects of the long-term effects of radiation is that they are often invisible. Radiation exposure does not cause immediate pain or visible injury, making it difficult to know when or how much exposure has occurred. The delayed onset of many radiation-related illnesses means that survivors may not experience symptoms until years after the initial exposure, making it harder to connect their health problems to the event. This delayed effect is why monitoring radiation levels and understanding the risks associated with even low doses of radiation is so critical for long-term survival.

For preppers and those planning to survive a nuclear event, the key to minimizing the long-term health effects of radiation exposure is to reduce exposure as much as possible from the outset. This means staying in a well-constructed fallout shelter for as long as necessary to avoid contact with radioactive particles, properly cleaning any contaminated items or surfaces, and ensuring that food and water supplies are free from contamination. Medical preparedness is also essential, as early detection and treatment of radiation-related illnesses can improve long-term outcomes.

In conclusion, the long-term effects of radiation exposure on health are profound and far-reaching. From an increased risk of cancer and cardiovascular disease to genetic mutations and psychological trauma, the impact of radiation exposure can last for decades. Understanding these risks is crucial for anyone preparing to survive a nuclear event, as the effects of radiation are not limited to the immediate aftermath but can continue to shape health and wellbeing long into the future. Preparedness, knowledge, and vigilance are the best defenses against the long-term dangers posed by radiation.

The Science behind Nuclear Weapons: A Breakdown

Nuclear weapons are among the most powerful and destructive technologies ever created by humans, capable of causing unparalleled devastation in a matter of seconds. The science behind these weapons is both complex and fascinating, as it taps into the fundamental forces of nature at the atomic level. To truly understand the power of nuclear weapons—and how to prepare for the consequences of their use—it is essential to break down the scientific principles that govern their operation, including nuclear fission, nuclear fusion, and the mechanics of detonation.

At the core of nuclear weapons is the process of releasing the immense energy stored within the nuclei of atoms. This energy is released through two types of nuclear reactions: fission and fusion. These reactions are what make nuclear weapons so much more powerful than conventional explosives, which rely on chemical reactions to release energy.

Nuclear Fission: The Foundation of Atomic Bombs

The first type of nuclear reaction used in nuclear weapons is fission, which occurs when the nucleus of a heavy atom, such as uranium-235 or plutonium-239, is split into two smaller nuclei. This splitting releases a tremendous amount of energy, as well as additional neutrons that can go on to split other atomic nuclei, creating a chain reaction. It is this self-sustaining chain reaction that makes nuclear weapons so powerful.

In a nuclear bomb, the process of fission begins when a sufficient amount of fissile material is brought together to form a critical mass. Critical mass refers to the minimum amount of fissile material needed to sustain a chain reaction. If the amount of material is too small, the neutrons produced by fission will escape without causing further reactions, and the chain reaction will fizzle out. But when the material reaches critical mass, the chain reaction accelerates, releasing vast amounts of energy in the form of heat, light, and radiation.

In the case of an atomic bomb, like the ones dropped on Hiroshima and Nagasaki, two methods are commonly used to achieve critical mass. The first is the "gun-type" design, in which two sub-critical masses of uranium-235 are rapidly brought together by conventional explosives. This method was used in the bomb dropped on Hiroshima, code-named "Little Boy." When the two masses collide, they form a supercritical mass, triggering a rapid fission chain reaction.

The second method is the "implosion" design, used in the bomb dropped on Nagasaki, known as "Fat Man." In this design, a sphere of plutonium-239 is surrounded by conventional explosives arranged in such a way that they compress the plutonium inward, increasing its density and bringing it to critical mass. The implosion method is more efficient than the gun-type design and is used in most modern nuclear weapons.

When fission occurs, a small amount of the mass of the atom is converted into energy, as described by Albert Einstein's famous equation, $\mathbf{E=mc^2}$. In this equation, "E" represents energy, "m" is mass, and "c" is the speed of light.

This equation illustrates how a small amount of mass can be converted into an enormous amount of energy. In the case of nuclear weapons, just a few kilograms of fissile material can produce the explosive energy equivalent to thousands or even millions of tons of TNT.

Nuclear Fusion: The Power behind Hydrogen Bombs

While fission forms the basis of atomic bombs, fusion is the reaction that powers hydrogen bombs, also known as thermonuclear weapons. Fusion occurs when the nuclei of light atoms, such as isotopes of hydrogen (deuterium and tritium), are forced together under extreme pressure and temperature, causing them to combine into a single, heavier nucleus. This process releases even more energy than fission, making fusion weapons far more powerful.

Fusion reactions are the same processes that power the sun and other stars, where intense heat and pressure cause hydrogen atoms to fuse, releasing energy in the form of light and heat. In a hydrogen bomb, this reaction is triggered by using a fission bomb as a "primary" stage to create the necessary conditions for fusion. When the fission bomb detonates, it generates the extreme heat and pressure needed to initiate fusion in a separate section of the bomb, known as the "secondary" stage.

The design of a thermonuclear weapon typically consists of two main stages: the primary (fission) and the secondary (fusion). When the primary stage detonates, it compresses the secondary stage, which contains fusion fuel (such as deuterium and tritium) along with more fissionable material. This compression generates temperatures in the millions of degrees, igniting the fusion reaction. The fusion reaction then releases a massive burst of energy, which can also cause additional fission reactions in the surrounding material, further amplifying the explosion.

The result of this fission-fusion-fission process is a weapon that can be hundreds or even thousands of times more powerful than an atomic bomb. The first successful test of a hydrogen bomb, conducted by the United States in 1952 and code-named "Ivy Mike," demonstrated the destructive potential of thermonuclear weapons. While the atomic bomb dropped on Hiroshima had a yield of about 15 kilotons of TNT, Ivy Mike produced a yield of 10 megatons—over 600 times more powerful.

The Mechanics of a Nuclear Explosion

When a nuclear bomb detonates, it releases energy in several forms: heat, blast, and radiation. These three effects occur almost simultaneously but have different impacts on the environment and human health.

The initial burst of energy is released as intense heat, creating a fireball that can reach temperatures of millions of degrees. The heat generated by the explosion can ignite fires over a wide area, causing secondary damage far beyond the blast zone. The fireball also emits a blinding flash of light that can cause temporary or permanent blindness to anyone looking directly at it, even from miles away.

Next comes the blast wave, which travels outward at supersonic speeds, flattening buildings, trees, and anything else in its path. The pressure from the blast can cause immediate destruction over several miles, depending on the size of the weapon. People caught in the blast zone may suffer fatal injuries from the force of the explosion, as well as from flying debris.

Following the heat and blast, the detonation releases a burst of ionizing radiation, including gamma rays and neutrons. This radiation can penetrate deep into the body, damaging cells and tissues and leading to radiation sickness, as discussed earlier. Fallout, or the radioactive particles that are carried into the atmosphere by the explosion, is another major concern. These particles can spread over vast distances, contaminating soil, water, and air, and causing long-term health effects such as cancer.

The Role of Isotopes and Fallout

One of the key scientific components of nuclear weapons is the production of radioactive isotopes, or unstable forms of elements that release radiation as they decay. During both fission and fusion reactions, a variety of radioactive isotopes are produced, including cesium-137, strontium-90, and iodine-131. These isotopes pose significant health risks to humans and the environment because they can be absorbed into the body through food, water, or inhalation.

Cesium-137, for example, mimics potassium and can be absorbed into muscle tissue, where it continues to release radiation. Strontium-90 behaves like calcium and is deposited in bones, increasing the risk of bone cancer and leukemia. Iodine-131 is particularly dangerous to the thyroid gland, and exposure to this isotope has been linked to an increased risk of thyroid cancer, especially in children.

The dispersal of these isotopes through fallout is a major concern after a nuclear explosion. Fallout occurs when radioactive particles from the explosion are carried by the wind and settle over a wide area, contaminating everything they touch. The spread of fallout depends on a number of factors, including the altitude of the explosion, weather patterns, and geography. In a ground detonation, fallout can be particularly severe, as the explosion lifts large amounts of radioactive material from the ground into the atmosphere.

Nuclear Weapons and the Future

The science behind nuclear weapons has continued to evolve since the development of the first atomic bomb. Today, nuclear-armed states possess a range of weapons with varying yields and delivery systems, including intercontinental ballistic missiles (ICBMs) and submarine-launched ballistic missiles (SLBMs). Advances in missile technology and miniaturization have made it possible to deploy nuclear weapons with greater precision and flexibility, increasing the risk of their use in modern warfare.

Understanding the science behind nuclear weapons is crucial for anyone preparing for the possibility of a nuclear event. The sheer destructive power of these weapons, combined with the long-term environmental and health effects of radiation, makes them one of the most dangerous threats to humanity. By comprehending the principles of nuclear fission, fusion, and the mechanics of a nuclear explosion, preppers can better prepare for the challenges they may face in a post-nuclear world.

Understanding Fallout: What Happens After the Blast

After a nuclear detonation, the immediate destruction caused by the blast and intense heat is only the beginning of the danger. The aftermath brings an invisible but deadly threat: fallout. Fallout refers to the radioactive particles that are released into the atmosphere during a nuclear explosion and then fall back to the ground, contaminating everything they touch. These particles pose a serious risk to human health, the environment, and the long-term survivability of anyone exposed. Understanding how fallout forms, how it spreads, and how to protect yourself from its dangers is essential for surviving in the aftermath of a nuclear attack.

The Formation of Fallout

Fallout is a direct result of the nuclear fission process that occurs during a detonation. When a nuclear bomb explodes, it releases an immense amount of energy that vaporizes everything within the immediate vicinity, including soil, buildings, and other materials. These vaporized materials, along with the radioactive by-products of the nuclear reaction, are lifted high into the atmosphere by the force of the explosion. As the fireball from the explosion rises, it pulls up debris and radioactive particles, forming the characteristic mushroom cloud seen in nuclear explosions.

The particles that make up fallout are a mix of radioactive isotopes created during the fission or fusion process, as well as materials from the ground and structures that have been irradiated. These isotopes include highly dangerous substances such as cesium-137, strontium-90, and iodine-131, all of which emit radiation as they decay. Fallout particles can range in size from tiny, dust-like particles to larger chunks of material. The smaller particles, in particular, are especially hazardous because they can travel long distances and be inhaled or ingested by humans and animals.

The Spread of Fallout

Once the mushroom cloud forms, the radioactive particles within it begin to cool and solidify, falling back to Earth over the course of hours, days, or even weeks, depending on their size and the altitude they reached. This process is where the term "fallout" originates, as the radioactive particles literally fall from the sky.

The spread of fallout is influenced by several factors, including weather conditions, wind speed, the altitude of the explosion, and the size of the bomb. For example, an airburst detonation, where the bomb explodes above the ground, tends to produce less local fallout but can spread radioactive particles over a wider area. A ground burst, where the explosion occurs at or near the ground, results in much heavier local fallout as the explosion lifts large amounts of irradiated earth and debris into the atmosphere. Ground bursts are especially dangerous for creating fallout because they produce larger quantities of radioactive particles that settle more quickly, leading to higher levels of radiation in the immediate area.

Wind plays a significant role in how far and how fast fallout spreads. High-altitude winds can carry radioactive particles hundreds or even thousands of miles from the blast site. Fallout patterns can be unpredictable, with weather systems causing it to fall in areas far removed from the original detonation. In some cases, rain can cause fallout to settle more quickly in a particular area, creating what is known as "hot spots"—areas where radiation levels are particularly high due to concentrated fallout. This uneven distribution means that even areas seemingly far from the blast zone can become heavily contaminated.

Radioactive Decay and Half-Life

One of the key characteristics of fallout is that it contains radioactive isotopes, which emit ionizing radiation as they decay. The rate at which these isotopes decay is measured by their half-life—the time it takes for half of the radioactive atoms in a substance to decay into a more stable form. Different isotopes have different half-lives, and understanding these can help determine how long fallout will remain dangerous.

For example, iodine-131 has a relatively short half-life of about eight days. This means that the radiation from iodine-131 decreases significantly after a few weeks, but it also poses an immediate threat to health, especially the thyroid gland. Cesium-137 and strontium-90, on the other hand, have half-lives of about 30 years, meaning they remain hazardous for much longer periods and can contaminate the environment for decades. These longer-lived isotopes are particularly concerning for food and water supplies, as they can be absorbed into plants, animals, and water sources, leading to prolonged exposure for humans.

While the initial fallout from a nuclear explosion is extremely dangerous, the intensity of radiation decreases over time due to radioactive decay. A common rule of thumb is the "7-10 rule," which states that for every sevenfold increase in time after the detonation, radiation levels decrease by a factor of 10. For example, after 7 hours, radiation levels will be 10% of what they were immediately after the explosion, and after 49 hours (7×7), they will be 1%. This decline in radiation means that fallout is most dangerous in the first few days after a nuclear detonation, but areas with heavy fallout could remain hazardous for months or even years.

Health Risks of Fallout

The primary danger of fallout comes from the radiation it emits, which can penetrate living tissue and damage cells, leading to a range of health problems. The type of radiation released by fallout includes alpha particles, beta particles, and gamma rays. Alpha particles are relatively large and can be stopped by skin or clothing, but they are extremely dangerous if inhaled or ingested, as they can cause severe damage to internal organs. Beta particles are smaller and can penetrate the skin, causing burns or radiation sickness if exposure is prolonged. Gamma rays are the most penetrating form of radiation and can pass through the body, damaging cells and DNA deep within tissues.

The health effects of fallout exposure can vary depending on the dose of radiation received, the duration of exposure, and the part of the body affected. Some of the immediate health effects include:

Radiation Sickness: High doses of radiation can lead to acute radiation syndrome (ARS), also known as radiation sickness. Symptoms include nausea, vomiting, diarrhea, fatigue, and, in severe cases, internal bleeding, infections, and organ failure. Radiation sickness can occur within hours or days of exposure, depending on the level of radiation.

Burns: Exposure to beta radiation can cause radiation burns on the skin, which can range from mild irritation to severe tissue damage.

Thyroid Damage: Inhalation or ingestion of iodine-131 can lead to radiation accumulation in the thyroid gland, increasing the risk of thyroid cancer, particularly in children and adolescents. Taking potassium iodide (KI) pills immediately after exposure can help block the absorption of radioactive iodine by the thyroid.

Long-Term Cancer Risks: Even low levels of fallout exposure can increase the risk of developing cancers, particularly of the thyroid, lungs, and digestive system. This risk persists for years or even decades after exposure.

Internal Contamination: Ingesting food or water contaminated with fallout particles can lead to internal radiation exposure, which is difficult to treat and poses long-term health risks.

Protecting Yourself from Fallout

Surviving in the aftermath of a nuclear explosion requires understanding how to minimize exposure to fallout. The key to protection is time, distance, and shielding.

Time: The most dangerous period for fallout exposure is within the first 48 hours after a nuclear detonation. During this time, radiation levels are highest, and it is crucial to stay sheltered indoors, away from any outside contamination. As radiation levels drop over time due to radioactive decay, it becomes safer to venture outside, but extreme caution is still needed, especially in heavily contaminated areas.

Distance: The further you are from the site of the explosion, the less likely you are to be affected by fallout. Fallout spreads out as it travels, so increasing your distance from the blast zone can greatly reduce your exposure. In addition, staying indoors, particularly in a well-sealed fallout shelter, can help keep you away from radioactive particles in the air.

Shielding: The thicker and denser the materials between you and the fallout, the better protected you will be. Fallout shelters are typically constructed with thick walls made of concrete, lead, or earth to provide maximum shielding from radiation. If a fallout shelter is not available, basements or rooms in the center of a building can offer some protection from radiation.

In addition to finding shelter, it is essential to avoid any contact with fallout. This means removing contaminated clothing and thoroughly washing exposed skin with soap and water to remove radioactive particles. Water supplies should also be treated with caution, as fallout can contaminate rivers, lakes, and groundwater. Only drink water from sealed or protected sources, and avoid consuming any food that may have been exposed to fallout.

Fallout and the Long-Term Environment

Beyond the immediate health risks, fallout poses long-term environmental challenges. Radioactive isotopes in the soil and water can contaminate food supplies, making agriculture and livestock farming difficult or dangerous. Cesium-137 and strontium-90, in particular, can be absorbed by plants and animals, leading to bioaccumulation in the food chain. This means that even years after a nuclear detonation, people could still be exposed to radiation through contaminated food and water.

Fallout can also render large areas of land uninhabitable for extended periods. The exclusion zones around Chernobyl and Fukushima are examples of the long-lasting impact of radioactive contamination. In a post-nuclear scenario, survivors may need to relocate to areas with lower radiation levels and find alternative means of food production until the environment begins to recover.

Conclusion

Understanding fallout is critical for surviving in a post-nuclear environment. Fallout spreads over vast areas, bringing deadly radioactive particles with it that can cause both immediate and long-term health problems. Knowing how fallout forms, how it spreads, and how to protect yourself is essential for minimizing exposure and ensuring your survival after a nuclear detonation. By taking the necessary precautions—finding shelter, limiting exposure, and avoiding contaminated food and water—you can reduce the risk of radiation sickness and other fallout-related illnesses, increasing your chances of survival in the days, months

Dangerous Contaminants in Nuclear Fallout: What to Avoid

When a nuclear weapon detonates, it not only releases an immense amount of energy but also disperses radioactive particles into the environment, creating fallout. This fallout contains a range of dangerous contaminants that pose significant risks to human health and the environment. These contaminants, primarily in the form of radioactive isotopes, are hazardous because they emit ionizing radiation, which can damage living tissue, disrupt cellular function, and lead to both immediate and long-term health issues. Understanding the key radioactive isotopes found in nuclear fallout and how to avoid exposure is crucial for survival in a post-nuclear environment.

Radioactive Isotopes in Fallout

Nuclear fallout contains a variety of radioactive isotopes, also known as radionuclides, that are created during the fission or fusion reactions in a nuclear detonation. These isotopes decay over time, emitting radiation in the form of alpha particles, beta particles, and gamma rays. Some of the most dangerous and commonly found isotopes in nuclear fallout include:

Iodine-131 (I-131)

Iodine-131 is a particularly dangerous contaminant in fallout because it accumulates in the thyroid gland when inhaled or ingested. It emits beta and gamma radiation, which can damage thyroid tissue and increase the risk of thyroid cancer, especially in children. Iodine-131 has a relatively short half-life of about eight days, meaning its radiation decreases quickly over a few weeks, but its immediate threat is severe.

What to Avoid:

Iodine-131 is primarily absorbed through inhalation or ingestion of contaminated food and water. After a nuclear event, avoid consuming any fresh milk, vegetables, or water that could be contaminated. Potassium iodide (KI) pills, taken shortly after exposure, can block the thyroid from absorbing radioactive iodine, reducing the risk of damage.

Cesium-137 (Cs-137)

Cesium-137 is one of the most long-lasting contaminants in fallout, with a half-life of about 30 years. It is a gamma emitter and can be absorbed by the body in a way similar to potassium, meaning it can accumulate in muscle tissue. Exposure to cesium-137 increases the risk of cancer and other health problems over time. It is particularly dangerous because it can contaminate soil and water, making it a long-term environmental hazard.

What to Avoid:

Cesium-137 can contaminate food and water sources. Avoid consuming food grown in contaminated soil or water, and ensure that any drinking water is from a clean, protected source. In heavily contaminated areas, farming and food production can remain unsafe for years or even decades.

Strontium-90 (Sr-90)

Strontium-90 is another long-lived radioactive isotope found in fallout, with a half-life of about 28 years. It behaves similarly to calcium and is absorbed into bones and teeth, where it can remain for long periods. The accumulation of strontium-90 in the body increases the risk of bone cancer and leukemia. Like cesium-137, it is a major contaminant of soil and water, affecting the food supply for extended periods.

What to Avoid:

Avoid consuming dairy products, leafy greens, and root vegetables grown in contaminated soil, as strontium-90 tends to accumulate in these products. In areas affected by fallout, be cautious of bone-in meats, as strontium can concentrate in the bones of animals.

Plutonium-239 (Pu-239)

Plutonium-239 is a highly toxic and long-lived isotope with a half-life of 24,000 years. Although it is primarily an alpha emitter, which means its radiation cannot penetrate the skin, it is extremely dangerous if inhaled or ingested. Plutonium exposure can cause lung cancer, liver damage, and bone cancer. While not as widespread in fallout as other isotopes, plutonium contamination can occur if a nuclear bomb with a plutonium core is used.

What to Avoid:

Plutonium is most dangerous when inhaled as dust or particles. In areas with fallout, it is essential to avoid breathing in dust from the ground or contaminated surfaces. Wearing a mask or using an air filtration system can help reduce the risk of inhalation. Plutonium can also contaminate water sources, so ensure that drinking water is properly filtered or sourced from a clean area.

Tritium (H-3)

Tritium is a radioactive isotope of hydrogen used in some nuclear weapons, particularly in thermonuclear or hydrogen bombs. It is a beta emitter and has a relatively short half-life of 12 years. Tritium can contaminate water sources and is easily absorbed into the body if consumed in drinking water. While its radiation is not as intense as other isotopes, prolonged exposure to tritium can still pose health risks.

What to Avoid:

Contaminated water is the primary source of tritium exposure. In the aftermath of a nuclear event, it is crucial to avoid drinking untreated water from natural sources like rivers, lakes, or rainwater, as these can be contaminated with tritium.

Carbon-14 (C-14)

Carbon-14 is a naturally occurring isotope, but nuclear fallout can significantly increase its levels in the environment. It has a half-life of about 5,730 years and is a beta emitter. Although the levels of carbon-14 produced by fallout are relatively small compared to other isotopes, it can be incorporated into living organisms, affecting the carbon cycle. Long-term exposure can contribute to the risk of cancer, particularly in cases where food and water sources are contaminated.

What to Avoid:

Like other radioactive isotopes, carbon-14 can enter the food chain. Avoid consuming food or water from contaminated areas, particularly long-lived plants or animals that have absorbed carbon-14.

Pathways of Contamination

The main ways these radioactive isotopes enter the human body are through inhalation, ingestion, and external exposure. After a nuclear explosion, fallout particles are dispersed into the atmosphere and settle over a wide area, contaminating the air, water, soil, and food supply. Understanding how to limit exposure is critical for survival.

Inhalation: Fallout particles can be inhaled when they are suspended in the air. This is particularly dangerous in areas with high levels of fallout or where radioactive dust has been disturbed. Breathing in radioactive particles can lead to lung damage, internal radiation exposure, and an increased risk of cancer. Wearing a mask or respirator in contaminated areas is essential for reducing the risk of inhalation.

Ingestion: Fallout particles can contaminate food and water, leading to internal exposure when consumed. Contaminated crops, livestock, and water sources are major risks in areas affected by fallout. Drinking only from sealed or purified water sources and avoiding food that may have been exposed to fallout are vital precautions.

External Exposure: Fallout particles can settle on skin, hair, and clothing, causing external exposure to radiation. While alpha particles cannot penetrate the skin, beta and gamma radiation can cause burns or penetrate deeper into the body. Washing thoroughly with soap and water and changing into clean clothing can help remove fallout particles and reduce exposure.

Protecting Yourself from Fallout Contaminants

The best way to protect yourself from the dangerous contaminants in nuclear fallout is to limit exposure as much as possible. Here are some key strategies for avoiding radioactive contamination:

Shelter in Place: Immediately after a nuclear detonation, seek shelter indoors, preferably in a basement or a room with thick walls. Fallout is most dangerous in the first few hours to days after a blast, so staying inside can significantly reduce exposure. Ensure that windows and doors are sealed to prevent radioactive particles from entering.

Water Filtration: Use water from stored supplies or filtration systems designed to remove radioactive contaminants. Simple boiling will not remove radioactive particles, so it's essential to have proper filtration equipment. If possible, use sealed bottles of water for drinking and cooking.

Food Precautions: Avoid eating fresh produce, milk, or meat from areas affected by fallout. Stick to canned or stored food that was sealed before the detonation. If fresh food must be consumed, peel vegetables and fruits, and thoroughly wash them to remove any surface contamination.

Personal Decontamination: After exposure to fallout, remove contaminated clothing and wash thoroughly with soap and water to remove radioactive particles from the skin. If possible, use uncontaminated water for this purpose.

Air Filtration: If you are in an area where fallout is present, use air filtration systems or wear masks to reduce inhalation of radioactive particles. N95 masks or similar can help filter out larger particles, but specialized filters may be required for more thorough protection.

Potassium Iodide (KI): If radioactive iodine is present in the fallout, taking potassium iodide (KI) pills can help protect the thyroid from absorbing the radioactive iodine. However, KI does not protect against other types of radiation, so it should be used as part of a broader fallout protection plan.

The dangerous contaminants in nuclear fallout—such as iodine-131, cesium-137, strontium-90, plutonium-239, and others—pose serious threats to human health and the environment. These radioactive isotopes can spread over vast areas, contaminating air, water, food, and soil, and their effects can last for years or even decades. Avoiding exposure to these contaminants is crucial for survival after a nuclear event. By understanding the pathways of contamination and taking steps to protect yourself from inhalation, ingestion, and external exposure, you can significantly reduce the risks associated with fallout.

Poison in the Air: Isotopes and Elements that Threaten Survival

When a nuclear weapon detonates, the explosion generates not only immense heat and a blast wave but also releases a variety of dangerous radioactive isotopes and elements into the air. These particles, which form part of the nuclear fallout, are carried by wind and atmospheric currents, creating a toxic environment that poses significant risks to human survival. Inhaling these radioactive particles can lead to serious health consequences, including acute radiation sickness, long-term cancers, and genetic damage. Understanding the specific isotopes and elements that contaminate the air after a nuclear explosion, and how they threaten survival, is crucial for minimizing exposure and ensuring safety.

The Formation of Airborne Radioactive Isotopes

In the moments following a nuclear explosion, the extreme heat generated by the detonation vaporizes materials near the blast site, including buildings, soil, water, and any organic matter. The energy released also triggers nuclear reactions within the fissile material of the bomb—typically uranium-235 or plutonium-239—producing a host of radioactive isotopes as byproducts. These isotopes are lifted into the atmosphere by the force of the explosion, forming the characteristic mushroom cloud seen in nuclear detonations.

As the radioactive cloud ascends, it spreads over large areas, with some particles settling close to the explosion site and others carried by wind currents over long distances. The airborne particles can remain suspended for days, slowly falling back to Earth as fallout. However, during this time, they pose a significant threat to anyone in the fallout zone, as breathing in these contaminants can lead to severe internal radiation exposure.

Key Radioactive Isotopes and Elements in Fallout

Several specific isotopes and elements are particularly dangerous when inhaled. Each one emits radiation that can damage cells, tissues, and organs, leading to both short- and long-term health problems. The most harmful isotopes commonly found in nuclear fallout include:

Iodine-131 (I-131)

Iodine-131 is one of the most hazardous radioactive isotopes released in nuclear fallout, especially because it is easily inhaled and absorbed into the body. Once inside the body, iodine-131 tends to accumulate in the thyroid gland, where it emits beta and gamma radiation. The thyroid's role in regulating metabolism makes it highly sensitive to radiation, and damage to this gland can result in a significantly increased risk of thyroid cancer, particularly in children and young adults.

Threat to Survival:

Inhalation of iodine-131 can cause rapid thyroid damage, leading to long-term health problems or death if exposure is high enough. Immediate protection includes taking potassium iodide (KI) pills, which saturate the thyroid with non-radioactive iodine, preventing the absorption of iodine-131. However, this needs to be done quickly, as iodine-131 has a half-life of about eight days, meaning it decays relatively quickly, but its effects can be immediate.

Cesium-137 (Cs-137)

Cesium-137 is a long-lived isotope, with a half-life of about 30 years, meaning it remains in the environment and poses a threat to human health for decades after a nuclear event. This isotope is a gamma emitter and can spread widely through the air, contaminating vast areas. Once inhaled, cesium-137 is absorbed into muscle tissues, where it continues to emit radiation, leading to an increased risk of cancer over time.

Threat to Survival:

The inhalation of cesium-137 can lead to long-term internal contamination, affecting the body's organs and tissues. As a persistent contaminant, it can cause chronic exposure to radiation, raising the risk of cancers, particularly in the muscles and soft tissues. Additionally, cesium-137's ability to contaminate food and water supplies makes it a widespread environmental hazard.

Strontium-90 (Sr-90)

Strontium-90 behaves similarly to calcium, meaning it is absorbed by bones and teeth when ingested or inhaled. With a half-life of approximately 28 years, strontium-90 poses long-term risks, particularly to children and adolescents whose bones are still developing. Once inside the body, it emits beta radiation, which can damage bone marrow and lead to bone cancer or leukemia.

Threat to Survival:

Inhalation of strontium-90 poses a significant threat because it becomes incorporated into bone tissue, where it continues to emit harmful radiation for many years. This internal exposure can weaken the bones and immune system, leading to fatal conditions like leukemia or bone cancer over time. Protecting against inhalation or ingestion of strontium-90 is crucial in the aftermath of a nuclear event.

Plutonium-239 (Pu-239)

Plutonium-239 is one of the most toxic radioactive elements found in nuclear fallout, particularly because of its long half-life of 24,000 years. Although it primarily emits alpha particles, which cannot penetrate the skin, plutonium is extremely dangerous when inhaled. Inhalation of plutonium-239 particles can lodge in the lungs, where it continues to emit radiation, leading to lung cancer or other respiratory diseases over time.

Threat to Survival:

Inhalation of even tiny amounts of plutonium-239 can be fatal in the long term. Once lodged in the lungs, plutonium is nearly impossible to remove, and its alpha radiation will damage the surrounding tissue, leading to cancer. Strict avoidance of areas contaminated with plutonium dust or particles is necessary for survival in a fallout zone.

Tritium (H-3)

Tritium is a radioactive isotope of hydrogen, often found in the form of water molecules (tritiated water). Tritium emits low-energy beta radiation and has a half-life of about 12 years. While not as immediately dangerous as other isotopes, inhalation or ingestion of tritiated water can lead to internal exposure, as the body treats it like regular water. Once inside, tritium emits radiation that can damage cells and increase the risk of cancer.

Threat to Survival:

Tritium poses a relatively low risk compared to other isotopes, but long-term exposure can still be harmful, especially if contaminated water is consumed over time. In the aftermath of a nuclear event, ensuring that water supplies are free of tritium contamination is essential.

Radon-222 (Rn-222)

Radon-222 is a naturally occurring radioactive gas that can become more prevalent in the aftermath of a nuclear explosion due to the disturbance of the Earth's crust. It is a decay product of uranium and can accumulate in enclosed spaces, such as basements or shelters. Radon gas is odorless and invisible but poses a significant risk when inhaled, as it can cause lung cancer.

Threat to Survival:

Radon exposure is dangerous because it can accumulate without detection, and inhaling radon gas increases the risk of lung cancer, particularly in poorly ventilated shelters. In a post-nuclear environment, monitoring for radon levels in fallout shelters or underground spaces is essential.

Health Effects of Inhaled Fallout

Inhaling radioactive isotopes presents immediate and long-term health risks, ranging from radiation sickness to life-threatening cancers. Some of the key health effects include:

Acute Radiation Syndrome (ARS): Inhalation of high doses of radioactive particles can lead to ARS, which manifests as nausea, vomiting, fatigue, and in severe cases, organ failure. ARS occurs when the body is overwhelmed by ionizing radiation, and without treatment, it can be fatal within days or weeks.

Lung Damage: Radioactive particles inhaled into the lungs can lead to long-term respiratory damage, including lung cancer. Alpha emitters like plutonium-239 are particularly dangerous because they directly damage lung tissue, causing mutations that can lead to cancer.

Thyroid Cancer: Inhalation of iodine-131 leads to an accumulation of radiation in the thyroid gland, increasing the risk of thyroid cancer, especially in children and young adults.

Bone and Blood Disorders: Strontium-90, once inhaled, integrates into bone tissue, where it damages the bone marrow and increases the risk of leukemia and other blood disorders.

Protecting Yourself from Airborne Fallout

Given the severity of the threat posed by inhaling radioactive isotopes, taking steps to protect yourself from airborne fallout is critical for survival in the aftermath of a nuclear event. Here are essential measures to minimize the risk:

Shelter in Place: The best protection against inhaling fallout is to stay indoors, ideally in a sealed, well-constructed shelter. Fallout particles settle quickly after an explosion, so limiting your exposure to outdoor air during the first 24 to 48 hours is crucial.

Air Filtration: In shelters or enclosed spaces, use air filtration systems designed to remove radioactive particles. If these are unavailable, improvised methods like sealing windows and using wet cloths over ventilation points can help reduce contamination.

Wear Respiratory Protection: If you must go outside or are in an area where fallout is present, wear a mask or respirator capable of filtering out fine particles. An N95 mask or similar can reduce the inhalation of larger fallout particles, though specialized filters may be required for complete protection.

Limit Time Outdoors: Fallout is most dangerous in the hours and days immediately following a nuclear detonation. Limiting your time outdoors and avoiding exposure to dust, ash, or debris can help reduce your risk of inhaling radioactive particles.

Radioactive isotopes and elements in nuclear fallout pose a serious threat to human survival, particularly when inhaled. Contaminants like iodine-131, cesium-137, strontium-90, plutonium-239, and tritium can cause immediate and long-term health problems, ranging from acute radiation sickness to various cancers. Understanding the risks associated with these airborne contaminants and taking steps to minimize exposure is essential for surviving in a fallout zone.

By sheltering in place, using air filtration systems, and wearing appropriate respiratory protection, you can significantly reduce your exposure to the dangerous isotopes present in fallout. Being aware of the environment around you—whether it's the air, water, or surfaces—will also help in making safer choices as you navigate the post-nuclear landscape.

Additional Steps for Protection:

Decontamination: After being exposed to fallout, it's critical to decontaminate yourself as soon as possible. This involves removing contaminated clothing and thoroughly washing your skin and hair with soap and clean water to remove any radioactive particles that may have settled on your body. If water is scarce, use wipes or dry decontamination methods to remove as much particulate matter as possible.

Monitor Radiation Levels: If available, a Geiger counter or dosimeter can help you monitor radiation levels in the air, on surfaces, and in water. This equipment can give you a better understanding of which areas are more dangerous and allow you to make informed decisions about where to go and what to avoid.

Potassium Iodide (KI) Pills: As previously mentioned, taking potassium iodide (KI) pills immediately after exposure to iodine-131 can block the thyroid from absorbing radioactive iodine, reducing the risk of thyroid cancer. However, these pills only protect the thyroid and do not shield the body from other forms of radiation. Therefore, they should be used as part of a broader fallout protection strategy, not as a sole solution.

Avoid Stirring Up Dust: Fallout can settle on surfaces, and any activity that stirs up dust—such as walking through contaminated areas or moving debris—can re-suspend radioactive particles into the air. Be mindful of your actions in areas where fallout may have settled, and avoid disturbing dust whenever possible.

Long-Term Considerations:

The presence of long-lived isotopes like cesium-137 and strontium-90 means that the environment could remain hazardous for years or even decades. Contaminated soil, food supplies, and water will continue to pose risks, making it necessary to adopt long-term strategies for survival.

Water Safety: Ensure that drinking water is sourced from uncontaminated supplies, such as sealed bottles or water treated with reliable filtration systems designed to remove radioactive particles. Boiling water will not remove radioactive isotopes, so proper filtration is essential.

Food Sources: Foraging or consuming locally grown food may be dangerous if fallout has contaminated the area. Stick to stored, sealed, or canned food for the immediate aftermath of a nuclear event. Over time, growing your own food in areas with minimal contamination, or in greenhouses that protect against fallout, may become necessary.

Movement and Relocation: If the area you are in has been heavily contaminated with fallout, consider relocating to a safer zone if possible. Monitoring local radiation levels can help guide decisions on when and where it is safe to move.

The threat of airborne radioactive isotopes following a nuclear explosion is one of the most serious challenges survivors will face. The inhalation of dangerous elements like iodine-131, cesium-137, strontium-90, and plutonium-239 can lead to both immediate and long-term health consequences, including acute radiation sickness, cancer, and organ damage. Understanding the risks posed by these contaminants, knowing how to minimize exposure, and taking proactive steps like sheltering, wearing protective equipment, and decontaminating yourself are key to surviving the fallout.

By staying informed about the dangers in your environment and acting swiftly to protect yourself from airborne contaminants, you increase your chances of survival in the highly dangerous aftermath of a nuclear detonation. Time, distance, and shielding remain the essential principles for staying safe, but preparation and awareness are just as critical to overcoming the poison in the air.

The Dangers of Fallout Dispersal: How Wind and Weather Play a Role

The dispersal of nuclear fallout—the radioactive particles that fall back to Earth after a nuclear explosion—plays a critical role in determining how far and wide contamination spreads. Fallout dispersal is influenced by several environmental factors, with wind and weather being among the most significant. Understanding how these factors affect the movement and deposition of fallout can help you prepare for survival, assess potential risks, and determine the best strategies for protection in the aftermath of a nuclear event.

How Fallout Disperses

When a nuclear weapon is detonated, a massive amount of energy is released, creating a fireball that vaporizes everything in the immediate vicinity. The intense heat of the explosion lifts vaporized material, radioactive particles, and debris into the atmosphere, where they form a mushroom cloud. As the cloud rises, the particles begin to cool and solidify, eventually falling back to Earth as fallout. The size and type of fallout particles vary, and this influences how far they travel and how quickly they settle.

Heavy particles, such as larger pieces of debris, tend to fall back to the ground quickly, within a few miles of the detonation site.

Smaller particles and dust, which are more dangerous due to their radioactive nature, can remain suspended in the atmosphere for hours, days, or even weeks. These particles are carried by wind currents, potentially spreading fallout across hundreds or thousands of miles.

The dispersal of fallout is not uniform, and wind patterns, precipitation, and atmospheric conditions play a major role in determining where and when the radioactive particles will fall to Earth.

The Role of Wind in Fallout Dispersal

Wind is one of the primary drivers of fallout dispersal. In the immediate aftermath of a nuclear detonation, local wind patterns determine the direction in which the fallout cloud travels. Wind can carry radioactive particles over vast distances, creating fallout zones that may extend hundreds or even thousands of miles away from the original blast site.

Local Wind Patterns

Surface Winds: These winds near the ground play a key role in dispersing fallout in the hours immediately following a nuclear explosion. The direction and speed of surface winds can spread fallout particles in a specific direction, creating a "fallout plume" that extends downwind from the detonation site. This plume is where the concentration of fallout is likely to be highest, and areas within the plume are at the greatest risk of contamination.

Upper-Level Winds: Fallout particles that are lifted high into the atmosphere are affected by upper-level wind currents, such as the **jet stream**, which can transport radioactive particles over much greater distances than surface winds. Jet streams, which flow at high altitudes and speeds, can carry fallout across continents and oceans, leading to contamination in regions far removed from the detonation.

Long-Distance Fallout

When fallout particles are carried into the upper atmosphere, they can be transported over long distances before settling back to Earth. For example, after the Chernobyl nuclear disaster in 1986, radioactive fallout was detected as far away as Scandinavia, the United Kingdom, and North America. Similarly, during atmospheric nuclear testing in the 1950s and 1960s, fallout from tests conducted in the Pacific spread across the globe, with radioactive particles detected thousands of miles from the test sites.

The global transport of fallout depends on the altitude of the explosion (airburst vs. ground burst) and the strength of the winds at different atmospheric levels. **Airbursts**, which occur when a nuclear weapon is detonated above the ground, tend to create less local fallout but result in wider dispersal due to particles being lofted higher into the atmosphere. **Ground bursts**, where the explosion occurs at or near the surface, generate much more fallout in the immediate area, as the explosion vaporizes soil and debris, creating larger radioactive particles that settle quickly.

The Role of Weather in Fallout Dispersal

In addition to wind, weather conditions like precipitation, temperature, and atmospheric pressure can have a significant impact on how fallout spreads and where it eventually settles.

Precipitation (Rain, Snow, and Fog)

Rain plays a particularly crucial role in fallout dispersal, as it can cause radioactive particles to fall to the ground much faster than they would in dry conditions. When fallout particles mix with rain, they are brought down in what is known as "radioactive rain" or "black rain." This rain is highly contaminated and can create dangerous hot spots where radiation levels are significantly higher than surrounding areas.

Snow can similarly accelerate fallout deposition. In cold climates, fallout particles can become trapped in snow, which may remain radioactive for long periods, especially in areas where snow accumulates for extended periods during winter.

Fog and mist can cause fallout particles to condense and fall to the ground more rapidly, leading to localized contamination, particularly in low-lying areas.

Temperature and Atmospheric Conditions

Temperature inversions, where a layer of warm air traps cooler air near the surface, can affect fallout patterns by slowing the dispersion of particles. During an inversion, fallout may remain concentrated in the lower atmosphere and settle over a smaller area than it would in normal conditions. This can create highly concentrated fallout zones near the detonation site or in specific downwind areas.

High humidity can cause fallout particles to adhere to water vapor in the air, increasing the likelihood that they will settle more quickly in moist conditions. This can lead to contamination over a wider area but with less concentration than in dry conditions.

Thunderstorms and High Winds

Severe weather events like thunderstorms or hurricanes can influence the spread of fallout in unpredictable ways. Strong winds associated with these storms can carry fallout in multiple directions or distribute it unevenly. In addition, thunderstorms can generate **lightning**, which may ionize the air and affect the behavior of radioactive particles. Precipitation from storms will also cause fallout to settle faster, often in the form of concentrated "hot spots" where rain or snow brings down large amounts of radioactive material in localized areas.

Hot Spots and Fallout Concentration

Not all areas downwind of a nuclear explosion will receive the same level of fallout contamination. Due to variations in wind speed, direction, and precipitation, certain areas may become "hot spots" where radioactive particles accumulate in higher concentrations.

Hot spots can be particularly dangerous because they may appear in seemingly random locations, far from the blast site, but contain dangerously high levels of radiation. These areas often occur after rainfall or snowfall, which concentrates fallout in small, localized regions. Hot spots are particularly concerning because they can create sudden and intense radiation exposure for anyone passing through or living in those areas. For preppers and survivors, being aware of the possibility of hot spots and monitoring radiation levels in your area is essential for avoiding high-risk zones.

Predicting Fallout Dispersal

Predicting the exact dispersal of fallout in the aftermath of a nuclear explosion is challenging due to the many variables involved, including the size of the explosion, the altitude of the detonation, the type of nuclear device, and local weather conditions. However, there are some general guidelines that can help assess potential fallout risks:

Downwind areas from the blast site are most at risk, particularly within a few hundred miles.

Precipitation patterns can create higher concentrations of fallout in areas that receive rain or snow shortly after the explosion.

Sheltering in place during the initial fallout period (24–48 hours) and monitoring local weather reports and wind patterns can help determine when it may be safer to move or relocate.

In some cases, governments or organizations may provide fallout prediction maps or guidance on safe zones based on the prevailing wind and weather conditions. In the absence of such information, it is essential to rely on tools like Geiger counters to assess radiation levels in your immediate environment.

Protecting Yourself from Fallout Dispersal

Given the unpredictable nature of fallout dispersal, taking immediate and informed action is crucial to minimizing your exposure to radioactive particles:

Shelter in Place: In the hours following a nuclear explosion, staying indoors is the safest option, especially if you are downwind of the blast. A basement or an interior room with no windows is ideal for reducing exposure to airborne fallout particles.

Monitor the Weather: Be aware of wind direction and precipitation. If rain or snow is predicted, stay indoors, as these conditions can accelerate fallout deposition and create dangerous hot spots.

Time, Distance, and Shielding: Increase your distance from the blast site and affected areas as soon as it is safe to do so. Thick walls, concrete, and other dense materials provide effective shielding against radiation. Avoid traveling through areas where fallout may have concentrated due to wind or precipitation.

Radiation Monitoring: Use a Geiger counter or dosimeter to monitor radiation levels in your environment. This will help you identify potential hot spots and avoid areas with dangerous radiation levels.

Cover Up and Use Respirators: If you must go outside, wear protective clothing, such as long sleeves, gloves, and a mask or respirator, to reduce your risk of inhaling or coming into contact with fallout particles.

Wind and weather are powerful forces that can dramatically influence the spread of nuclear fallout, creating contamination zones far beyond the initial blast area. Understanding how wind currents, precipitation, and atmospheric conditions contribute to fallout dispersal is critical for assessing the risks and taking protective measures in the aftermath of a nuclear explosion. By sheltering in place, monitoring weather patterns, and using radiation detection tools, you can minimize your exposure and improve your chances of surviving the dangers of fallout dispersal.

Determining the Safe Duration: How Long to Stay in a Fallout Shelter

In the aftermath of a nuclear explosion, one of the most critical decisions for survival is determining how long to stay in a fallout shelter. The key to this decision lies in understanding the behavior of radioactive fallout and how quickly it decays over time. Fallout shelters provide protection from the immediate dangers of radioactive particles that can contaminate the air, water, and soil. However, once inside, it's essential to remain sheltered long enough for radiation levels outside to drop to a relatively safe level before venturing out.

This chapter will guide you through the factors that influence the safe duration to stay in a fallout shelter, the science behind radiation decay, and strategies for determining when it's safe to leave.

The Immediate Fallout Period

Immediately following a nuclear explosion, radioactive particles are dispersed into the atmosphere, where they begin to settle back to Earth as fallout. Fallout contains a mix of dangerous radioactive isotopes, including iodine-131, cesium-137, and strontium-90, which emit harmful radiation in the form of alpha, beta, and gamma rays. Exposure to high levels of radiation can cause acute radiation sickness, cancer, and other life-threatening health effects.

The initial hours and days after the explosion are the most dangerous because radiation levels are at their highest. This period is known as the **"immediate fallout period"**, and during this time, it is critical to remain in a well-constructed fallout shelter to minimize exposure. Shelters offer protection by shielding you from the ionizing radiation that can penetrate buildings and enter through the air.

Radioactive Decay and the "7-10 Rule"

Radioactive fallout decays rapidly after the detonation, meaning that radiation levels decrease over time. One of the most important principles for understanding this decay is the **"7-10 rule."** This rule provides a simple way to estimate how much radiation levels will drop over time:

7-10 Rule: For every 7-fold increase in time after the explosion, the radiation level decreases by a factor of 10.

For example:

After 7 hours, radiation levels will be 10% of what they were immediately following the explosion.

After 49 hours (7×7), radiation levels will drop to 1% of their initial intensity.

After 2 weeks (approximately 14 days), radiation levels will be about 0.1% of the original level.

The 7-10 rule reflects the exponential nature of radioactive decay. While radiation levels drop quickly in the first few hours and days, they continue to decrease at a slower rate over time. This means that the first few days in the fallout shelter are the most critical, as the radiation outside is still dangerous.

Factors Influencing How Long to Stay in a Fallout Shelter

Several factors influence how long you need to remain in a fallout shelter. While the 7-10 rule provides a basic guideline, additional considerations include the size of the nuclear explosion, your proximity to the blast site, local weather patterns, and the type of shelter you have.

Proximity to the Blast Site

The closer you are to the detonation, the higher the initial radiation levels will be. If you are within a few miles of the explosion, the fallout will be most intense in the hours and days immediately following the blast. In this case, you may need to remain in the shelter for several days or even weeks before it is safe to leave.

For those located further away from the detonation (50+ miles), the fallout will be less concentrated, and radiation levels will drop more quickly. In these cases, the duration of sheltering may be shorter, but it is still critical to stay inside for at least the first 48 hours.

Type and Quality of Shelter

The effectiveness of your fallout shelter in protecting you from radiation will also influence how long you need to stay inside. A well-constructed shelter with thick walls made of concrete, lead, or earth provides the most protection by significantly reducing the amount of radiation that penetrates the structure. Shelters buried underground or in basements offer even greater shielding, as they block more radiation.

If you are in a less secure shelter—such as a home without proper shielding—radiation may be able to penetrate more easily, and you may need to extend your time inside to ensure your safety. Reinforcing your shelter with additional barriers, such as sandbags or metal sheeting, can improve protection.

Weather and Wind Patterns

Weather conditions can play a role in how long you need to remain in the shelter. Fallout particles are carried by wind currents, and local weather patterns will influence how quickly fallout settles to the ground. Rain or snow can accelerate the deposition of fallout, creating localized hot spots where radiation levels are higher. In these cases, it may take longer for radiation levels to drop to safe levels.

If fallout is carried by the wind into your area days after the explosion, radiation levels may spike, even if you were initially far from the blast. Monitoring weather reports and wind patterns can help you gauge when it is safest to emerge from the shelter.

Determining When It's Safe to Leave

There are several ways to determine when it is safe to leave your fallout shelter. While the 7-10 rule provides a general timeline, it's important to have tools and strategies for assessing radiation levels in your area. Here are some methods for making an informed decision:

Radiation Detection Equipment

The most reliable way to assess radiation levels is with a radiation detector, such as a Geiger counter or dosimeter. These devices measure the amount of ionizing radiation in the environment and provide real-time data on exposure levels. Ideally, you should have one of these devices in your shelter to monitor the levels of radiation outside. The goal is to wait until radiation levels have dropped to a level that is considered safe for short-term exposure.

Safe Level: A commonly accepted safe radiation level for short-term exposure is 0.5 to 1 roentgen per hour (R/hr). If levels are above this, you should remain in the shelter longer.

Stay Informed with Local Authorities

In the event of a nuclear detonation, local and national authorities may provide updates on radiation levels and advice on when it is safe to leave shelters. Stay informed by listening to emergency broadcasts on a battery-powered or hand-crank radio. Government agencies may issue warnings about fallout zones, radiation levels, and evacuation routes. If you are able to receive this information, follow official guidance closely.

The "Core Survival" Timeline

For those without access to radiation detection equipment, following a general timeline based on the 7-10 rule is still an effective strategy. Here's a breakdown of a basic timeline for sheltering after a nuclear explosion:

First 24 Hours: Radiation levels are at their highest. Stay sheltered in a protected area, ideally underground or in a reinforced room. Do not go outside.

24–48 Hours: Radiation levels begin to decrease, but are still dangerous. Continue to shelter in place, especially if you are in a high-risk zone near the blast.

48–72 Hours: Radiation levels may have dropped enough to assess conditions outside, but extreme caution is still necessary. If you must go outside briefly, wear protective clothing and minimize exposure time.

72 Hours to 1 Week: Depending on your proximity to the blast and the effectiveness of your shelter, radiation levels should have decreased significantly. After one week, radiation may be at a level that allows limited time outside, but you should still limit exposure and wear protective gear.

1 Week to 2 Weeks: Radiation levels will likely be low enough to allow for more extended outdoor activity, but avoid spending too much time in areas that could still be contaminated, especially if you are in a high-risk zone.

Beyond 2 Weeks: Radiation levels will continue to decrease, and you may be able to safely leave the shelter for longer periods, but continue to monitor the situation.

Venturing Out for Short Periods

After several days in the shelter, if you need to leave for essential tasks (such as retrieving water or supplies), limit your time outside to the shortest duration possible. Wear protective clothing, including masks, gloves, and long sleeves, to reduce exposure to fallout particles. Return to the shelter as quickly as possible and decontaminate yourself by washing off any potential fallout particles with soap and water.

Determining how long to stay in a fallout shelter is one of the most important decisions for survival in the aftermath of a nuclear event. The first 48 hours are the most critical, as radiation levels are at their highest during this time. The longer you remain sheltered, the safer you will be, as radiation levels decay rapidly over time, following the 7-10 rule. Using radiation detection equipment, staying informed about local conditions, and following general timelines can help you assess when it is safe to emerge.

While the instinct may be to leave the shelter as soon as possible, patience is essential for minimizing your exposure to deadly radiation. By understanding the science of fallout decay and taking informed steps to protect yourself, you can increase your chances of survival in the days and weeks following a nuclear explosion.

Contaminants in the Air and Soil: How to Assess the Threat

In the aftermath of a nuclear explosion, radioactive contaminants in the air and soil pose serious threats to human health and the environment. Understanding how to assess the presence and concentration of these contaminants is crucial for survival, particularly if you plan to stay in or return to an area affected by fallout. These radioactive particles, carried by wind and weather patterns, settle in the environment, contaminating the air we breathe, the soil we grow food in, and the water we drink.

This chapter will cover the key radioactive contaminants found in air and soil after a nuclear event, methods for assessing the threat they pose, and practical steps you can take to protect yourself and your surroundings.

Radioactive Contaminants in Air and Soil

The contaminants found in fallout are a mix of radioactive isotopes, each with different half-lives, levels of radiation emission, and biological impacts. The most dangerous isotopes are those that emit ionizing radiation in the form of alpha particles, beta particles, and gamma rays, which can damage cells and tissues in the body.

Key radioactive contaminants include:

Iodine-131 (I-131)

Half-Life: ~8 days

Primary Danger: Inhalation or ingestion

Health Risks: Accumulates in the thyroid, causing thyroid cancer or damage, particularly in children.

Iodine-131 is one of the first contaminants to become airborne after a nuclear event. It poses an immediate threat because it is rapidly absorbed by the body through inhalation or ingestion. Its relatively short half-life means it decays quickly, but in the days following the explosion, it is one of the most hazardous substances in the air.

Cesium-137 (Cs-137)

Half-Life: ~30 years

Primary Danger: Inhalation, ingestion, and external exposure

Health Risks: Absorbed by muscle tissue, increases the risk of cancer.

Cesium-137 is a gamma emitter and can travel long distances in the air. Once it settles in the soil, it can contaminate food and water supplies for decades, making it one of the most persistent and widespread contaminants in nuclear fallout.

Strontium-90 (Sr-90)

Half-Life: ~28 years

Primary Danger: Inhalation and ingestion

Health Risks: Mimics calcium and is absorbed into bones, increasing the risk of bone cancer and leukemia.

Strontium-90, like cesium-137, is a long-lived contaminant that can persist in the environment for decades. It poses a particular risk because it becomes incorporated into the bone structure, where it continues to emit beta radiation, damaging bone marrow and increasing cancer risk.

Plutonium-239 (Pu-239)

Half-Life: ~24,000 years

Primary Danger: Inhalation

Health Risks: Lodges in lung tissue, causing lung cancer and other respiratory diseases.

While plutonium-239 is primarily an alpha emitter (which cannot penetrate the skin), it is extremely toxic when inhaled. Even minute amounts of plutonium dust can lodge in the lungs and cause long-term damage.

Methods for Assessing the Threat

Assessing the presence and concentration of radioactive contaminants in the air and soil is essential for determining the safety of an environment after a nuclear explosion. Here are the most common methods for assessing these threats:

Radiation Detection Devices

The most effective way to assess radiation levels is by using radiation detection equipment. There are several types of devices that can help you measure the presence of radioactive contaminants in your environment:

Geiger Counters: These devices detect ionizing radiation (alpha, beta, and gamma radiation) and provide real-time readings of radiation levels in your immediate surroundings. A Geiger counter will give you a reading in counts per minute (CPM) or roentgens per hour (R/hr), allowing you to assess the radiation levels in the air or on surfaces.

Dosimeters: A dosimeter measures the cumulative amount of radiation exposure over time. This is useful for tracking your long-term exposure, especially if you are living or working in an area with low-level contamination.

Alpha and Beta Particle Detectors: These detectors are specialized for identifying specific types of radiation, such as alpha or beta particles. They are useful for assessing contamination in soil or surfaces, as these particles can be harmful when inhaled or ingested but do not penetrate deeply into the body.

Gamma Spectrometers: These are more advanced instruments used to identify specific radioactive isotopes by measuring the gamma radiation they emit. While not commonly used by individuals, they can provide valuable data on the specific types of contamination present in your environment.

Soil and Water Testing

Testing soil and water for radioactive contamination is crucial if you plan to grow food or use local water sources after a nuclear event. Even if radiation levels in the air have dropped, contaminated soil and water can pose long-term health risks.

Soil Testing Kits: Soil testing kits designed to detect radiation can help you determine whether it is safe to plant crops in a particular area. Cesium-137 and strontium-90 are the main contaminants to watch for in soil, as they can be absorbed by plants and enter the food chain.

Water Testing: Water sources can become contaminated by fallout particles that settle into lakes, rivers, or groundwater. Specialized water testing kits can help identify radioactive contaminants in drinking water. Contaminated water can carry cesium-137, strontium-90, and tritium (a radioactive isotope of hydrogen) into your body.

Visual and Environmental Clues

In addition to using radiation detection devices, you can also look for environmental clues to assess the safety of an area. These clues may not provide precise data but can help you identify areas that may be heavily contaminated:

Black Rain: After a nuclear explosion, you may observe "black rain," which is rainwater mixed with fallout particles. This rain is a sign of concentrated contamination, and areas where it falls are likely to have elevated radiation levels.

Soil Discoloration: Fallout particles can cause changes in soil color, particularly near the blast zone. Dark, ashy soil or unusual deposits of dust may indicate areas with high radiation levels.

Plant and Animal Behavior: In heavily contaminated areas, plants may exhibit stunted growth or unusual discoloration. Similarly, animals in the area may become sick or behave erratically. While these signs are not definitive, they may indicate environmental contamination.

Understanding Safe Exposure Levels

Radiation exposure is measured in units of **roentgens (R)**, **rem (Roentgen Equivalent Man)**, or **sieverts (Sv)**, with 1 Sv equaling 100 rem. These units indicate how much radiation is absorbed by the body and the potential biological effects.

0–50 millirems (mrem) per year: Typical background radiation exposure from natural sources.

50–500 mrem: Considered safe for short-term exposure. This level is unlikely to cause immediate health problems but should be avoided over long periods.

500 mrem to 1 rem: Exposure at this level can lead to health problems if sustained over a longer period. Short-term exposure should be minimized.

Above 1 rem: Exposure at this level can cause radiation sickness, increase the risk of cancer, and lead to other long-term health effects.

It's important to remember that any exposure to radiation carries some risk. While the human body can tolerate low levels of radiation over time, minimizing exposure is essential, especially in a contaminated environment.

Protecting Yourself from Contaminants in Air and Soil

Once you've assessed the contamination levels in your environment, taking protective measures is essential to reduce exposure to radioactive particles.

Air Protection

Stay Indoors: In the initial hours and days after a nuclear explosion, staying inside a fallout shelter or a well-sealed building is the best way to avoid inhaling contaminated air.

Air Filters: If you must ventilate your shelter, use air filters designed to trap radioactive particles. High-efficiency particulate air (HEPA) filters can help remove fine particles from the air.

Respirators and Masks: If you need to go outside, wear a high-quality respirator or mask (such as an N95 mask) to reduce the risk of inhaling fallout particles.

Soil and Water Protection

Avoid Contaminated Soil: If soil contamination is present, avoid disturbing the ground, as this can release radioactive particles into the air. Do not consume food grown in contaminated soil unless you are certain it is safe.

Water Filtration: Water filtration systems designed to remove radioactive particles are essential for ensuring safe drinking water. Boiling water does not remove radioactive isotopes, so use filters capable of removing particles like cesium-137 and strontium-90.

Decontamination

Wash Regularly: After being outside in contaminated areas, remove your clothing and wash your skin thoroughly with soap and clean water. This will help remove any radioactive dust or particles from your body.

Clothing and Equipment: If your clothing or gear becomes contaminated, isolate it in sealed bags and avoid bringing it into living areas. Clothing contaminated with fallout can continue to emit radiation for extended periods.

Assessing the threat of radioactive contaminants in the air and soil is crucial for making informed decisions about your safety in a post-nuclear environment. With the right tools—such as radiation detectors, soil and water testing kits, and visual clues—you can determine whether an area is safe for habitation, food production, or other activities. However, assessing the environment goes beyond detection—it's about understanding the potential health risks, taking appropriate actions to minimize exposure, and making decisions that prioritize long-term survival in a contaminated world.

Summary of Key Protective Steps

Use Radiation Detection Devices: Geiger counters, dosimeters, and other radiation measurement tools are critical for assessing the levels of contamination in both the air and soil. Monitoring radiation levels frequently can help you decide when it is safe to move outside, gather resources, or begin to rehabilitate the land for farming.

Test Soil and Water: Before using local soil for gardening or drinking from natural water sources, make sure they are free from dangerous levels of radioactive isotopes like cesium-137, strontium-90, and iodine-131. Water filtration systems designed to remove radioactive particles are necessary to prevent contamination through ingestion.

Follow the 7-10 Rule: Use the 7-10 rule to estimate the decay rate of radiation in your area. While radiation decreases rapidly in the days immediately following a nuclear event, some radioactive elements will persist for much longer periods, so patience and continuous monitoring are vital.

Stay Indoors and Use Air Filters: In the initial days after a nuclear explosion, staying inside a fallout shelter or a sealed building will protect you from inhaling radioactive dust. If you need ventilation, use HEPA filters to reduce the presence of fallout particles in the air.

Use Protective Gear Outdoors: If you must venture outside, wear appropriate protective clothing, including masks or respirators, to minimize your exposure to radioactive contaminants in the air and soil.

Practice Decontamination: After coming into contact with contaminated environments, decontaminate by removing clothes and washing skin thoroughly with soap and clean water. Isolate contaminated items and avoid bringing them back into safe areas.

Long-Term Considerations

The presence of long-lived radioactive isotopes like cesium-137 and strontium-90 means that some areas may remain hazardous for years or even decades. Long-term survival strategies involve creating self-sufficient systems in less contaminated zones, using safe water sources, and growing food in controlled or greenhouse environments until the land becomes viable again.

For those living in or near fallout zones, the threat from radioactive contamination is a prolonged one. Constant vigilance, proper planning, and the ability to adapt to the environment as it slowly recovers will be key to long-term survival.

Contaminants in the air and soil after a nuclear event present significant challenges, but with knowledge and the right tools, you can assess the risks and make informed decisions to protect yourself and your family. The combination of radiation detection, testing, and protective measures will allow you to navigate the complex dangers of radioactive fallout, ensuring that you avoid unnecessary exposure while preparing for a future in which the environment slowly becomes safer.

By remaining aware of the hazards, taking proactive steps to limit your exposure, and monitoring the changing levels of contamination, you can survive and ultimately thrive in a world impacted by nuclear fallout.

Testing for Nuclear Contamination: Essential Tools and Techniques

In the wake of a nuclear event, the environment becomes contaminated with radioactive particles that pose serious health risks. The ability to accurately test for nuclear contamination is crucial for survival, as it allows you to assess radiation levels in the air, soil, water, and on surfaces. This chapter focuses on the essential tools and techniques needed to test for nuclear contamination, helping you make informed decisions about safety, resources, and movement in a post-nuclear world.

Understanding Nuclear Contamination

Nuclear contamination occurs when radioactive isotopes, such as iodine-131, cesium-137, and strontium-90, are released into the environment. These isotopes emit ionizing radiation in the form of alpha, beta, and gamma rays, which can cause both immediate and long-term health problems, including radiation sickness and cancer.

To minimize exposure, it's essential to regularly test the environment for radiation levels. This involves detecting both external radiation (from fallout settling on surfaces and in the air) and internal radiation (from contaminated food, water, and inhaled particles).

Essential Tools for Testing Nuclear Contamination

There are several reliable tools you can use to detect and measure radiation. Each tool serves a different function, but together they provide a comprehensive approach to assessing contamination in the environment.

Geiger Counters

A Geiger counter is the most commonly used device for detecting and measuring ionizing radiation. It is a handheld tool that provides real-time readings of radiation levels, making it invaluable for testing both air and surfaces.

How It Works: A Geiger counter detects radiation by using a gas-filled tube that becomes ionized when exposed to radiation. The device measures alpha, beta, and gamma radiation and gives a reading in counts per minute (CPM) or micro-sieverts per hour (μSv/h), indicating the intensity of radiation.

When to Use It: Use a Geiger counter to monitor radiation levels in your immediate environment, especially after venturing outdoors or into potentially contaminated areas. You can also scan surfaces like clothing, food, or shelter walls to ensure they are free from dangerous contamination.

Tips for Use:

Always zero your Geiger counter before taking readings.

Move the device slowly over surfaces to detect any lingering radiation.

Use it regularly to track changes in environmental radiation levels over time.

Dosimeters

A dosimeter measures cumulative radiation exposure over time, rather than providing real-time readings like a Geiger counter. It's an essential tool for tracking long-term exposure to low-level radiation, especially for individuals living in areas where contamination is present but not immediately life-threatening.

How It Works: Dosimeters come in both analog and digital forms and measure the amount of radiation absorbed by the body over a set period. The device typically shows the total exposure in rem (roentgen equivalent man) or sieverts (Sv).

When to Use It: Use a dosimeter if you are living or working in a fallout zone where long-term exposure is a concern. It helps ensure that your cumulative exposure does not exceed safe levels, which could lead to health problems over time.

Tips for Use:

Wear your dosimeter at all times to accurately measure cumulative exposure.

Reset the dosimeter periodically to track specific periods of exposure, such as before and after going outside.

Alpha and Beta Particle Detectors

Alpha and beta particles, while less penetrating than gamma radiation, can be hazardous when inhaled, ingested, or absorbed through the skin. Specialized detectors for alpha and beta radiation are useful for pinpointing contamination on surfaces like food, water, and soil.

How It Works: Alpha and beta particle detectors are similar to Geiger counters but are designed specifically to detect these less-penetrating forms of radiation. They provide readings that can help you determine whether certain areas, objects, or consumables have been contaminated by fallout.

When to Use It: These detectors are especially useful for testing food and water for contamination. Use them to check surfaces that may have come into contact with fallout particles.

Tips for Use:

Make sure to calibrate the detector for the specific type of radiation (alpha or beta) you are testing for.

When testing food or water, ensure that all surfaces are dry to get accurate readings.

Gamma Spectrometers

A gamma spectrometer is a more advanced tool used to identify specific radioactive isotopes by analyzing the gamma radiation they emit. It provides a detailed breakdown of which isotopes are present and their relative concentrations.

How It Works: Gamma spectrometers detect gamma rays and use sophisticated software to determine the specific isotopes present in a sample. The device provides information on the types of contaminants, such as cesium-137 or iodine-131.

When to Use It: Use a gamma spectrometer when you need precise information about the type and quantity of radioactive materials in your environment, food, or water.

Tips for Use:

These devices are more complex and may require training to operate effectively.

Use this tool when you need to identify the specific risks in a contaminated environment, especially for long-term planning.

Soil and Water Testing Kits

Specialized kits are available for testing soil and water for radioactive contamination. These kits often include reagents or testing strips that change color or provide digital readings when radiation is detected.

How It Works: Soil and water testing kits are designed to detect radioactive particles in the environment. Some kits can detect specific isotopes like cesium-137 and strontium-90, which are common in nuclear fallout.

When to Use It: Use soil and water testing kits before using local resources for agriculture or drinking. These kits are essential for ensuring that crops, livestock, and water sources are safe for consumption.

Tips for Use:

Follow the kit's instructions carefully to ensure accurate results.

Regularly test water sources and soil in different areas to monitor changes in contamination levels over time.

Techniques for Testing Contaminated Areas

In addition to using radiation detection tools, there are specific techniques you should follow to ensure accurate and reliable assessments of nuclear contamination.

Regular Monitoring

The levels of radiation in the environment can change over time, particularly in the days and weeks following a nuclear event. Regularly monitor the air, soil, and water to detect any fluctuations in radiation levels. Use your Geiger counter or dosimeter frequently to track changes and identify areas of increased risk.

Surface Testing

When testing surfaces for contamination, move your detection device slowly and systematically over the area. Pay attention to high-risk surfaces, such as clothing, floors, doorways, and windows, where fallout particles may accumulate. In outdoor environments, focus on soil, leaves, and structures that may have collected fallout dust.

Sampling

For soil and water testing, collect samples from multiple locations to get an accurate assessment of contamination levels. In the case of soil, take samples from different depths to determine whether radioactive particles have penetrated the ground. For water testing, sample from both surface water and groundwater sources to assess potential contamination in drinking supplies.

Testing Food

Food, particularly fresh produce, livestock, and foraged items, can become contaminated by radioactive fallout settling on leaves, soil, or water sources. Use your alpha and beta particle detectors or spectrometers to test food before consumption. Focus on items that were exposed to fallout, and consider peeling or washing fruits and vegetables to reduce surface contamination.

Interpreting the Results

Once you've gathered data from your tests, you need to understand what the readings mean for your safety. Here's a basic guide to interpreting radiation readings:

0.05 to 0.2 μSv/h: This range is considered background radiation and is typically safe.

0.2 to 1 μSv/h: Slightly elevated radiation levels. Short-term exposure is generally safe, but prolonged exposure should be minimized.

1 to 10 μSv/h: These levels are considered high and should be avoided for prolonged periods. Limit your exposure by staying indoors or wearing protective gear.

10+ μSv/h: Dangerous levels of radiation. Seek shelter immediately, as prolonged exposure can lead to acute radiation sickness.

For dosimeters, measure your cumulative exposure to ensure that it does not exceed safe levels. The maximum recommended exposure for non-emergency workers is typically 100 millirems (1 mSv) per year. In a nuclear fallout scenario, it's critical to minimize exposure by staying sheltered and limiting time spent in contaminated areas.

Testing for nuclear contamination is one of the most important skills you can have in a post-nuclear environment. By using essential tools like Geiger counters, dosimeters, and soil and water testing kits, you can accurately assess the levels of radiation in your surroundings and make informed decisions about your safety. Regular monitoring, surface testing, and proper interpretation of results will allow you to minimize your exposure to radioactive fallout, protect your resources, and increase your chances of survival in a contaminated world.

Understanding how to test for contamination is not just about survival in the short term—it's about ensuring your long-term health and safety as the environment slowly recovers from the effects of a nuclear event.

Protecting Your Family from Fallout: Cleaning and Decontamination

After a nuclear explosion, fallout—a mixture of radioactive particles released into the atmosphere—presents a significant and immediate danger to human health. Fallout can settle on clothing, skin, and surfaces, contaminating everything it touches. One of the most important survival steps is learning how to protect your family from fallout through proper cleaning and decontamination. This chapter will cover effective techniques for decontaminating people, clothing, and your environment, as well as best practices for limiting exposure to radioactive particles.

Why Fallout is Dangerous

Fallout contains radioactive isotopes, such as cesium-137, iodine-131, and strontium-90, which emit ionizing radiation. When fallout particles come into contact with your skin, clothing, or enter your body through inhalation or ingestion, they can cause both immediate and long-term health effects, such as radiation burns, radiation sickness, and an increased risk of cancer.

Decontamination is the process of removing radioactive particles from your body, clothing, and surroundings to reduce your exposure to radiation. Timely and proper decontamination can greatly reduce the health risks associated with fallout.

Immediate Steps to Protect Your Family from Fallout

Before beginning the decontamination process, take these immediate steps to minimize exposure to fallout:

Shelter in Place: If you know that a nuclear explosion has occurred, the first priority is to get indoors as quickly as possible. The best protection is a basement or an interior room without windows. Seal windows, doors, and any other openings to prevent fallout particles from entering.

Turn Off Ventilation Systems: If you are sheltering in a building, turn off HVAC systems, fans, or any devices that might draw contaminated air into the space. This helps keep fallout particles outside.

Remove Outer Clothing: Once inside, remove outer layers of clothing immediately, as these may have collected fallout particles. Clothing can retain up to 90% of the radioactive material that has settled on it, so removing it quickly can significantly reduce exposure.

Store Contaminated Clothing: Place contaminated clothing in a plastic bag, seal it tightly, and store it in an area away from living spaces. This prevents further spread of radioactive particles.

Cleaning and Decontaminating People

The first priority in decontamination is to remove radioactive particles from your skin and hair. Follow these steps to properly clean yourself and your family:

Remove Outer Clothing and Accessories

As mentioned earlier, removing contaminated clothing is one of the most effective ways to reduce radiation exposure. Have family members remove their shoes, jackets, hats, and other outer garments as soon as they are sheltered indoors.

Tips: Handle contaminated clothing carefully to avoid spreading fallout particles. Do not shake clothing, as this can release particles into the air.

Shower with Soap and Water

Once outer clothing is removed, wash thoroughly with soap and lukewarm water. Avoid using hot water, as it can open pores and increase the absorption of radioactive particles into the skin. If a shower is not available, use wet wipes or a damp cloth to clean the skin.

Steps:

Begin by rinsing the entire body with water.

Lather up with soap and gently scrub the skin, focusing on areas that were exposed (hands, face, neck, and hair).

Rinse thoroughly to remove any remaining soap and particles.

Avoid using harsh scrubbing motions, as this can irritate the skin and make it more vulnerable to radiation.

Hair Cleaning: Wash your hair with regular shampoo, avoiding conditioner. Conditioner can bind radioactive particles to your hair, making them harder to remove.

Avoid Harsh Products

Do not use creams, oils, or lotions on your skin after decontamination. These substances can trap radioactive particles on your skin and make it more difficult to remove them during the cleaning process.

Nail and Eye Decontamination

Particles may settle under your nails or in your eyes. To clean under your nails, scrub gently with a nail brush and soap. If you feel irritation in your eyes, rinse them with clean water or a saline solution.

Tips: Use an eyewash or sterile saline solution to rinse out any fallout particles that may have gotten into your eyes. Avoid rubbing your eyes.

Clean Children and Pets

Children and pets are also vulnerable to fallout. Gently wash their skin with soap and water, being careful not to use hot water. For pets, ensure they are washed thoroughly, particularly around the paws and face where particles may have collected. Avoid letting pets lick themselves until they have been fully decontaminated.

Decontaminating Clothing and Gear

Fallout particles can cling to clothing and personal items, so it's important to either remove these items or clean them thoroughly. Follow these steps to decontaminate clothing and gear:

Separate Contaminated Clothing

Any clothing that was worn outside during the fallout should be considered contaminated. Immediately place these items in plastic bags and seal them to prevent the spread of radioactive particles. Store the bags in a location away from living areas, such as a garage or outdoor shed.

Washing Contaminated Clothing

If the contamination levels are low and the fallout is not severe, it may be possible to wash contaminated clothing. Use the following method to clean clothing:

Wash the clothing in cold water to avoid further setting the particles into the fabric.

Use a strong detergent to help remove the particles from the fibers.

Wash clothing separately from uncontaminated garments.

After washing, rinse the machine with clean water before using it again.

Note: In cases of heavy contamination, it's often best to discard the clothing rather than attempting to clean it, as radioactive particles can become embedded in fabric.

Decontaminating Gear and Personal Items

Backpacks, tools, and other personal gear that were exposed to fallout can also be decontaminated using soap and water. For hard surfaces, wipe them down with a damp cloth or disinfectant wipes to remove any fallout particles.

Steps:

Wipe down the entire surface of each item with soap and water.

Rinse thoroughly to remove any residue.

For electronics or delicate items, use a clean, damp cloth to gently wipe away particles without damaging the device.

Cleaning and Decontaminating Your Environment

In addition to decontaminating people and clothing, it's essential to clean the space you're living in to remove any fallout particles that may have entered your home or shelter. Here's how to effectively decontaminate your living environment:

Seal the Shelter

If you have not already done so, seal windows, doors, and other openings with plastic sheeting and duct tape. This will help prevent additional fallout from entering your home. Pay attention to vents, cracks, and any other potential entry points for contaminated air.

Clean Floors and Surfaces

If fallout particles have made their way into your home, it's important to clean all surfaces, particularly floors where particles can settle. Use damp cloths or mops to wipe down surfaces. Dry sweeping or dusting can stir up radioactive particles, making them airborne and more likely to be inhaled.

Tips:

Use disposable cloths or paper towels for cleaning, and discard them after use.

Mop floors with soap and water, then rinse the mop thoroughly between uses.

Vacuuming

If you need to vacuum, use a vacuum cleaner with a HEPA filter to trap radioactive particles. Standard vacuum cleaners without HEPA filters can release particles back into the air, increasing the risk of inhalation.

Regular Cleaning and Monitoring

Continue to clean regularly to reduce the buildup of radioactive particles. Monitor radiation levels in your home using a Geiger counter or radiation detector to ensure that your environment remains safe.

Waste Disposal

Contaminated clothing, cleaning supplies, and other items need to be disposed of properly to prevent further spread of radiation. Follow these steps to safely dispose of contaminated materials:

Bag and Seal Contaminated Items: Place contaminated items in sealed plastic bags. Double-bagging may be necessary for heavily contaminated materials.

Store Safely: Store these bags in a designated area away from living spaces. Do not burn or bury contaminated items, as this can release radioactive particles into the air or groundwater.

Follow Local Guidelines: In a widespread nuclear event, authorities may issue specific guidelines for the disposal of contaminated waste. Follow these instructions to minimize environmental damage and ensure the safety of others.

Protecting Against Inhalation of Fallout

In addition to cleaning and decontaminating your body, clothing, and environment, protecting your family from inhaling fallout particles is crucial. Here's how to reduce the risk of inhalation:

Wear Masks: If you need to go outside or into contaminated areas, wear an N95 mask or respirator to prevent the inhalation of radioactive dust.

Use Air Filters: Install air filtration systems in your home or shelter that can remove fallout particles from the air. HEPA filters are particularly effective for this purpose.

Limit Outdoor Exposure: Avoid going outside unless absolutely necessary, especially in the first 24 to 48 hours after a nuclear event when fallout levels are highest.

Protecting your family from fallout through proper cleaning and decontamination is one of the most critical steps to surviving a nuclear event. By understanding the dangers of fallout, immediately sheltering in place, and following effective decontamination procedures, you can minimize radiation exposure and reduce the risk of long-term health effects.

Proper decontamination involves cleaning your body, clothing, and living environment to remove radioactive particles. Regular monitoring of radiation levels, along with regular cleaning and decontamination efforts, will help ensure that your living space remains as safe as possible in the days and weeks following a nuclear event. Here's a summary of the most important steps to take when protecting your family from fallout:

Summary of Key Decontamination Practices

Shelter Immediately: As soon as you are aware of a nuclear explosion, get indoors and seal your environment to prevent fallout from entering. Turn off ventilation systems and close all openings.

Remove and Store Contaminated Clothing: Remove outer layers of clothing and store them in sealed plastic bags to minimize further contamination.

Wash with Soap and Water: Thoroughly clean your body with soap and lukewarm water, taking care not to use harsh scrubbing or hot water, which can increase skin absorption of radioactive particles.

Avoid Lotions or Conditioners: Do not use creams, conditioners, or other products that can trap radioactive particles on your skin or hair.

Clean Your Living Space: Wipe down surfaces with damp cloths, mop floors with soap and water, and vacuum with a HEPA filter. Avoid dry sweeping or dusting to prevent fallout from becoming airborne.

Decontaminate Gear and Objects: Clean personal items, gear, and surfaces exposed to fallout using soap and water. For hard-to-clean items, consider storing them away from living areas until they can be safely handled.

Dispose of Contaminated Waste Properly: Place contaminated clothing, cleaning materials, and other waste in sealed plastic bags, and follow local guidelines for safe disposal when available.

Use Masks and Air Filters: Wear an N95 mask or respirator when outside to prevent inhaling fallout particles. Use HEPA filters in your home to clean the air and prevent contamination.

Ongoing Protection

Even after the initial decontamination, it's important to continue taking protective measures:

Regularly Clean and Monitor: Continue cleaning your living environment and monitor radiation levels using a Geiger counter or dosimeter. Clean regularly to ensure that fallout particles are removed from surfaces and to prevent buildup.

Limit Exposure: Avoid unnecessary trips outside until you are certain that radiation levels have decreased significantly. If you must go outside, wear protective clothing and minimize your exposure.

Plan for Long-Term Contamination: Radioactive particles like cesium-137 and strontium-90 can persist in the environment for years. Decontamination will be an ongoing process, so understanding and adapting to this reality is essential for long-term survival.

By following these decontamination practices, you can protect your family from the harmful effects of fallout and reduce the risk of radiation exposure. The process of decontaminating people, clothing, and living spaces requires consistent effort and attention to detail, but it's a critical part of surviving in the aftermath of a nuclear event.

The key to success is acting quickly, staying informed, and remaining vigilant. With proper decontamination, regular monitoring, and effective protective measures, you can minimize the dangers of fallout and create a safer environment for yourself and your family during the critical post-nuclear period.

Fallout Poisoning: Symptoms, Treatment, and Prevention

Fallout poisoning, also known as radiation sickness or acute radiation syndrome (ARS), occurs when a person is exposed to high levels of ionizing radiation, such as that found in nuclear fallout. The radiation from fallout can damage cells and tissues, leading to severe health consequences. Understanding the symptoms, treatment options, and prevention strategies for fallout poisoning is crucial for protecting yourself and your family in the aftermath of a nuclear event.

This chapter will explore the symptoms of fallout poisoning, the available treatments, and how to prevent exposure to dangerous levels of radiation.

Understanding Fallout Poisoning

Fallout poisoning happens when radioactive particles from nuclear fallout enter the body, either through inhalation, ingestion, or direct exposure to the skin. These particles emit radiation, which damages the cells in your body by breaking chemical bonds in DNA and proteins. The severity of radiation sickness depends on the dose of radiation, the duration of exposure, and the individual's proximity to the radiation source.

Exposure to large amounts of radiation over a short period can lead to acute radiation sickness, while long-term, low-level exposure can increase the risk of cancer and other chronic health conditions.

Symptoms of Fallout Poisoning

The symptoms of radiation poisoning can vary depending on the level of exposure. They are generally divided into four stages: **prodromal stage**, **latent stage**, **manifest illness stage**, and **recovery or death**. The severity and progression of symptoms will depend on how much radiation the body has absorbed.

Prodromal Stage (Initial Symptoms)

This stage occurs within hours to days after exposure to a high dose of radiation. The symptoms are often similar to those of other illnesses, making it difficult to recognize radiation poisoning in the early stages.

Nausea and vomiting: Often the first symptoms to appear, usually within hours of exposure.

Diarrhea: May occur alongside nausea and vomiting, leading to dehydration.

Headache: Accompanied by dizziness or confusion.

Fever: A result of the body's initial response to radiation exposure.

The severity of these symptoms can provide a clue to the amount of radiation exposure. For example, if nausea and vomiting occur within an hour of exposure, it indicates a high dose of radiation.

Latent Stage (Temporary Symptom Relief)

After the initial symptoms, there may be a period where the affected person feels better, sometimes lasting hours to weeks. During this stage, the body's cells continue to be damaged, and internal organs may begin to fail, even though the outward symptoms seem to improve.

Manifest Illness Stage

This stage marks the return of more severe symptoms, as radiation damages the bone marrow, digestive system, skin, and nervous system. Depending on the radiation dose, this stage can be life-threatening.

Severe nausea and vomiting: Returns after the latent stage.

Loss of appetite: Accompanied by weight loss and weakness.

Bleeding and bruising: Caused by damage to the bone marrow, which reduces the body's ability to produce blood cells.

Hair loss: A result of radiation damaging the hair follicles.

Infections: The immune system is weakened due to damage to the bone marrow, making the body more vulnerable to infections.

Confusion and disorientation: Damage to the brain and nervous system can lead to neurological symptoms.

Shock and organ failure: In extreme cases, multiple organ systems may begin to fail, leading to shock.

Recovery or Death

The final stage of fallout poisoning depends on the level of exposure and the person's overall health. For those who survive, recovery may take weeks to years, with lingering health problems such as an increased risk of cancer or chronic illness. In cases of extreme radiation exposure, death may occur within days to weeks, often due to organ failure or infections.

Treatment for Fallout Poisoning

The treatment for radiation poisoning focuses on managing symptoms, preventing infection, and minimizing the damage caused by radiation exposure. While there is no cure for radiation poisoning, prompt medical attention can improve the chances of survival.

Decontamination

The first step in treatment is decontaminating the affected person to prevent further exposure to radioactive particles. This includes removing contaminated clothing, washing the skin with soap and water, and rinsing the eyes and mouth if they have been exposed.

Remove and bag contaminated clothing: Clothing can carry fallout particles that continue to emit radiation. Place contaminated clothing in sealed bags and keep it away from living areas.

Shower or wash thoroughly: Use lukewarm water and soap to wash the body, paying special attention to exposed skin, hair, and nails.

Supportive Care

Supportive care involves treating the symptoms of radiation sickness and helping the body recover from the damage caused by radiation.

Fluids and electrolytes: Dehydration is a common problem in radiation poisoning, especially if vomiting or diarrhoea is severe. Intravenous (IV) fluids may be needed to restore hydration and electrolyte balance.

Blood transfusions: If radiation has damaged the bone marrow, the body may not be able to produce enough red blood cells, white blood cells, or platelets. Blood transfusions can help restore these vital components of the blood.

Antibiotics: Because radiation weakens the immune system, the risk of infections is high. Antibiotics can be given to treat or prevent infections.

Medications

There are several medications that can help reduce the effects of radiation poisoning or limit the body's absorption of radioactive materials.

Potassium Iodide (KI): One of the most well-known treatments for radiation exposure, potassium iodide helps protect the thyroid gland from absorbing radioactive iodine-131. It is most effective when taken within hours of exposure and can reduce the risk of thyroid cancer.

Prussian Blue: This medication binds to cesium-137 and thallium in the intestines, helping to speed up their removal from the body. It reduces the amount of time these radioactive materials stay in the body, limiting radiation exposure.

Filgrastim (Neupogen): This drug stimulates the production of white blood cells in the bone marrow, helping to reduce the risk of infection in people with radiation poisoning.

Calcium and Zinc DTPA: These agents bind to radioactive particles like plutonium, americium, and curium, helping the body eliminate them more quickly.

Long-Term Care

For survivors of fallout poisoning, long-term medical care may be necessary to monitor and manage the delayed effects of radiation exposure. These effects include an increased risk of cancer (particularly thyroid cancer, leukemia, and bone cancer) and chronic conditions such as cardiovascular disease.

Preventing Fallout Poisoning

The best way to prevent fallout poisoning is to avoid exposure to radioactive fallout in the first place. Here are some key strategies for minimizing exposure:

Sheltering in Place

After a nuclear explosion, sheltering in place is one of the most effective ways to avoid exposure to fallout. Seek shelter indoors, ideally in a basement or an interior room without windows, and seal off any openings to prevent radioactive dust from entering.

Stay indoors for at least 24–48 hours, as this is when fallout levels are highest. Waiting out this critical period reduces the risk of radiation poisoning.

Avoid Contaminated Areas

If you must go outside, wear protective clothing, including long sleeves, pants, gloves, and a mask or respirator to prevent inhaling fallout particles. Limit your time outdoors and avoid touching surfaces that may be contaminated.

Use Potassium Iodide (KI)

If there is a risk of exposure to radioactive iodine-131, taking potassium iodide (KI) pills can protect your thyroid gland. However, KI only protects against radioactive iodine and does not prevent radiation sickness from other radioactive isotopes like cesium-137 or strontium-90.

Monitor Radiation Levels

Use a Geiger counter or dosimeter to monitor radiation levels in your environment. Knowing when radiation levels have dropped can help you make informed decisions about when it is safe to venture outside or when to seek medical help.

Clean and Decontaminate

After exposure to fallout, immediately remove contaminated clothing and wash your body with soap and water. Clean surfaces and objects in your home that may have come into contact with fallout particles.

Fallout poisoning is a serious condition that can result from exposure to radioactive particles in nuclear fallout. Recognizing the symptoms of radiation sickness and seeking prompt treatment can make a significant difference in survival. The key to reducing the risk of fallout poisoning is to minimize exposure to radioactive particles through proper sheltering, decontamination, and the use of protective measures like potassium iodide. By staying informed and taking preventive steps, you can protect yourself and your family from the dangers of radiation in the aftermath of a nuclear event.

Detoxifying Your Environment: Steps for Safe Living

In the aftermath of a nuclear event, ensuring that your environment is safe from radioactive contamination is critical for long-term survival. Fallout from a nuclear explosion can contaminate air, soil, water, and surfaces, making it unsafe for daily living. Detoxifying your environment involves removing or reducing radioactive particles and taking steps to ensure that your home and surrounding area are safe for habitation.

This chapter will outline the steps you need to take to detoxify your environment, including cleaning strategies, monitoring for contamination, and long-term solutions for safe living in a post-nuclear world.

Understanding Fallout and Environmental Contamination

Radioactive fallout consists of fine particles that settle back to Earth after a nuclear explosion. These particles can contaminate surfaces, soil, water, and even the air you breathe. The radioactive isotopes in fallout, such as iodine-131, cesium-137, and strontium-90, emit radiation that can pose significant health risks if not properly addressed.

Detoxifying your environment is a multi-step process aimed at reducing or eliminating these harmful particles from your living area, minimizing exposure to radiation, and ensuring a healthier environment for long-term survival.

Immediate Steps for Decontaminating Your Home

After a nuclear event, the first priority is to ensure that your home or shelter is as free from radioactive contamination as possible. Begin by addressing the immediate risks posed by fallout particles that may have entered your living space.

Seal Your Shelter

If you are sheltering in place during the fallout period, start by sealing off your home to prevent further contamination. Use plastic sheeting, duct tape, and other materials to block windows, doors, and vents where fallout particles might enter.

Tip: Cover air vents and any gaps around doors or windows with plastic to prevent fallout dust from entering your living area. Keep this sealed until you are sure it is safe to allow airflow again.

Clean Surfaces with Damp Cloths

Radioactive particles can settle on surfaces, such as floors, countertops, furniture, and walls. Dry sweeping or dusting will stir up these particles, making them airborne and more likely to be inhaled. Instead, use damp cloths, sponges, or mops to clean surfaces.

Step-by-step:

Wet a clean cloth or sponge with soapy water.

Wipe down all exposed surfaces, paying special attention to high-traffic areas where people may have brought in contaminated dust.

For floors, use a mop with soapy water to clean thoroughly.

Rinse the mop or cloth frequently in clean water and replace the water when it becomes visibly dirty.

Tip: Use disposable cloths or towels if possible, and discard them safely after use.

Vacuum with a HEPA Filter

If you need to vacuum carpets or rugs, use a vacuum cleaner equipped with a high-efficiency particulate air (HEPA) filter. Standard vacuums can release radioactive particles back into the air, but a HEPA filter can trap small particles and prevent them from recirculating.

Tip: Avoid using standard vacuum cleaners, as they can spread fallout particles into the air. Stick to HEPA-filtered vacuums for more thorough decontamination.

Ventilation and Air Filtration

Once you are confident that the initial fallout period has passed (typically 24 to 48 hours), and radiation levels have dropped, consider introducing ventilation to your living space. However, ensure that air entering your home is filtered to prevent bringing in contaminated particles.

Use HEPA Air Filters: Install HEPA air purifiers in your home to filter out radioactive particles that may still be in the air. These filters are highly effective at removing fine dust and airborne contaminants.

Tip: Check the filters regularly and replace them as needed to maintain their effectiveness.

Cleaning and Detoxifying Outdoor Areas

The outdoor environment is likely to be contaminated with fallout particles that have settled on the ground, in the soil, and on plants. Decontaminating the outside areas around your home is essential for safe outdoor activities and food production.

Remove Fallout from Outdoor Surfaces

Radioactive particles can settle on outdoor surfaces like walkways, patios, and vehicles. Cleaning these surfaces is crucial to reduce the risk of tracking fallout into your home.

Hose Down Surfaces: Use a garden hose to wash away fallout particles from paved surfaces like driveways, sidewalks, and patios. Ensure that the water is directed away from your home to prevent it from contaminating other areas.

Tip: Avoid dry sweeping or using blowers, as these can lift radioactive dust back into the air.

Cover or Replace Contaminated Soil

Soil contamination poses a long-term threat, especially if you plan to grow food. Cesium-137 and strontium-90 are two of the most persistent contaminants in soil, as they can remain radioactive for decades. To mitigate the effects of contaminated soil:

Cover the Soil: If decontaminating large areas of soil is impractical, consider covering the ground with a tarp, plastic sheeting, or a layer of uncontaminated soil to reduce the radiation levels at the surface.

Remove Topsoil: In small areas, such as gardens, removing the top few inches of soil where fallout has settled can reduce contamination. Dispose of the contaminated soil safely, according to local guidelines for hazardous waste disposal.

Plant Grass or Ground Cover: In some cases, planting grass or other ground cover plants can help stabilize the soil and reduce the spread of radioactive dust. However, avoid consuming any plants grown in contaminated soil until it is safe.

Testing Soil for Contamination

Before growing food in an area that may have been contaminated, it's important to test the soil for radioactive isotopes. Soil testing kits or professional services can detect levels of cesium-137, strontium-90, and other radioactive particles.

Tip: If the soil tests show high levels of contamination, consider using raised garden beds with uncontaminated soil brought in from safe sources. This allows you to grow food without the risk of exposure to radiation.

Water Decontamination and Filtration

Water supplies can also become contaminated by fallout, particularly if fallout particles have settled into lakes, rivers, or groundwater. Drinking contaminated water can lead to internal radiation exposure, so it's essential to ensure that your water is safe.

Avoid Open Water Sources

In the immediate aftermath of a nuclear event, avoid drinking from open water sources like rivers, lakes, ponds, and streams, as fallout particles may have settled in these bodies of water. Instead, rely on stored or bottled water that was sealed before the event.

Filter Water for Safe Consumption

If bottled water or stored water is not available, use water filtration systems designed to remove radioactive particles. Standard water filters will not remove radioactive contamination, so it's important to use advanced filtration methods.

Use Reverse Osmosis Filters: Reverse osmosis systems are highly effective at removing a wide range of contaminants, including radioactive particles, from water. These systems force water through a semipermeable membrane that filters out harmful substances.

Boiling Water: Boiling water will not remove radioactive contaminants, but it can kill biological pathogens. Always use a filtration system in addition to boiling if the water source is suspected to be contaminated.

Water Testing Kits: Use water testing kits to check for the presence of radioactive particles in your water supply before using it for drinking, cooking, or bathing.

Long-Term Solutions for Safe Living

While immediate decontamination is crucial after a nuclear event, long-term strategies for reducing environmental radiation are essential for safe living in the months and years that follow. Here are some long-term measures you can take to detoxify your environment.

Geiger Counters and Radiation Detectors

Regularly monitor radiation levels in your home and surrounding areas using a Geiger counter or dosimeter. This will help you track changes in radiation levels over time and ensure that your environment remains safe.

Tip: Check radiation levels before engaging in outdoor activities like gardening, building, or gathering natural resources.

Raised Gardens and Greenhouses

If soil contamination is a concern, consider building raised garden beds or using greenhouses to grow food in uncontaminated soil. Raised beds allow you to control the soil quality, and greenhouses provide additional protection against radioactive fallout.

Tip: Import uncontaminated soil for raised beds and regularly test it to ensure it remains safe for food production.

Air Quality Monitoring

Even long after the initial fallout period, radioactive dust may still linger in the environment. Using air quality monitors and purifiers in your home can help detect and remove any remaining airborne contaminants.

Tip: Continue using HEPA air purifiers in enclosed spaces to maintain a safe indoor environment.

Personal Protective Equipment (PPE)

If you need to work in contaminated areas or handle materials that may contain fallout particles, wear protective clothing, including gloves, masks, and goggles. This will help reduce your exposure to lingering radioactive dust.

Detoxifying your environment after a nuclear event is an ongoing process that requires vigilance and careful attention to cleaning, testing, and maintaining safety protocols. By taking immediate steps to decontaminate your home, monitor radiation levels, and clean outdoor areas, you can reduce the risks posed by radioactive fallout.

In the long term, using tools like Geiger counters, water filtration systems, and soil testing kits will help you maintain a safe living environment. Growing food in controlled conditions, using clean water sources, and continually monitoring your surroundings will allow you and your family to live safely in a post-nuclear world.

When is it Safe? Assessing the Outside World After Fallout

After a nuclear event, one of the most critical questions for survivors is, "When is it safe to go outside?" Assessing the safety of the outside world after fallout involves understanding how radioactive particles behave, monitoring radiation levels, and knowing what signs to look for when deciding whether it's safe to venture outdoors. Radiation from fallout can pose significant risks, especially in the immediate aftermath of a nuclear explosion. However, as time passes, radiation levels decrease, and with the right tools and knowledge, you can assess when it's safe to leave the shelter and begin moving about your environment.

This chapter will guide you through the process of evaluating the safety of the outside world, how to monitor radiation, and what precautions to take before resuming outdoor activities.

Understanding Fallout and Radiation Decay

When a nuclear explosion occurs, radioactive fallout is carried into the atmosphere and eventually settles back to Earth, contaminating everything it touches. Fallout consists of particles from the bomb itself, as well as debris from vaporized buildings, soil, and other materials that become irradiated during the explosion. These radioactive particles emit ionizing radiation, which can be extremely dangerous in high doses.

The intensity of radiation decreases over time due to **radioactive decay**. One of the most important principles in radiation safety is the **"7-10 rule"**, which states that for every sevenfold increase in time, radiation levels decrease by a factor of 10:

After **7 hours**, radiation levels will be 10% of what they were immediately after the explosion.

After **49 hours** (7×7), radiation levels will have decreased to 1%.

After **2 weeks**, levels will be approximately 0.1% of the original level.

While radiation levels drop significantly after the first few days, it's important to remember that dangerous levels of radiation can persist for weeks or even months, particularly near the blast site or in areas with high fallout concentrations.

Key Factors in Assessing When It's Safe to Go Outside

Several factors influence how soon it will be safe to go outside after a nuclear event. Understanding these factors will help you make an informed decision about when to leave the shelter and begin assessing the world around you.

Proximity to the Blast Site

The closer you are to the site of the explosion, the higher the initial levels of radiation will be. Areas within a few miles of the blast are likely to experience intense radiation and heavy fallout, making it unsafe to go outside for an extended period. If you are located further from the blast, radiation levels may decrease more quickly, allowing for earlier movement outdoors.

Close to the blast: In areas near the detonation, you may need to stay indoors for several days or weeks before it's safe to go outside.

Farther from the blast: Those farther away from the explosion may be able to leave the shelter after 24–48 hours, depending on radiation levels.

Type of Nuclear Explosion

The type of nuclear explosion—**airburst** or **groundburst**—also affects how long fallout will remain dangerous. In an airburst, the explosion occurs above the ground, reducing the amount of local fallout but spreading radiation over a wider area. A groundburst, on the other hand, produces more local fallout because the explosion lifts large amounts of radioactive debris into the air, making the immediate area around the blast more dangerous for a longer period.

Weather and Wind Conditions

Weather patterns, including wind speed and direction, can spread fallout over a wide area, creating "hot spots" of concentrated radiation. Rain or snow can cause radioactive particles to fall more quickly, leading to localized areas of higher contamination. If you are in an area downwind from the blast or where precipitation occurred shortly after the explosion, radiation levels may be higher, and it may take longer for the area to become safe.

Type of Shelter

The effectiveness of your shelter in protecting you from radiation will influence how long you need to remain inside. A well-built fallout shelter with thick walls made of concrete, lead, or earth provides excellent protection from radiation and allows you to wait out the most dangerous period safely. However, if your shelter is less secure (such as a home without proper shielding), you may need to stay inside longer or move to a better-protected location.

Monitoring Radiation Levels

The most reliable way to determine when it's safe to go outside is by monitoring radiation levels with detection devices. These tools allow you to measure the intensity of radiation in your environment and provide real-time data on whether it is safe to leave the shelter.

Geiger Counter

A Geiger counter is the most commonly used device for measuring radiation levels. It detects ionizing radiation (alpha, beta, and gamma rays) and gives a reading in counts per minute (CPM) or microsieverts per hour (μSv/h). These readings help you assess the immediate level of radiation in your surroundings.

Safe levels: Radiation levels below 0.5–1 μSv/h are generally considered safe for short-term outdoor activity. If levels are above this threshold, it's best to remain indoors or limit exposure.

Dosimeter

A dosimeter measures cumulative radiation exposure over time. It is especially useful if you need to venture outside briefly or are exposed to low-level radiation for extended periods. Wearing a dosimeter allows you to track your total radiation exposure and ensure it does not exceed safe limits.

Safe exposure limits: The general safe limit for radiation exposure is 100 millirems (1 mSv) per year. In an emergency situation, short-term exposure to higher levels may be necessary, but you should monitor your total exposure closely to avoid long-term health risks.

Radiation Maps

In some cases, authorities may provide radiation maps or reports showing the spread of fallout and areas where radiation levels are particularly high. These maps can help you assess whether your location is in a danger zone or if it's safe to venture outside. If you have access to emergency broadcasts or updates, use this information to guide your decisions.

Signs to Look For in the Environment

In addition to using radiation detection tools, you can look for physical signs in the environment that indicate whether it's safe to go outside.

Visible Fallout

After a nuclear explosion, fallout particles may be visible as fine dust or ash on surfaces like cars, rooftops, plants, and the ground. If you can see this dust, it's a sign that fallout has settled in your area, and you should take extra precautions before going outside. If possible, avoid contact with any surface that may have fallout particles on it.

Tip: If you must go outside, wear protective clothing and a mask to prevent inhalation or direct contact with fallout dust.

Plant and Animal Life

In heavily contaminated areas, plant and animal life may show signs of distress. Stunted or discolored plants, animals acting lethargic or sick, or dead wildlife can all indicate that the area is heavily contaminated with radiation. If you notice these signs, radiation levels may still be too high for safe outdoor activity.

Tip: Avoid consuming any plants, fruits, or animals from the environment until you are sure the area is safe and free of contamination.

"Hot Spots"

Fallout does not settle evenly across an area, and localized "hot spots" of high radiation can form, particularly in areas where fallout has been concentrated by rain or wind. These hot spots can be difficult to detect without a Geiger counter, so it's essential to scan the area before spending extended time outdoors.

Precautions before Going Outside

Even when radiation levels have decreased, it's important to take precautions before venturing outdoors. Here's how to minimize your exposure and ensure your safety:

Wear Protective Clothing

If you need to go outside, wear protective clothing that covers your skin as much as possible. Long sleeves, pants, gloves, and boots can help shield your body from fallout particles. Wear a mask or respirator to avoid inhaling contaminated dust.

Limit Time Outdoors

If radiation levels are still elevated but it's necessary to go outside (for example, to gather supplies or check on neighbors), limit your time outdoors. The longer you are exposed, the higher your risk of radiation sickness, so keep your trips short and return to shelter as quickly as possible.

Avoid Contaminated Areas

Avoid areas where fallout is visible or where radiation levels are higher, such as near buildings that may have concentrated fallout on their roofs or in low-lying areas where water may have carried radioactive particles.

Decontaminate Upon Return

After returning indoors, immediately remove your outer clothing and place it in a sealed plastic bag. Wash your body thoroughly with soap and water to remove any fallout particles that may have settled on your skin or hair.

Long-Term Considerations

While radiation levels will decrease over time, certain radioactive isotopes, such as cesium-137 and strontium-90, can remain dangerous for years. These isotopes can contaminate soil, water, and food supplies, posing long-term health risks. If you are living in a fallout zone for an extended period, it's important to continue monitoring radiation levels and take steps to minimize your exposure.

Tip: Regularly check radiation levels in your environment using a Geiger counter or dosimeter, and avoid consuming food or water from potentially contaminated sources.

Determining when it's safe to go outside after a nuclear event requires careful assessment of radiation levels, environmental factors, and the risks posed by fallout. By monitoring radiation with tools like Geiger counters and dosimeters, staying informed about weather and fallout patterns, and taking appropriate precautions before going outdoors, you can minimize your exposure to dangerous radiation.

Remember that even after the initial fallout period has passed, it is important to remain vigilant and cautious, as radioactive contamination can persist in the environment for weeks, months, or even years, depending on the severity of the fallout. By continuing to monitor radiation levels and adopting long-term safety measures, you can protect yourself and your family from the lingering dangers of fallout.

Summary of Key Steps to Assess Outside Safety

Monitor Radiation Levels: Use a Geiger counter or dosimeter to regularly check the radiation levels outside your shelter. Wait until levels drop below 0.5–1 µSv/h before considering extended time outdoors.

Stay Informed: Keep track of radiation reports from authorities or emergency broadcasts, and look for radiation maps that indicate safe and dangerous zones.

Observe Environmental Signs: Look for visible fallout, signs of plant and animal distress, and avoid areas where fallout dust or particles are present.

Wear Protective Gear: Always wear protective clothing, masks, and gloves when venturing outside, especially in the days and weeks following a nuclear event.

Limit Outdoor Exposure: Minimize the time spent outside and avoid unnecessary contact with contaminated surfaces. Be sure to decontaminate yourself and your clothing upon returning indoors.

Avoid Consuming Local Resources: Do not eat plants, fruits, or animals from the environment until you are certain they are free of contamination. Use water filtration systems for safe drinking water.

Long-Term Safety

As time passes and radiation levels drop, your ability to safely move about the environment will increase. However, some areas may remain hazardous for long periods due to persistent radioactive isotopes like cesium-137 and strontium-90. Long-term radiation monitoring and careful planning are essential for rebuilding your life after a nuclear event.

Consider Relocation: If you live near the site of the explosion or in an area with heavy fallout, consider relocating to a safer zone if possible. This can significantly reduce long-term health risks from radiation exposure.

Use Clean Soil and Water: If you plan to grow food or collect water, ensure that you are using uncontaminated soil and water sources. Testing these resources for radiation is critical before consumption.

Continue Monitoring: Even after the immediate crisis has passed, regularly use radiation detectors to assess the safety of your surroundings. Radiation levels can fluctuate, and some areas may remain dangerous for extended periods.

Surviving a nuclear event requires not only immediate shelter and protection but also ongoing vigilance in assessing when it is safe to re-enter the outside world. By carefully monitoring radiation levels, observing environmental cues, and taking appropriate protective measures, you can minimize your exposure to fallout and make informed decisions about your safety.

With proper tools, knowledge, and planning, you can assess when it's safe to go outside and begin the process of rebuilding your life in a post-nuclear environment. The key to long-term survival is patience, caution, and continual monitoring of the risks posed by radioactive fallout, ensuring that you and your family remain safe as you navigate the aftermath of such an event.

Building the Ideal Fallout Shelter for Long-Term Survival

In the event of a nuclear catastrophe, having a well-designed fallout shelter is essential for long-term survival. A fallout shelter provides protection from the dangerous radioactive particles released into the environment after a nuclear explosion. While immediate sheltering can help you survive the initial fallout, a shelter designed for long-term survival is critical if you plan to remain in a contaminated area for weeks, months, or even longer. This chapter will guide you through the key considerations and steps to build an ideal fallout shelter that can sustain you and your family for an extended period.

The Importance of a Fallout Shelter

A fallout shelter shields you from the most dangerous effects of nuclear fallout—radioactive particles that can settle in the air, soil, and water after a nuclear explosion. These particles emit ionizing radiation, which can cause severe health issues, including radiation sickness, cancer, and even death.

The primary goal of a fallout shelter is to protect against **gamma radiation**, which can penetrate walls, and to reduce exposure to **beta and alpha particles**, which are less penetrating but still harmful when inhaled or ingested. A well-built shelter also provides a controlled environment where you can store supplies, access clean water, and maintain safe air quality while the radiation levels outside decrease over time.

Key Features of the Ideal Fallout Shelter

To ensure long-term survival, your fallout shelter should have certain key features that provide protection, comfort, and sustainability. These features include:

Location and Placement

Structural Design

Radiation Shielding

Ventilation and Air Filtration

Water Supply and Filtration

Food Storage and Preparation

Power Generation

Medical Supplies and First Aid

Waste Management and Sanitation

Communication and Information

Location and Placement

The location of your fallout shelter plays a significant role in its effectiveness. Ideally, your shelter should be placed in an area that is easy to access during an emergency but far enough from potential blast zones or major urban areas that could be targeted in a nuclear attack.

Underground Shelters: The most effective fallout shelters are built underground, as the Earth provides natural radiation shielding. An underground shelter offers better protection from gamma radiation and can help regulate temperature. Basements, converted bunkers, or purpose-built underground shelters are all good options.

Above-Ground Shelters: If building underground is not possible, above-ground shelters can still provide protection, but they require thicker walls and additional shielding to protect against radiation. These shelters should be built in a sturdy location, such as within a reinforced concrete or steel structure.

Structural Design

A fallout shelter must be strong enough to withstand the pressure of the blast wave (if close to the explosion) and the weight of earth or materials used for shielding. The structure must also be resistant to fire, debris, and other hazards associated with nuclear fallout.

Materials: Concrete, brick, and steel are ideal materials for building a fallout shelter due to their strength and ability to absorb radiation. Use reinforced concrete or thick steel walls to provide the necessary structural integrity and radiation shielding.

Thickness of Walls: The walls of your shelter should be thick enough to absorb gamma radiation. A general guideline is that 3 feet (1 meter) of earth, 2 feet (60 cm) of concrete, or 6 inches (15 cm) of lead provides adequate protection from gamma rays.

Radiation Shielding

The most important feature of a fallout shelter is its ability to shield you from radiation. The level of radiation inside the shelter should be reduced to a fraction of the levels outside.

Materials for Shielding: To block gamma radiation effectively, you need dense materials like concrete, lead, or earth. The thicker the material, the better the protection. In general:

6 inches of lead or steel can reduce gamma radiation by 50%.

2 feet of concrete or brick provides substantial protection.

3 to 4 feet of packed earth offers significant shielding and can be used if you are building a partially underground shelter.

Design for Shielding: Make sure the shelter is fully enclosed, with no gaps or thin sections that allow radiation to enter. The roof and doors should be made from the same shielding materials as the walls.

Ventilation and Air Filtration

Air quality inside a fallout shelter is critical for long-term survival. You must ensure that the air you breathe is free of radioactive particles, carbon dioxide, and other harmful gases.

Air Filtration Systems: Use a HEPA filter system to remove radioactive dust and particles from the air. In more advanced shelters, an NBC (nuclear, biological, and chemical) filtration system can provide clean air by filtering out hazardous substances.

Ventilation: Your shelter needs a reliable source of fresh air. Install air intake and exhaust vents, but ensure they are equipped with filters to block fallout particles. To protect against external contamination, vents should be designed with blast doors or covers that can be sealed if needed.

Manual Ventilation: In case of a power outage, have a manual ventilation system, such as hand-crank fans, to circulate air through the shelter. Stale air can cause health problems, so maintaining good airflow is essential.

Water Supply and Filtration

A reliable water supply is crucial for long-term survival. You will need enough water for drinking, cooking, and hygiene, and it must be free of radioactive contamination.

Water Storage: Store a minimum of 1 gallon (3.8 liters) of water per person per day. For long-term survival, aim to store at least 2 to 3 months' worth of water. Use food-grade containers and keep them sealed.

Water Filtration: If you need to source water from outside the shelter, install a high-quality water filtration system capable of removing radioactive particles and other contaminants. Reverse osmosis filters are highly effective for this purpose, but ensure they are designed for emergency use.

Food Storage and Preparation

To survive long-term, you need an adequate supply of non-perishable food. Focus on calorie-dense and nutrient-rich foods that have a long shelf life.

Types of Food: Canned goods, freeze-dried foods, dehydrated meals, and sealed packaged foods are excellent for long-term storage. Store foods high in protein, fats, and carbohydrates to provide sustained energy.

Cooking: A small, safe cooking method, such as a propane stove or solar oven, is useful for heating food in the shelter. Ensure that your shelter has proper ventilation if you plan to cook inside.

Power Generation

Having a power source in your shelter ensures that you can run essential equipment, such as air filters, communication devices, and lighting.

Generators: A small, portable generator (diesel or propane) can power essential devices. Be sure to store enough fuel for long-term use and have a ventilation system for generator exhaust.

Solar Power: Solar panels, combined with a battery storage system, can provide a renewable energy source without the need for fuel. This is an excellent option for long-term use, provided you can safely access sunlight.

Battery Backup: Keep a supply of rechargeable batteries and a charging system to power small devices, such as radios and lights.

Medical Supplies and First Aid

A well-stocked first aid kit and a supply of essential medications are critical for addressing medical needs in a fallout shelter.

First Aid Kit: Include bandages, antiseptics, pain relievers, and other basic first aid items. Be prepared for common injuries, such as cuts, burns, and sprains.

Medications: Stock up on prescription medications needed by family members, as well as potassium iodide (KI) tablets to protect against radioactive iodine exposure.

Radiation Treatments: If available, stock medications like Prussian blue (to treat cesium exposure) and calcium or zinc DTPA (to treat plutonium and americium exposure).

Waste Management and Sanitation

Waste disposal is a major challenge in a fallout shelter, especially if you are staying for an extended period. Poor sanitation can lead to the spread of disease.

Toilets: Use portable toilets, composting toilets, or bucket systems with plastic liners. Ensure you have plenty of liners, disinfectants, and odor-control products.

Waste Disposal: Store human waste and garbage in sealed containers, keeping them away from living areas. If possible, create a designated space for waste disposal outside the main shelter.

Hygiene: Maintain hygiene with wet wipes, hand sanitizer, and soap. Cleanliness is critical for preventing illness in a confined space.

Communication and Information

Staying informed about the situation outside is crucial for long-term survival in a fallout shelter. You will need ways to receive updates and communicate with the outside world.

Radios: A battery-powered or hand-crank emergency radio can provide vital information about radiation levels, rescue operations, and other developments. Ensure it has access to local and national emergency broadcast channels.

Two-Way Communication: Consider having a two-way radio or satellite phone to communicate with others in your community or with emergency services.

Backup Power: Keep spare batteries or a hand-crank charging system to ensure that your communication devices remain functional.

Building the ideal fallout shelter for long-term survival involves careful planning and preparation. By focusing on radiation shielding, ventilation, water, food, and sanitation, you can create a safe and sustainable living environment that protects you and your family from the dangers of nuclear fallout. The key to a successful shelter is not only its construction but also ensuring that it is well-stocked, properly maintained, and equipped to support long-term survival. Whether you're preparing for days, weeks, or even months inside, the functionality and sustainability of your fallout shelter will be vital to your health and well-being.

Here are some key takeaways for building the ideal fallout shelter:

Summary of Key Points

Location and Placement: Opt for an underground or well-shielded above-ground shelter, strategically placed to offer protection from fallout and, if possible, away from blast zones.

Structural Design: Build using strong, dense materials such as concrete, steel, or brick. Ensure the walls, roof, and doors are thick enough to block gamma radiation effectively.

Radiation Shielding: Ensure proper shielding by using materials like lead, concrete, or earth. Thickness is crucial: the more substantial the material, the more protection it offers from radiation.

Ventilation and Air Filtration: Use HEPA or NBC filtration systems to provide clean air, and ensure proper ventilation to prevent the buildup of carbon dioxide and other gases inside the shelter.

Water Supply and Filtration: Store enough clean water for each person and include filtration systems like reverse osmosis to purify any outside water you collect.

Food Storage and Preparation: Stockpile non-perishable, calorie-dense food that can last months or longer. Consider safe cooking methods that require minimal energy and can function in confined spaces.

Power Generation: Plan for power needs with generators, solar panels, and battery backup systems. Ensure you have the means to recharge essential devices and run critical systems like air filters.

Medical Supplies: Prepare a well-stocked first aid kit and keep essential medications and radiation treatments on hand. This includes basic medical supplies and specialized treatments like potassium iodide tablets.

Waste Management and Sanitation: Set up proper waste disposal systems, such as portable or composting toilets, and maintain cleanliness to avoid the spread of disease.

Communication and Information: Equip your shelter with emergency radios and communication devices. Ensure that you can receive updates from authorities and maintain contact with the outside world.

Planning for the Long Term

Long-term survival in a fallout shelter requires careful attention to resources, hygiene, and mental health. The shelter must be designed not only to protect against radiation but also to provide a livable environment for extended periods.

Resource Management

Ration your food and water supplies carefully to ensure they last as long as possible. Keep track of usage and replenish resources when it's safe to do so.

Mental Health and Comfort

Spending weeks or months in a confined space can take a toll on mental health. Ensure your shelter includes items for comfort and entertainment, such as books, games, or music, to help pass the time and reduce stress.

Routine Maintenance

Regularly inspect your shelter for any signs of wear, damage, or leaks in ventilation or shielding systems. Ensure that air filtration, water filtration, and power generation systems are working properly at all times.

Monitoring Outside Conditions

Continue to monitor radiation levels outside using a Geiger counter or dosimeter. Once levels drop to safe levels, begin planning for life outside the shelter, including potential relocation or rehabilitation of your home and land.

A well-designed fallout shelter can mean the difference between life and death in the event of a nuclear disaster. By investing time and resources into building a shelter that offers adequate radiation protection, clean air, water, food, and medical supplies, you are taking essential steps to safeguard yourself and your family from the long-term effects of nuclear fallout. Remember that the shelter's effectiveness depends not only on its construction but also on how well it is stocked and maintained over time.

Above Ground vs. Below Ground Shelters: Which is Right for You?

Choosing between an above-ground and a below-ground shelter for nuclear fallout protection is a critical decision that depends on a variety of factors, including your location, budget, available space, and the type of threat you're preparing for. Each shelter type offers distinct advantages and disadvantages, which need to be weighed carefully to determine which is best suited for your survival needs. This chapter will explore the key differences between above-ground and below-ground shelters, highlighting their pros and cons, as well as the situations in which each type might be most appropriate.

Understanding the Differences

The primary goal of a fallout shelter is to protect you from the harmful effects of radioactive fallout, including gamma radiation, radioactive dust, and particles. The design and placement of the shelter can affect its ability to provide adequate shielding, regulate temperature, and maintain livability for extended periods.

Below-ground shelters are built underground or partially buried beneath the earth, relying on the natural shielding properties of the soil and the ground itself to block radiation.

Above-ground shelters are built on the surface, using man-made materials like concrete, steel, or lead to create a barrier that provides similar radiation protection.

Below-Ground Shelters: The Advantages

Superior Radiation Protection

One of the primary advantages of a below-ground shelter is its superior radiation protection. The Earth provides an excellent natural shield against gamma radiation, which is the most penetrating and dangerous form of radiation from nuclear fallout. Being underground or surrounded by dense soil significantly reduces your exposure to radiation, making it a safer option for long-term survival.

Earth as a Shield: As little as 3 feet (1 meter) of packed earth can block 90% of gamma radiation. Deeper shelters offer even greater protection.

Temperature Regulation

Below-ground shelters benefit from the Earth's natural insulation, which helps regulate temperature. This can be especially advantageous in extreme climates, as the ground temperature remains relatively constant, making it easier to maintain a livable environment without excessive energy use.

Cool in the Summer, Warm in the Winter: The Earth's stable temperature moderates the shelter's internal environment, reducing the need for artificial heating or cooling.

Protection from Blast and Debris

If the nuclear explosion occurs relatively close to your location, a below-ground shelter offers better protection from the blast wave, flying debris, and other immediate effects of the explosion. Being underground reduces the risk of structural damage to your shelter from above-ground forces, making it more resilient.

Stealth and Concealment

A below-ground shelter can be hidden from view, providing an extra layer of protection in situations where privacy or security is a concern. In the aftermath of a nuclear event, looting or desperate individuals may pose a threat, and an underground shelter is less likely to attract attention.

Below-Ground Shelters: The Disadvantages

Cost and Labor-Intensive Construction

Building a below-ground shelter requires more resources, labor, and planning compared to an above-ground shelter. Excavation, reinforcement, waterproofing, and ventilation systems all add to the complexity and cost of the project. If you're on a tight budget or have limited space, constructing an underground shelter may not be feasible.

Excavation Costs: Digging out the required space, especially in areas with difficult terrain or bedrock, can significantly increase construction costs.

Risk of Flooding

Below-ground shelters are more vulnerable to water infiltration, particularly in areas with a high water table or prone to flooding. Without proper waterproofing and drainage systems, water can seep into the shelter, compromising its structural integrity and safety.

Waterproofing and Drainage: Ensuring the shelter remains dry requires careful planning and potentially costly waterproofing measures.

Air Circulation and Ventilation

Proper ventilation is crucial for any fallout shelter, but it can be more challenging to maintain in a below-ground shelter. Air must be actively pumped or circulated, requiring reliable filtration systems and backup power to ensure clean, breathable air. If these systems fail, the shelter can become uninhabitable.

Mechanical Ventilation: Below-ground shelters often require mechanical ventilation systems, which rely on electricity and maintenance to function properly.

Above-Ground Shelters: The Advantages

Easier and Less Expensive to Build

Above-ground shelters are generally easier and less expensive to construct than below-ground options. You don't need to worry about excavation or dealing with soil conditions, which can significantly reduce labor and material costs. Additionally, above-ground shelters can be built using prefabricated materials or converted from existing structures.

Modular or Prefabricated Options: Prefabricated shelters are available for quick installation and often require less specialized labor compared to underground shelters.

Accessibility

Above-ground shelters are more easily accessible, particularly for individuals with mobility issues or in situations where rapid entry and exit are necessary. Without the need to descend into a dug-out space, you can move in and out of the shelter more easily in an emergency.

Quick Entry and Exit: In a fast-developing crisis, the ability to quickly enter and seal the shelter could make a significant difference in survival.

Customizable Features

Because you're not constrained by the need to dig or reinforce underground structures, above-ground shelters allow for more customization. You can easily add features like additional rooms, larger living areas, windows (with proper shielding), and other amenities that make long-term living more comfortable.

Comfortable Living Space: Above-ground shelters can offer more room for storage, comfort, and even luxuries like natural lighting, which can be added with radiation-proof glass.

Above-Ground Shelters: The Disadvantages

Less Natural Radiation Protection

Above-ground shelters do not benefit from the Earth's natural shielding, so they require much thicker walls to achieve the same level of radiation protection as a below-ground shelter. This means more materials—such as concrete, steel, or lead—are needed to block gamma rays, which can add to the cost and complexity of the build.

Material-Heavy Design: To achieve adequate shielding, walls may need to be 2–3 feet thick, which increases the amount of materials needed for construction.

Vulnerability to Blast Waves

In the event of a nearby nuclear explosion, above-ground shelters are more vulnerable to the blast wave, flying debris, and fire. While they can be reinforced to withstand some of these forces, they don't offer the same level of protection as an underground shelter when it comes to shielding from physical destruction.

Structural Damage Risk: The blast wave from a nuclear explosion can cause significant damage to above-ground structures, especially if not properly reinforced.

Temperature Control

Above-ground shelters are more affected by external weather conditions, making it harder to regulate temperature without relying on energy-intensive heating and cooling systems. This is especially challenging in extreme climates where temperature fluctuations could make long-term survival more difficult.

Energy Requirements: You may need to rely on generators or alternative energy sources to maintain a liveable temperature inside an above-ground shelter.

Which Shelter is Right for You?

Deciding whether an above-ground or below-ground shelter is the best option depends on your specific circumstances, including location, budget, and the type of threats you are preparing for. Here are a few key factors to consider:

Geographic and Environmental Conditions

Flood-Prone Areas: If you live in a region prone to flooding or with a high water table, an above-ground shelter may be a safer option. Below-ground shelters in these areas run the risk of flooding or water damage.

Rural vs. Urban Locations: In rural areas with more space, a below-ground shelter is often more feasible. Urban areas, however, may have limited options for digging, making an above-ground shelter a more practical choice.

Budget and Resources

Low Budget: If you're working with limited financial resources, an above-ground shelter is likely to be more affordable. Prefabricated shelters or retrofitting an existing structure can save on costs.

Higher Budget: If you have the resources to invest in a long-term solution, a below-ground shelter offers superior radiation protection and more stable environmental conditions.

Expected Duration of Stay

Short-Term Survival: If you are preparing for a relatively short-term stay, such as a few days to weeks, an above-ground shelter may suffice, provided it offers adequate radiation protection.

Long-Term Survival: For long-term survival (months or longer), a below-ground shelter is preferable due to its superior radiation shielding, temperature regulation, and potential for sustainable living conditions.

Mobility and Accessibility

Elderly or Disabled: If mobility is a concern, an above-ground shelter may be easier to access and maintain. Without the need to navigate stairs or ladders, these shelters offer quicker and safer entry in emergencies.

Both above-ground and below-ground shelters offer distinct advantages and challenges when it comes to nuclear fallout protection. The decision on which type of shelter is right for you depends on your location, budget, and long-term survival needs. Below-ground shelters offer superior radiation shielding and temperature control, making them ideal for long-term survival. However, they can be more costly and difficult to construct. Above-ground shelters are easier to build, more affordable, and offer greater accessibility, but they require thicker walls and more materials to achieve the same level of protection.

Ultimately, the choice between an above-ground or below-ground shelter should be based on your specific circumstances, the risks you are preparing for, and your ability to invest in building the most suitable option for your needs. Proper planning, construction, and maintenance are essential for ensuring your shelter offers the best possible protection in the event of a nuclear disaster. Whether you opt for an above-ground or below-ground shelter, the key to survival lies in thoughtful preparation, adequate resources, and a solid understanding of the risks and protective measures necessary for long-term safety.

Final Considerations for Shelter Construction

Regardless of which type of shelter you choose, there are several important factors to keep in mind to maximize the effectiveness of your fallout shelter:

Adequate Shielding

Both above-ground and below-ground shelters need sufficient shielding from gamma radiation. Whether using earth, concrete, steel, or lead, the thickness and density of your materials are crucial in blocking harmful radiation. Be sure to consult experts or reliable guidelines to ensure your shelter meets recommended protection standards.

Ventilation and Air Filtration

Good air quality is essential for long-term survival. Make sure your shelter has a well-designed air filtration system to remove radioactive particles from the air. For below-ground shelters, this is especially important, as ventilation must be actively managed to avoid stale air and carbon dioxide buildup.

Waterproofing and Drainage

For below-ground shelters, proper waterproofing is a critical concern, especially in areas with a high water table or the potential for flooding. Installing drainage systems, sump pumps, and waterproof linings can help keep your shelter dry and functional.

Energy Independence

Consider how you will generate power for lighting, ventilation, and essential systems. Backup generators, solar panels, and battery storage can ensure your shelter remains operational during extended periods without external power.

Food and Water Supplies

Your shelter must be stocked with enough food and water to last for an extended period. Store non-perishable, calorie-dense foods and clean water, and ensure that you have water filtration systems in place in case you need to access external water sources.

Sanitation and Hygiene

Long-term living in a confined space requires careful attention to waste management and sanitation. Include portable toilets, waste storage systems, and cleaning supplies to maintain hygiene and prevent the spread of disease.

Mental Health and Comfort

Being in a fallout shelter for extended periods can take a toll on mental health. Include items that provide comfort and help pass the time, such as books, games, radios, and communication devices. Staying informed about the outside world through emergency broadcasts or two-way communication can also provide psychological comfort.

Preparing for the Future

As you plan for your fallout shelter, always consider the future. The ability to leave the shelter safely after a nuclear event will depend on several factors, including radiation levels, environmental conditions, and access to clean resources. Whether you choose an above-ground or below-ground shelter, building with long-term sustainability in mind will give you the best chance of survival.

The key to deciding between an above-ground and a below-ground shelter ultimately lies in your unique situation. By evaluating the factors discussed in this chapter—radiation protection, cost, location, accessibility, and long-term survival needs—you can make an informed decision that best fits your circumstances. Whatever option you choose, the goal remains the same: to protect you and your family from the devastating effects of nuclear fallout and ensure you have the resources to endure the aftermath of a nuclear event.

The Best Materials for Building a Safe Nuclear Shelter

When constructing a nuclear fallout shelter, selecting the right materials is crucial to ensure protection from radiation, blast waves, and other dangers associated with a nuclear event. The materials you use must provide adequate shielding from ionizing radiation, offer structural integrity, and help create a livable environment for long-term survival. This chapter will explore the best materials for building a safe nuclear shelter, considering factors such as radiation protection, durability, cost, and ease of construction.

Understanding the Threat: Types of Radiation

Before choosing materials, it's essential to understand the types of radiation your shelter needs to protect against:

Gamma radiation: The most penetrating form of radiation, capable of passing through walls and even several feet of solid materials. Gamma radiation requires dense materials for effective shielding.

Beta radiation: Less penetrating than gamma radiation, beta particles can be blocked by relatively thin materials like aluminium or glass but can still pose a threat when inhaled or ingested.

Alpha radiation: Alpha particles are the least penetrating but highly dangerous if inhaled or ingested. They can be stopped by a sheet of paper or the outer layer of human skin, but controlling airborne contamination is vital.

For your shelter, the primary concern is **gamma radiation**, as it is the hardest to block and can cause the most damage if exposure levels are high. Let's now explore the best materials for blocking this radiation and ensuring your shelter is structurally sound.

Best Materials for Radiation Shielding

Concrete

Concrete is one of the most commonly used materials for building fallout shelters due to its high density, affordability, and ease of use. It provides excellent protection against gamma radiation, and its availability makes it accessible for many types of construction.

Advantages:

Concrete is cost-effective and widely available.

It provides excellent radiation shielding when used in sufficient thickness (typically 2–3 feet for optimal protection from gamma radiation).

Concrete structures are durable and can withstand external forces, such as blast waves and debris.

Disadvantages:

Concrete can be heavy, making it harder to work with without proper equipment.

It may require steel reinforcement (rebar) to ensure structural integrity over long periods.

Best Use: Thick concrete walls, floors, and ceilings in both above-ground and below-ground shelters. A 2-foot-thick (60 cm) concrete wall can provide significant radiation protection.

Steel

Steel is another excellent material for radiation shielding, offering strength, durability, and the ability to absorb radiation, especially in above-ground shelters where structural reinforcement is needed. Steel is particularly effective when combined with other materials like concrete or lead.

Advantages:

Steel provides strong structural support, making it ideal for reinforcing shelter walls, doors, and roofs.

It has high tensile strength, making it resistant to physical damage from blast waves.

While steel is not as effective at shielding gamma radiation as lead, it can still offer protection when used in thick layers or combined with other materials.

Disadvantages:

Steel is more expensive than concrete and requires specialized skills and equipment for installation.

It offers less radiation protection on its own compared to denser materials like lead or concrete.

Best Use: Reinforcing shelter walls, doors, and roofing, particularly in above-ground shelters. Steel panels or frames can be used to add strength and protection to concrete structures.

Lead

Lead is one of the best materials for shielding against radiation due to its high density. Even relatively thin layers of lead can block a significant amount of gamma radiation. However, it is also one of the most expensive and heavy materials, making it more challenging to use in large quantities.

Advantages:

Lead is highly effective at blocking gamma radiation, requiring less thickness than other materials to provide the same level of protection.

It is useful in situations where space is limited, such as reinforcing doors or windows.

Disadvantages:

Lead is expensive, heavy, and difficult to work with, especially in large quantities.

It may require support structures to handle the weight, particularly in roofing or ceiling applications.

Best Use: Lead is ideal for shielding doors, windows, and areas that require high radiation protection but have space constraints. A 1-inch layer of lead can provide similar protection to several feet of concrete.

Earth and Packed Soil

Earth and packed soil offer excellent protection from radiation when used as a natural barrier, especially in below-ground shelters. Earth absorbs gamma radiation effectively and can be used to create a relatively inexpensive but highly protective shelter.

Advantages:

Earth is abundant, making it cost-effective and easily accessible for most shelter construction.

3 feet (1 meter) of packed earth can provide substantial protection from radiation.

Earth is ideal for use in below-ground shelters where it also provides natural insulation for temperature regulation.

Disadvantages:

Earth requires excavation and proper reinforcement to prevent cave-ins, which can be labor-intensive and time-consuming.

Moisture from the earth can seep into the shelter without proper waterproofing.

Best Use: Below-ground shelters or earth-bermed walls and roofs for above-ground shelters. Earth can also be used as additional shielding around concrete structures.

Brick and Stone

Brick and stone offer a natural alternative to concrete, providing good radiation protection while also being structurally sound. These materials are less common for modern fallout shelters but can still be effective in certain situations.

Advantages:

Brick and stone are dense materials that provide radiation shielding when used in thick layers.

They can be aesthetically pleasing for above-ground shelters and provide a traditional look.

They are more fire-resistant than some other materials.

Disadvantages:

Brick and stone walls need to be much thicker than concrete to provide the same level of protection (approximately 2–3 feet thick).

These materials can be more expensive and labor-intensive to work with than concrete.

Best Use: Brick and stone are suitable for thick, solid walls in both above-ground and below-ground shelters. They can also be used for structural reinforcement and decorative finishes.

Sandbags

Sandbags filled with soil or sand are a simple and inexpensive method for adding radiation shielding to an existing structure or temporary shelter. They can be stacked to form walls and barriers, providing additional protection against fallout.

Advantages:

Sandbags are easy to transport, fill, and arrange, making them ideal for emergency situations or as temporary shielding.

They are relatively inexpensive and can be filled with locally available materials.

Sandbags can be used to reinforce doors, windows, and other vulnerable areas.

Disadvantages:

Sandbags provide less structural support than concrete or steel.

They can deteriorate over time, especially when exposed to moisture, and may require replacement.

Best Use: Sandbags are excellent for quickly reinforcing shelters, filling gaps, and adding extra layers of radiation protection in both above-ground and below-ground shelters.

Other Essential Materials for Shelter Construction

In addition to materials for radiation shielding, you'll need other building materials to create a safe, comfortable, and sustainable shelter:

Insulation: Proper insulation is essential for temperature regulation. Use foam insulation, mineral wool, or fiberglass to keep the shelter warm in winter and cool in summer.

Waterproofing Materials: For below-ground shelters, waterproofing is critical. Use moisture barriers like plastic sheeting, rubber membranes, or bitumen to prevent water from seeping into the shelter.

Ventilation and Air Filtration Systems: Ensure clean air with proper ventilation systems equipped with HEPA filters or NBC (nuclear, biological, chemical) filtration systems. Install blast valves and air intake systems to protect against contamination.

Reinforcement Bars (Rebar): For concrete construction, steel reinforcement bars (rebar) are essential for structural integrity, ensuring the shelter can withstand external forces like blast waves or shifting soil.

Doors and Windows: Use steel-reinforced or lead-lined doors to block radiation and protect against physical breaches. Windows should be covered or built with radiation-resistant materials, such as leaded glass.

The materials you choose for your fallout shelter play a vital role in its effectiveness at protecting you from radiation and ensuring long-term survival. Dense materials like concrete, steel, and lead offer the best protection against gamma radiation, while earth and sandbags provide cost-effective shielding for below-ground or temporary shelters. Each material has its strengths and weaknesses, so your choice will depend on factors like location, budget, and the type of shelter you plan to build.

Ultimately, combining these materials to reinforce structural integrity, block radiation, and maintain a livable environment will give you the best chance of surviving a nuclear event. Proper planning, resource management, and attention to detail in material selection will help you build a safe and reliable shelter for you and your family.

Designs for Long-Term Survival: Blueprinting Your Safe Haven

Designing a fallout shelter for long-term survival involves more than just building a basic structure. To ensure you and your family can live comfortably and safely for an extended period, your shelter must be well-planned, functional, and equipped with essential systems. A proper blueprint should consider factors like space utilization, storage capacity, air and water systems, power generation, waste management, and overall liveability.

In this chapter, we'll explore how to create a comprehensive design for your fallout shelter, focusing on practical layout strategies, system integration, and critical considerations for ensuring long-term survival.

Core Components of a Fallout Shelter for Long-Term Survival

A fallout shelter for long-term survival requires specific components to ensure safety, comfort, and sustainability. These include:

Radiation Shielding

Space Allocation and Layout

Air and Ventilation Systems

Water Storage and Filtration

Food Storage and Preparation

Waste Management

Power Generation

Medical Facilities

Communication and Monitoring Systems

Mental Health and Comfort Considerations

Radiation Shielding: Designing for Safety

The first priority when designing your shelter is to ensure it provides adequate protection from radiation. The walls, floors, and ceilings of your shelter must be thick and dense enough to block gamma radiation, the most dangerous form of fallout radiation.

Materials: Use dense materials like concrete, steel, lead, or packed earth for radiation shielding. As discussed in the previous chapter, the thickness of the material will depend on the type of shelter and the materials used. For example:

2–3 feet of concrete or packed earth

1 foot of steel-reinforced concrete with lead lining for above-ground shelters

Double Wall Construction: Consider designing the walls with an air gap between two layers of concrete or steel for additional insulation and radiation protection. This adds an extra layer of defense and helps regulate temperature.

Space Allocation and Layout: Maximizing Efficiency

Effective space allocation is key to long-term survival in a confined shelter. Your blueprint should optimize available space for essential activities, storage, and comfort. Consider dividing your shelter into the following sections:

Living Quarters: Design separate sleeping areas for each person, allowing for privacy and comfort. Bunk beds or foldable cots can maximize space. Ensure there is enough room for personal belongings and comfort items.

Storage Rooms: Allocate a significant portion of the shelter for food, water, medical supplies, and tools. Use shelving units, cabinets, and bins to organize and store items efficiently. Include extra space for long-term food storage and any items you may need to replenish or rotate over time.

Kitchen and Food Prep Area: Set aside a small area for food preparation and cooking. Use a compact kitchen setup with a portable stove, solar oven, or other energy-efficient cooking methods. Include enough counter space for meal preparation and cleaning.

Bathroom and Sanitation: Create a designated area for waste management and personal hygiene. If space allows, a small room with a composting toilet or chemical toilet can be installed for privacy. Store hygiene supplies, such as toilet paper, disinfectants, and soap, nearby.

Workstation/Communication Area: Designate a space for communication and monitoring equipment. This can include a small desk or work area for using radios, charging devices, or keeping logs of daily activities.

Common Area: If space permits, plan for a small common area where the family can gather to eat, play games, or socialize. Keeping morale high is important for long-term survival, so this space should be comfortable and welcoming.

Air and Ventilation Systems: Ensuring Breathable Air

Maintaining a fresh and clean air supply is crucial for long-term survival in a sealed shelter. Your design must include an air filtration and ventilation system to protect against radioactive particles and maintain air quality.

HEPA Filtration System: Install a HEPA or NBC (nuclear, biological, chemical) filtration system that can remove radioactive particles from the air. This is especially important if your shelter is exposed to fallout.

Air Intake and Exhaust Vents: Your shelter should have a dedicated air intake and exhaust vent system. Vents must be protected with blast valves or filters to prevent fallout from entering.

Manual Ventilation: Include a manual ventilation option, such as a hand-crank fan, in case power is lost. This will allow you to circulate air through the shelter when mechanical systems are unavailable.

Water Storage and Filtration: Accessing Clean Water

Water is one of the most critical resources in a fallout shelter. You must design your shelter to store enough clean water for drinking, cooking, and hygiene.

Water Storage Tanks: Allocate space for large, sealed water storage tanks. Aim to store at least 1 gallon of water per person per day. For long-term survival, plan for several months' worth of water, or include the capacity to collect and filter external water.

Water Filtration Systems: Install a high-quality water filtration system capable of removing radioactive particles. Reverse osmosis filters are effective for filtering contaminated water. If your shelter is above-ground, include a system for rainwater collection and filtration.

Backup Water Sources: If possible, design your shelter near a groundwater source, such as a well, and include filtration systems for safely accessing this water.

Food Storage and Preparation: Sustaining Nutrition

Your shelter design should account for long-term food storage, ensuring that you have enough supplies to last for months or longer. In addition, you'll need space for preparing and preserving food.

Non-Perishable Food Storage: Plan shelving and storage for canned goods, dried foods, freeze-dried meals, and other non-perishable items. Include food high in calories, protein, and nutrients.

Cooking Area: Design a compact kitchen with a propane stove, portable burner, or solar oven. Ensure proper ventilation if using fuel-based stoves. Include enough space for meal preparation and food hygiene.

Food Preservation: If space allows, include a method for preserving fresh food, such as a dehydrator or vacuum sealer.

Waste Management: Maintaining Sanitation

Long-term survival requires careful attention to sanitation. Waste management systems should be designed to handle human waste and garbage without compromising the safety or comfort of the shelter.

Toilets: Composting toilets, chemical toilets, or bucket systems with biodegradable liners can be used in small shelters. These systems require minimal water and can be safely managed with proper disposal.

Waste Storage: Design a secure area for storing waste, such as sealed bins or bags, to prevent contamination and odor. Include space for cleaning supplies and disinfectants to maintain hygiene.

Power Generation: Keeping Essential Systems Running

A reliable power source is essential for running air filtration systems, lighting, communication devices, and more. Your shelter design must include a backup power solution.

Generators: Install a small generator (propane, diesel, or gasoline) with proper ventilation for exhaust. Store enough fuel to last several weeks or months, depending on usage.

Solar Power: Consider solar panels with a battery storage system to provide renewable energy. If your shelter is above-ground or semi-underground, solar power can be an excellent long-term solution for keeping essential systems running.

Battery Backup: Design your shelter with enough space to store batteries, power packs, and chargers. Rechargeable batteries can power small devices and lighting for extended periods.

Medical Facilities: Handling Health Issues

Your shelter should be equipped with a dedicated area for storing medical supplies and administering first aid.

First Aid Station: Include a small area for a first aid station, with space for storing medical supplies, over-the-counter medications, prescription medications, and radiation treatments like potassium iodide.

Emergency Medical Equipment: Store basic medical tools, such as bandages, splints, antiseptics, and equipment to handle minor injuries or illnesses. If possible, include a manual or reference guide for medical procedures.

Communication and Monitoring Systems: Staying Informed

Communication is critical in a post-nuclear environment, allowing you to stay informed about radiation levels, rescue efforts, or other important developments.

Radios: Install a battery-powered or hand-crank radio to receive emergency broadcasts and weather updates. Include a space for two-way radios to communicate with others in your area.

Radiation Monitoring: Your shelter design should include a space for radiation monitoring equipment, such as a Geiger counter or dosimeter, to track radiation levels inside and outside the shelter.

Mental Health and Comfort: Supporting Well-Being

Surviving in a fallout shelter for an extended period can take a toll on mental health. Design your shelter with comfort and morale in mind.

Comfortable Living Spaces: Ensure that living spaces are comfortable, with enough bedding, blankets, and personal items to make the environment as pleasant as possible.

Entertainment: Plan for entertainment options like books, games, radios, or electronic devices to help pass the time and maintain mental health.

Lighting and Ambiance: Include sufficient lighting to create a livable environment. Use LED lights or battery-powered lamps to conserve energy, and ensure that the shelter feels as homelike as possible.

Designing a fallout shelter for long-term survival requires careful consideration of space, resources, and essential systems. Your shelter must provide adequate protection from radiation, ensure access to clean air and water, store enough food for extended periods, and maintain a safe, sanitary living environment. By carefully planning your shelter's layout and integrating key systems for long-term survival, you can create a safe haven that will support you and your family through the aftermath of a nuclear event.

Summary of Key Design Elements

Radiation Shielding: Ensure your shelter provides sufficient shielding from gamma radiation using materials like concrete, steel, lead, or packed earth. Proper wall thickness and structure are critical for long-term protection.

Space Allocation and Layout: Divide your shelter into functional spaces, including living quarters, food storage, water storage, bathroom and sanitation areas, and a common area for relaxation. Maximize the use of vertical space with shelving and bunk beds to conserve room.

Air and Ventilation Systems: Install HEPA or NBC filtration systems to ensure clean air inside the shelter. Design the shelter with both mechanical and manual ventilation systems to guarantee air circulation in the event of power loss.

Water Storage and Filtration: Store enough water to last for months and install filtration systems for external water sources, such as rainwater or groundwater. Plan for backup water sources and ensure storage tanks are properly sealed to prevent contamination.

Food Storage and Preparation: Design the shelter to accommodate long-term food storage with easy access to non-perishable items. Create a small kitchen or food prep area with compact cooking devices like solar ovens or portable stoves.

Waste Management: Install sanitation systems like composting or chemical toilets, and designate space for waste storage. Include hygiene supplies to maintain cleanliness in confined quarters.

Power Generation: Include a generator, solar panels, or battery backup systems to power essential shelter functions like air filtration, lighting, and communication devices. Ensure fuel or power supplies are stored safely and in sufficient quantities for long-term use.

Medical Facilities: Set up a first aid station and store a comprehensive supply of medical and radiation treatment supplies. Include space for administering care in case of injuries or illnesses during your time in the shelter.

Communication and Monitoring: Plan for communication systems to receive emergency updates and monitor radiation levels. Designate space for radios, Geiger counters, and other monitoring devices to stay informed and ensure safety.

Mental Health and Comfort: Design with comfort in mind, including areas for relaxation, entertainment, and personal space. Plan for adequate lighting and entertainment options to reduce stress and maintain morale during extended confinement.

Long-Term Survival Strategies

Beyond the physical structure of the shelter, planning for long-term sustainability is essential. Consider these additional strategies to make your shelter more livable for extended periods:

Resource Management: Carefully monitor food, water, and power usage to ensure resources last as long as possible. Develop a rationing plan to conserve supplies and avoid shortages.

Maintenance and Repairs: Include tools and supplies for making repairs to the shelter or its systems, such as air filters, water filtration, and power systems. Regular maintenance checks will help ensure everything functions properly over time.

Psychological Support: Keep the shelter environment as comfortable and normal as possible. Rotate tasks, schedule daily routines, and encourage social interaction to reduce the psychological strain of long-term confinement.

Designing and building a fallout shelter for long-term survival is a complex task that requires careful planning, resource allocation, and attention to detail. By focusing on radiation protection, space efficiency, and the integration of essential systems for air, water, food, and sanitation, you can create a shelter that will sustain life in the harsh conditions following a nuclear event. Ensuring the comfort and well-being of your family throughout the survival period is just as important as the technical aspects of the shelter. With a well-thought-out design, you can transform a basic fallout shelter into a safe, functional, and sustainable environment that offers protection and hope for the future.

Ensuring Clean Air in a Fallout Shelter: Filtration Systems

Ensuring clean air in a fallout shelter is one of the most critical aspects of long-term survival. After a nuclear event, the air outside will likely be contaminated with radioactive particles, making it hazardous to breathe without proper filtration. A well-designed air filtration system can protect you and your family from these dangers, as well as from other potential airborne threats like chemical or biological agents. In this chapter, we'll explore the various types of air filtration systems, how they work, and how to integrate them into your fallout shelter design to ensure a continuous supply of clean, breathable air.

The Importance of Clean Air

Nuclear fallout releases radioactive particles into the air, which can be inhaled, ingested, or absorbed through the skin, leading to radiation poisoning, respiratory issues, and other health problems. Fallout particles—composed of radioactive isotopes like cesium-137, iodine-131, and strontium-90—pose a significant risk to anyone exposed to them for prolonged periods. In addition to radioactive fallout, air inside a shelter can become stale and accumulate harmful gases like carbon dioxide, leading to asphyxiation if not properly ventilated.

To mitigate these risks, a reliable air filtration system is necessary to remove radioactive particles, circulate fresh air, and maintain a safe and livable environment in your shelter.

Types of Air Filtration Systems for Fallout Shelters

There are several types of air filtration systems available, each designed to address different kinds of contaminants. The most effective fallout shelter filtration systems focus on removing radioactive particles, while also filtering out other harmful contaminants like chemicals, dust, and biological agents. Here are the primary types of filtration systems suitable for fallout shelters:

HEPA (High-Efficiency Particulate Air) Filters

HEPA filters are widely recognized for their ability to trap small airborne particles, including radioactive fallout. These filters work by forcing air through a fine mesh that captures particles as small as 0.3 microns, which includes radioactive dust, allergens, and other harmful contaminants.

How HEPA Filters Work: HEPA filters are designed to remove particulate matter from the air by trapping particles within multiple layers of fine fibers. Air is pushed through the filter, where particles are captured, and clean air is released.

Effectiveness: HEPA filters are highly effective at removing up to 99.97% of airborne particles, including most radioactive dust. However, they do not filter out gases, so additional filtration may be needed for volatile organic compounds (VOCs) or chemical agents.

Maintenance: HEPA filters need to be replaced periodically to maintain their effectiveness. Ensure you have replacement filters stored in your shelter.

Best Use: Install HEPA filters in your shelter's air intake system to trap radioactive fallout particles and other contaminants from the external environment.

NBC (Nuclear, Biological, and Chemical) Filtration Systems

NBC filtration systems are designed to protect against nuclear, biological, and chemical threats, making them one of the most comprehensive air filtration solutions for a fallout shelter. These systems include a combination of particulate filters (such as HEPA) and activated carbon filters to neutralize gases and chemicals.

How NBC Filters Work: NBC systems typically combine multiple stages of filtration:

Particulate filtration (HEPA) to remove radioactive fallout, biological contaminants, and dust.

Activated carbon filtration to absorb chemical gases, such as chlorine, ammonia, and VOCs.

Pre-filters to capture larger particles before they reach the more delicate HEPA or carbon filters.

Effectiveness: NBC systems provide excellent all-around protection from radioactive particles, chemical agents, and biological threats. They are the most comprehensive air filtration systems available for fallout shelters.

Maintenance: NBC systems are more complex than standard air filters and require regular maintenance, including replacing filters and carbon beds. Ensure you have sufficient supplies for long-term use.

Best Use: NBC filtration systems are ideal for shelters where maximum protection is needed, such as those located near potential nuclear targets or in areas where chemical or biological threats may arise.

Activated Carbon Filters

Activated carbon filters are highly effective at removing gases and odors from the air, making them an important component of any comprehensive air filtration system. While they are not designed to filter radioactive particles, they work in conjunction with HEPA or NBC systems to absorb harmful gases and vapors.

How Activated Carbon Filters Work: Activated carbon filters use a bed of carbon granules or charcoal that adsorb gases, trapping chemical molecules as air passes through. These filters are particularly effective at neutralizing chemicals like chlorine, formaldehyde, and volatile organic compounds (VOCs).

Effectiveness: While not effective against particulate matter like radioactive fallout, activated carbon filters are excellent for removing toxic gases and odors that may accumulate in a sealed environment.

Maintenance: Activated carbon filters must be replaced periodically as they become saturated with contaminants. Monitor the filter's performance and have spares available.

Best Use: Use activated carbon filters in combination with HEPA or NBC systems to remove chemical contaminants and improve overall air quality in the shelter.

UV-C Light Air Purifiers

UV-C light air purifiers use ultraviolet light to neutralize bacteria, viruses, and other pathogens in the air. While not effective against radioactive particles, UV-C systems can help reduce the spread of airborne diseases in a closed environment, which can be a concern in long-term survival scenarios.

How UV-C Light Air Purifiers Work: These systems use short-wave ultraviolet light to destroy the DNA of microorganisms, preventing them from reproducing and causing infections. UV-C light can be used in conjunction with HEPA filters to enhance air quality.

Effectiveness: UV-C purifiers are effective against biological agents like bacteria, viruses, and mold spores but are not designed for filtering radioactive particles or gases.

Maintenance: UV-C lamps need to be replaced periodically, and the system should be cleaned to maintain its effectiveness.

Best Use: UV-C light air purifiers are best used in conjunction with a particulate filtration system to improve air quality and reduce the risk of airborne diseases inside the shelter.

Designing an Effective Filtration System for Your Shelter

When designing your shelter's air filtration system, it's important to create a setup that addresses both immediate fallout protection and long-term air quality. Here are key factors to consider when integrating an air filtration system into your shelter:

Air Intake and Exhaust Vents

Your shelter must be equipped with intake and exhaust vents to ensure proper air circulation. Without these, the air inside the shelter can become stale, leading to oxygen depletion or a buildup of carbon dioxide and other harmful gases.

Air Intake: The air intake system should pull in fresh air from outside the shelter. Ensure that this intake is equipped with a filtration system (HEPA or NBC) to remove radioactive particles and other contaminants.

Exhaust Vents: An exhaust vent system expels stale air from inside the shelter. To prevent contamination, the exhaust system should be located away from the air intake and designed to limit exposure to external fallout.

Blast Valves: Both intake and exhaust vents should be equipped with blast valves to prevent the shelter from being compromised in the event of a nearby explosion.

Manual Ventilation Systems

In the event of a power outage or system failure, your shelter should include a manual ventilation option. Hand-crank fans or manual air pumps can help circulate air and push it through the filtration system without the need for electricity.

Crank Ventilators: A hand-crank ventilator can provide airflow in and out of the shelter, ensuring that fresh air continues to circulate even if mechanical systems fail.

Backup Systems: Include backup air pumps or fans to ensure there's a secondary method for air filtration in the event of an extended power outage.

Redundancy and Backup Filters

To ensure long-term survivability, your shelter should be equipped with spare filters for each stage of the filtration system. Store extra HEPA filters, activated carbon beds, and any other necessary parts to ensure the system can be maintained over weeks or months.

Spare Filters: Keep multiple replacement filters for each type used in your system. Regularly check filters for buildup and replace them as needed.

Monitoring Air Quality

It's essential to monitor air quality in your shelter to ensure that your filtration system is functioning correctly. Install air quality monitors that measure carbon dioxide, oxygen levels, and any potentially hazardous gases.

Carbon Dioxide Monitors: These devices alert you if CO2 levels are rising, signaling that the shelter may not be properly ventilated.

Radiation Detectors: Use radiation detectors to check for any potential breaches in your filtration system. If radiation levels inside the shelter rise, you'll need to investigate the filtration system for potential failures.

Long-Term Maintenance and System Reliability

In a long-term survival scenario, keeping your air filtration system operational is vital. Here are key maintenance tips to ensure that your system remains reliable:

Regular Filter Changes: Track how often your filters need to be replaced based on usage. HEPA and carbon filters can become clogged with particles and lose their effectiveness over time.

Inspect and Clean Components: Periodically inspect your air intake, exhaust, and filtration systems for damage, blockages, or wear. Clean intake grates and replace any worn-out components as necessary.

Power Backup: Ensure you have backup power sources, such as batteries or a generator, to keep your filtration system running if the primary power source fails.

Ensuring clean air in a fallout shelter is critical for long-term survival. By installing an effective air filtration system—whether it's a HEPA filter, an NBC filtration system, or a combination of both—you can protect yourself and your family from the harmful effects of radioactive fallout, as well as other potential airborne threats. Clean air is essential not only for immediate protection from fallout but also for maintaining a liveable environment in the shelter over extended periods.

Summary of Key Steps to Ensure Clean Air in Your Fallout Shelter

Install a HEPA or NBC Filtration System: These systems are essential for removing radioactive particles and other airborne contaminants. NBC systems offer the most comprehensive protection, filtering out nuclear, biological, and chemical threats.

Use Activated Carbon Filters for Gas Filtration: If your shelter is at risk of chemical exposure, incorporate activated carbon filters to absorb toxic gases and neutralize odors.

Ensure Proper Ventilation: Your shelter must have a well-designed intake and exhaust system to maintain airflow and prevent the buildup of harmful gases like carbon dioxide. Install blast valves to protect against outside blasts and contamination.

Manual Ventilation Backup: Include a hand-crank ventilator or manual air pump as a backup to keep air circulating in case of power failure.

Monitor Air Quality: Use air quality monitors to track carbon dioxide and oxygen levels and ensure the air remains safe to breathe. Radiation detectors will alert you to any increases in radiation inside the shelter.

Keep Spare Filters and Backup Power: Regularly replace filters and store extras to ensure long-term functionality. Maintain backup power sources, such as batteries or generators, to keep your filtration system running if electricity fails.

Regular Maintenance: Inspect your filtration system frequently, clean air intake and exhaust vents, and replace worn or clogged filters to maintain clean, breathable air.

Final Considerations

Your air filtration system is the lifeline of your fallout shelter. By carefully planning and implementing an effective air filtration strategy, you can create a safe environment free of dangerous contaminants, allowing you to focus on other aspects of survival. Clean air, combined with proper ventilation and regular maintenance, will help ensure that you and your family can remain healthy and protected during and after a nuclear event. With these systems in place, your fallout shelter will be equipped to handle the challenges of long-term survival, giving you peace of mind knowing that the air you breathe is safe.

Clean Water Solutions for Nuclear Survival

Clean water is one of the most critical resources for long-term survival in a nuclear fallout situation. After a nuclear event, water sources can become contaminated with radioactive particles, chemicals, and other pollutants, making it essential to have reliable methods for storing, purifying, and filtering water. Ensuring access to clean, uncontaminated water is crucial for drinking, cooking, hygiene, and maintaining overall health. This chapter will explore various clean water solutions for nuclear survival, covering both short-term and long-term strategies for securing a safe water supply.

Understanding Water Contamination After a Nuclear Event

Nuclear explosions release radioactive fallout into the atmosphere, which eventually settles on the ground and contaminates water sources, including rivers, lakes, reservoirs, and even rainwater. The most dangerous contaminants include radioactive isotopes such as cesium-137, iodine-131, and strontium-90, all of which pose severe health risks when ingested.

In addition to fallout, other pollutants—such as heavy metals and chemical toxins—may enter the water supply due to the destruction of industrial areas, storage facilities, and infrastructure. In a nuclear survival scenario, clean water becomes scarce, and any available water must be purified and filtered to remove harmful contaminants.

Core Water Strategies for Nuclear Survival

Water Storage and Stockpiling

Water Filtration and Purification Systems

Rainwater Collection

Groundwater and Well Water

Emergency Water Sources

Long-Term Water Management

Water Storage and Stockpiling

The most reliable way to ensure access to clean water in the immediate aftermath of a nuclear event is to stockpile water before the disaster strikes. Properly stored water can last for months or even years, ensuring that you and your family have a safe supply for drinking, cooking, and hygiene.

How Much Water to Store: Aim to store at least 1 gallon (3.8 liters) of water per person per day. For long-term survival, it's recommended to stockpile at least 3 months' worth of water. For a family of four, this means storing at least 360 gallons (1,362 liters) for a 3-month period.

Storage Containers: Use food-grade, BPA-free plastic containers, stainless steel tanks, or glass containers to store water. Large water storage tanks, such as 55-gallon barrels, are ideal for long-term storage. Be sure to seal containers tightly to prevent contamination.

Water Treatment Before Storage: Treat stored water with unscented household bleach (5–6% concentration) to prevent bacterial growth. Use 8 drops of bleach per gallon of water and allow it to sit for 30 minutes before sealing.

Rotation: Rotate stored water every 6 to 12 months to ensure freshness, especially if it is stored in plastic containers. Keep stored water in a cool, dark place away from sunlight and chemicals.

Water Filtration and Purification Systems

Water filtration and purification systems are essential for removing radioactive particles, bacteria, viruses, and chemical contaminants from water. These systems can be used to treat water collected from external sources or stored in the shelter.

Portable Water Filters

Portable water filters are compact and effective for filtering water in emergency situations. They are ideal for short-term use or in situations where access to larger filtration systems is not possible.

Types of Portable Filters:

Ceramic Filters: These filters use a ceramic membrane to remove particles, bacteria, and some viruses from water. They are durable and reusable but may not remove chemical contaminants or radioactive isotopes.

Hollow Fiber Filters: These filters use a bundle of hollow fibers to trap bacteria and parasites, providing safe drinking water. However, they may not filter out chemicals or radioactive particles.

Activated Carbon Filters: These filters use activated carbon to remove chemicals, chlorine, and odors from water. They are useful for improving water taste but may need to be combined with other filtration systems to remove radioactive particles.

Gravity Water Filtration Systems

Gravity-based water filters are ideal for fallout shelters, as they require no electricity to operate. These systems use gravity to pass water through a series of filters, removing contaminants like bacteria, heavy metals, and radioactive particles.

Key Features:

Multi-stage Filtration: Gravity systems often use a combination of ceramic, carbon, and ion exchange filters to purify water.

Water Capacity: These systems can store several gallons of water at once, making them suitable for family use.

Reverse Osmosis Systems

Reverse osmosis (RO) systems are one of the most effective water filtration methods for removing a wide range of contaminants, including radioactive particles, heavy metals, and chemicals. RO systems work by forcing water through a semipermeable membrane, filtering out impurities and contaminants.

Advantages:

Highly Effective: RO systems remove up to 99% of contaminants, including cesium-137, strontium-90, and iodine-131.

Durable: With proper maintenance, an RO system can provide clean water for years.

Disadvantages:

Water Waste: RO systems can waste a significant amount of water during the filtration process. For every gallon of clean water produced, several gallons may be discarded.

Requires Power: RO systems typically require electricity to operate, making them less suitable for off-grid or emergency use unless you have a reliable power source.

Ultraviolet (UV) Water Purifiers

UV water purifiers use ultraviolet light to kill bacteria, viruses, and other microorganisms in water. While they are effective against biological contaminants, they do not remove radioactive particles or chemicals, so they should be used in conjunction with a filtration system.

Best Use: UV purifiers are ideal for treating stored water or water from external sources that may be contaminated with biological agents.

Rainwater Collection

Rainwater is one of the safest sources of water during a nuclear event if collected and filtered properly. However, fallout particles can settle on rooftops and other collection surfaces, so rainwater must be filtered before use.

Collection System: Set up a rainwater harvesting system with gutters, downspouts, and storage barrels. Ensure that the collection surface is clean, and use a fine mesh screen to filter out larger debris.

Filtration: Before drinking, rainwater must be passed through a filtration system, such as a gravity water filter or reverse osmosis unit, to remove any radioactive fallout particles.

Storage: Store rainwater in large, sealed containers in a cool, dark location. Use treated barrels to prevent bacterial growth.

Groundwater and Well Water

If you have access to a well, it can provide a reliable source of clean water. However, groundwater may become contaminated by fallout or chemicals, so it's important to test and treat well water regularly.

Testing Well Water: After a nuclear event, use a radiation detection kit to test well water for radioactive contamination. If radioactive particles are detected, a reverse osmosis or gravity filtration system can be used to purify the water.

Sealing the Well: To prevent fallout from entering the well, cover the wellhead with a protective seal or cap. Ensure that the area around the well is free of debris and runoff that could carry contaminants into the groundwater.

Emergency Water Sources

In a survival situation, it may become necessary to collect water from unconventional sources. Here are some emergency water sources and how to treat them:

Snow and Ice: If snow or ice is available, it can be melted and treated for drinking. Always filter and purify melted snow, as it may contain radioactive particles or pollutants.

Natural Water Bodies: Rivers, lakes, and ponds may be used as emergency water sources. These water bodies will likely be contaminated with fallout, so treat all collected water using a combination of filtration and purification methods, such as reverse osmosis or gravity filtration.

Distillation: If no clean water is available, distillation can be used to purify contaminated water. Distillation involves heating water to create steam, which is then collected and condensed back into liquid form, leaving contaminants behind.

Long-Term Water Management

Managing water supplies over the long term is crucial for survival. Here are some key strategies to ensure you have a sustainable water supply:

Conservation: Practice water conservation to stretch your stored water supply. Use water-efficient methods for hygiene and cooking, and reuse water when possible (for example, using leftover cooking water for cleaning).

Regular Maintenance: Maintain your filtration systems, water containers, and storage tanks. Check for leaks, replace filters as needed, and monitor water quality regularly.

Plan for Redundancy: Have multiple water sources and filtration methods available. If one system fails, you'll need a backup to ensure continuous access to clean water.

Clean water is essential for survival, especially in the aftermath of a nuclear event. By storing adequate amounts of water, using reliable filtration and purification systems, and exploring alternative water sources like rainwater and groundwater, you can ensure that your family has access to safe drinking water throughout a nuclear survival scenario. It's crucial to plan for both short-term and long-term water needs, incorporating multiple filtration methods and backup systems to guarantee a continuous supply of clean water. By taking these steps, you'll be better prepared to face the challenges of nuclear survival with confidence and security.

Storing and Growing Clean Food in a Fallout Environment

In a fallout survival scenario, food becomes a critical resource that must be carefully managed to ensure safety and sustainability. Storing and growing clean food in a fallout environment poses unique challenges, as nuclear fallout can contaminate the soil, water, and air, making traditional food sources potentially hazardous. This chapter will explore how to store clean food for long-term survival, as well as methods for safely growing food in a contaminated environment.

The Importance of Clean Food

Nuclear fallout can contaminate food through the deposition of radioactive particles on crops, soil, and water sources. Consuming contaminated food can expose you to harmful radioactive isotopes such as cesium-137, iodine-131, and strontium-90, which can lead to radiation poisoning and long-term health problems like cancer. To avoid these dangers, it is essential to store clean food and implement safe growing practices to ensure that your food supply remains uncontaminated.

Core Strategies for Food Storage and Growth in a Fallout Environment

Long-Term Food Storage

Safe Gardening and Food Production

Greenhouses and Indoor Farming

Hydroponics and Aquaponics Systems

Food Preservation Methods

Animal Husbandry in Fallout Conditions

Preventing Contamination in Food Supplies

Long-Term Food Storage

The safest way to ensure access to clean food after a nuclear event is to store a sufficient supply of non-perishable food items before the disaster occurs. Properly stored food can last for months or years, providing you with a reliable food source that is free from radioactive contamination.

Types of Food to Store

When preparing for a nuclear survival scenario, focus on storing non-perishable, calorie-dense foods that provide essential nutrients. Some ideal options include:

Canned Goods: Vegetables, fruits, meats, and soups in cans are ideal for long-term storage. Canned food can last several years when stored in a cool, dry place.

Freeze-Dried and Dehydrated Foods: These foods have had most of their moisture removed, giving them an extremely long shelf life. Freeze-dried meals, dried fruits, vegetables, and meats can last up to 25 years when stored in sealed containers.

Grains and Legumes: Rice, beans, lentils, oats, and other grains can be stored in airtight containers with oxygen absorbers to prevent spoilage. These foods are high in calories and provide valuable nutrition.

Powdered and Canned Milk: Dairy products can be stored in powdered or canned form to ensure a source of calcium and protein.

Nutrient-Rich Foods: Store multivitamins and nutrient supplements to compensate for any nutritional gaps in your diet, especially if fresh produce becomes unavailable.

Storage Conditions

To ensure the longevity of your stored food, it's important to create the right storage conditions:

Temperature Control: Store food in a cool, dry environment. Temperatures between 50°F and 70°F (10°C and 21°C) are ideal for maximizing the shelf life of non-perishable foods.

Airtight Containers: Use food-grade containers that are airtight to protect your stored food from moisture, pests, and air exposure. Mylar bags with oxygen absorbers, vacuum-sealed jars, and heavy-duty plastic bins are excellent options.

Rotation: Regularly rotate your stored food to ensure that you consume older items first. Use a first-in, first-out (FIFO) system to minimize the risk of spoilage.

Safe Gardening and Food Production

Growing your own food in a fallout environment can be challenging due to the contamination of soil, air, and water with radioactive particles. However, with proper precautions, it is possible to grow food safely.

Raised Garden Beds

Raised garden beds are one of the best methods for growing food in a contaminated environment. By creating a physical barrier between your plants and the contaminated soil, you can reduce the risk of radioactive particles affecting your crops.

Building Raised Beds: Construct raised garden beds using non-toxic, durable materials such as untreated wood, concrete blocks, or steel. The beds should be at least 12 inches (30 cm) deep to provide enough space for plant roots to grow.

Clean Soil: Fill the raised beds with clean, uncontaminated soil that has been stored indoors or sourced from a safe location. Avoid using soil from areas exposed to fallout.

Watering: Use filtered or treated water for irrigation. Avoid using rainwater or surface water unless it has been filtered to remove radioactive particles.

Protecting Plants from Fallout

If growing outdoors, take steps to protect your plants from fallout:

Row Covers and Tarps: Cover your plants with row covers or plastic tarps to prevent radioactive dust from settling on the crops. Ensure that the covers do not touch the plants directly to avoid contamination.

Windbreaks: Install windbreaks around your garden to reduce the amount of airborne fallout that reaches your plants. Trees, shrubs, or man-made barriers can help minimize the spread of radioactive particles.

Greenhouses and Indoor Farming

Greenhouses and indoor farming are excellent methods for growing clean food in a fallout environment. By creating a controlled environment, you can shield your crops from outside contamination while ensuring optimal growing conditions.

Greenhouse Farming

A greenhouse offers protection from fallout while allowing you to grow food year-round.

Materials: Build a greenhouse using durable, transparent materials such as polycarbonate or glass. Ensure that the structure is airtight to prevent fallout from entering.

Air Filtration: Install an air filtration system to remove airborne contaminants before they enter the greenhouse. Use HEPA filters or NBC (Nuclear, Biological, Chemical) filtration systems to keep the air clean.

Soil and Water Management: Use uncontaminated soil and filtered water to grow your plants. Raised beds, planters, or containers can be used to prevent direct contact with potentially contaminated ground.

Indoor Farming

Indoor farming allows you to grow food inside your shelter, fully protected from external contamination. While space may be limited, certain crops can be grown efficiently indoors using containers or hydroponic systems.

Lighting: Use grow lights (LED or fluorescent) to provide the necessary light spectrum for plant growth. Position the lights above the plants and adjust the intensity based on the type of crops you are growing.

Containers: Grow food in pots, containers, or raised trays filled with clean soil. Ensure the containers are large enough to support the root systems of your chosen plants.

Hydroponics and Aquaponics Systems

Hydroponics and aquaponics are innovative solutions for growing food indoors without soil, making them ideal for a fallout shelter environment. Both systems use nutrient-rich water to support plant growth and can be set up in a small space.

Hydroponics

Hydroponics involves growing plants in a soilless medium, such as gravel, sand, or perlite, while providing the plants with nutrient-rich water. This system allows for efficient water usage and rapid plant growth.

Benefits: Hydroponics is space-efficient, requires less water than traditional gardening, and eliminates the need for soil, reducing the risk of contamination.

System Setup: Set up hydroponic grow trays, nutrient reservoirs, and water pumps to circulate water through the system. Regularly monitor water quality and nutrient levels to ensure healthy plant growth.

Aquaponics

Aquaponics combines hydroponics with fish farming (aquaculture). The fish waste provides nutrients for the plants, while the plants filter and clean the water for the fish. This closed-loop system can provide both vegetables and fish for your diet.

Benefits: Aquaponics is highly sustainable and produces both plant-based food and protein (fish), making it ideal for long-term survival.

System Setup: Set up a tank for the fish, grow beds for the plants, and a filtration system to circulate the water. Choose fish species that thrive in a closed system, such as tilapia or catfish.

Food Preservation Methods

Preserving food is essential for ensuring a continuous supply during times when fresh food may be scarce. Use various preservation methods to extend the shelf life of your harvest or stored foods.

Canning

Canning is one of the best ways to preserve fruits, vegetables, and meats for long-term storage. The process involves sealing food in airtight jars and heating them to kill bacteria and other pathogens.

Water-Bath Canning: Suitable for high-acid foods like tomatoes, fruits, and pickles.

Pressure Canning: Necessary for low-acid foods like meats, beans, and vegetables to prevent botulism.

Dehydration

Dehydrating food removes moisture, preventing the growth of bacteria, mold, and yeast. Dehydrated foods are lightweight and have a long shelf life, making them ideal for storage.

Solar Dehydrators: Use solar dehydrators to dry fruits, vegetables, and herbs using the sun's heat.

Electric Dehydrators: These devices use low heat and fans to remove moisture from food.

Fermentation

Fermentation is an ancient method of preserving food that involves the conversion of sugars into alcohol or acids by bacteria or yeast. Fermented foods like sauerkraut, pickles, and yogurt can last for months and provide beneficial probiotics.

Animal Husbandry in Fallout Conditions

Raising small livestock such as chickens, rabbits, or fish can provide a valuable source of protein and other nutrients. However, you must ensure that the animals are kept in a clean environment and are not exposed to radioactive fallout, contaminated soil, or water. Animal husbandry in a fallout environment requires careful planning to ensure both the animals and their food sources remain uncontaminated.

Choosing the Right Livestock

Small livestock are easier to manage in a confined or controlled environment. Some ideal choices for raising in a fallout shelter or protected area include:

Chickens: Chickens provide eggs and meat and are relatively easy to care for. They do not require large spaces, and with proper shelter and food, they can thrive indoors or in a small outdoor pen that is protected from fallout.

Rabbits: Rabbits are excellent for meat production, as they reproduce quickly and can be raised in a small space. They are also efficient at converting plant material into protein.

Fish: Fish such as tilapia or catfish can be raised in aquaponic systems, providing both protein and a closed-loop system that supports plant growth. Fish farming requires clean water and a well-maintained filtration system.

Feeding Livestock

One of the biggest challenges of raising livestock in a fallout environment is ensuring that their food supply is clean and free from contamination. Consider the following strategies:

Stockpile Feed: Store enough clean, uncontaminated animal feed to last for several months. Ensure that the feed is stored in airtight containers in a cool, dry location to prevent spoilage.

Grow Feed Indoors: You can grow fodder indoors using hydroponic or soil-based systems. Grains like wheat or barley can be sprouted and used as feed for chickens, rabbits, or other small livestock.

Use of Food Scraps: If you're growing your own vegetables, consider using vegetable scraps or leftover produce to feed your livestock. Just ensure the food is clean and not exposed to any contaminants.

Housing and Protection

Livestock must be housed in a secure and clean environment to protect them from fallout. Here's how to safely raise livestock in a contaminated environment:

Indoor Pens: If space allows, raise animals indoors or in a greenhouse-type structure where they are shielded from external fallout. This ensures they are not exposed to contaminated soil, water, or air.

Outdoor Pens with Protection: If you have to keep animals outdoors, use covered pens with strong barriers like tarps, plastic sheeting, or other materials to protect against fallout particles. Ensure the pens are located away from areas where fallout may accumulate, such as downwind from the main shelter.

Preventing Contamination in Food Supplies

Whether you are growing food or storing it, preventing contamination is crucial to avoid exposure to radioactive fallout. Here are key practices for keeping food supplies clean:

Protect Food from Fallout

Cover Gardens: If you are growing food outdoors, cover your crops with tarps, plastic sheeting, or row covers to prevent fallout particles from settling on the plants.

Filter Water: Ensure that any water used for irrigation is filtered and free from contaminants. Use filtered rainwater, well water, or stored clean water for watering your crops.

Wash and Peel Produce

Before consuming any produce grown in a fallout environment, wash it thoroughly with clean, filtered water. If possible, peel fruits and vegetables to remove any surface contamination that may have settled on the skin. For leafy greens, remove the outer leaves before consumption.

Test for Contamination

If you have access to a radiation detector, test your crops, stored food, and water sources regularly for radioactive contamination. This is especially important for outdoor-grown produce, as fallout particles may still be present on the surface of the plants.

Storing and growing clean food in a fallout environment is essential for long-term survival. By stockpiling non-perishable food items, building raised beds or greenhouses for safe gardening, and using indoor farming techniques like hydroponics and aquaponics, you can maintain a sustainable and uncontaminated food supply.

Raising small livestock in a protected environment also adds a valuable protein source to your diet, but it requires careful planning to ensure their food and living conditions remain safe. Preventing contamination through proper food handling and protection practices is vital for minimizing the risks posed by radioactive fallout.

With proper preparation and a well-thought-out strategy, you can ensure access to clean, healthy food even in the harsh conditions following a nuclear event. By combining stored food with sustainable growing practices, you will be better equipped to support yourself and your family for the long term.

How to Protect Your Food Supply from Contamination

In the aftermath of a nuclear event, protecting your food supply from contamination is critical for ensuring long-term survival. Fallout from a nuclear explosion can contaminate the soil, water, and air, introducing radioactive particles into food sources. Consuming contaminated food can lead to radiation poisoning and other serious health issues. In this chapter, we will explore effective strategies to protect your food supply from contamination, focusing on prevention, storage methods, and safe handling practices.

Understanding the Threat of Fallout

Nuclear fallout consists of radioactive particles released into the atmosphere after a nuclear explosion. These particles eventually settle on the ground, contaminating crops, water sources, and surfaces. Fallout can carry isotopes such as cesium-137, iodine-131, and strontium-90, which pose long-term health risks when ingested through food or water. Fallout contamination can persist for months or years, so safeguarding your food supply is a vital part of surviving in a post-nuclear environment.

Key Strategies for Protecting Your Food Supply

Sealing and Storing Food Safely

Growing Food in a Controlled Environment

Preventing Fallout from Contaminating Outdoor Crops

Water Filtration for Clean Irrigation and Consumption

Proper Handling and Preparation of Food

Testing Food for Radiation

Preventing Secondary Contamination in Storage

Sealing and Storing Food Safely

The most reliable way to protect your food from fallout contamination is by sealing and storing it in airtight containers. Pre-packed, sealed food is less likely to be affected by fallout, and properly stored food can last for long periods.

Airtight Containers

To protect your food from radioactive particles, it's essential to store it in airtight, sealed containers. This prevents any airborne contaminants from reaching the food. Here's how to store food properly:

Mylar Bags with Oxygen Absorbers: Mylar bags are excellent for long-term food storage. When combined with oxygen absorbers, they create an airtight environment, protecting the food from oxidation, pests, and contamination.

Vacuum-Sealed Containers: Vacuum sealing food in bags or jars can extend shelf life by removing air and preventing the entry of radioactive particles. This is especially useful for grains, dried beans, and other dry goods.

Food-Grade Plastic Buckets: Store bulk food items, such as grains, rice, and flour, in food-grade plastic buckets with tight-fitting lids. Add a layer of protection by sealing the food in Mylar bags before placing them in the buckets.

Glass or Metal Containers: Use glass jars or metal containers for storing canned goods, nuts, and seeds. These materials provide a strong barrier against fallout particles and can be reused.

Long-Term Food Storage Locations

Food should be stored in a location that is secure from fallout and other contaminants. Ensure the storage area is cool, dry, and protected from environmental exposure.

Underground or Basement Storage: If possible, store food in an underground or basement shelter where it is shielded from fallout and extreme temperature fluctuations. The natural insulation of the earth can help maintain stable conditions for long-term storage.

Pantries or Sheltered Rooms: Use dedicated food storage areas inside your shelter or home that are well-ventilated, dry, and protected from external contamination.

Growing Food in a Controlled Environment

Growing food in a controlled environment, such as a greenhouse or indoors, helps protect crops from fallout exposure while providing a sustainable food source. These methods allow you to shield your crops from contaminated air, soil, and water.

Indoor Farming

Indoor farming is an ideal way to grow food in a fallout environment. By growing plants inside your home or shelter, you can maintain control over their exposure to contaminants.

Containers and Raised Beds: Use containers or raised beds with clean, uncontaminated soil to grow crops indoors. Ensure that the growing area is well-ventilated and uses clean water for irrigation.

Grow Lights: If growing indoors, use artificial lighting, such as LED grow lights, to provide the necessary light spectrum for plant growth. This allows you to cultivate crops year-round, even without access to natural sunlight.

Greenhouse Farming

A greenhouse can provide an enclosed space where crops are protected from radioactive fallout while still allowing natural sunlight for plant growth.

Sealing the Greenhouse: Ensure that the greenhouse is airtight and uses filtration systems to remove any radioactive particles from the air before it enters. Install plastic or glass panels to shield plants from external contamination.

Soil and Water Management: Use clean, uncontaminated soil and filtered water in your greenhouse to prevent crops from being exposed to fallout particles.

Preventing Fallout from Contaminating Outdoor Crops

If you must grow food outdoors, you need to take precautions to protect your crops from fallout particles. While growing crops outdoors is riskier, it can still be done safely with proper methods.

Covering Crops

Plastic Tarps and Row Covers: Use tarps or row covers to shield crops from fallout. These protective barriers prevent fallout particles from settling on the leaves, fruit, or soil. Ensure that the covers do not touch the plants directly, and remove them only when the area has been thoroughly decontaminated.

Windbreaks and Shields: Install barriers such as windbreaks, trees, or fences around your garden to reduce the movement of airborne fallout particles.

Watering with Filtered Water

Use filtered or treated water to irrigate outdoor crops, as rainwater and surface water sources may be contaminated with fallout. Consider using stored water or set up a filtration system to remove radioactive particles from the water before use.

Water Filtration for Clean Irrigation and Consumption

Clean water is essential not only for drinking but also for watering crops. Fallout particles can contaminate water sources, making it crucial to filter any water used in food production or consumption.

Reverse Osmosis Systems: Reverse osmosis is one of the most effective methods for removing radioactive particles from water. Install a reverse osmosis system to filter water for both irrigation and drinking.

Activated Carbon Filters: Use activated carbon filters to remove chemicals, heavy metals, and radioactive particles from water before using it for irrigation or consumption.

Boiling and Chemical Treatment: Boiling water or using chemical treatments (such as iodine or chlorine tablets) can help neutralize bacteria and pathogens, though it may not remove radioactive particles. Always filter water first before boiling.

Proper Handling and Preparation of Food

Even if your food is stored properly, fallout particles can still enter the shelter on clothing, skin, or equipment. Practicing safe food handling and preparation techniques is crucial to prevent contamination.

Washing and Peeling Produce

If you have grown food outdoors or suspect contamination, wash all produce thoroughly with clean, filtered water. For extra precaution, peel fruits and vegetables before consumption to remove any surface contaminants.

Scrub Hard-Skinned Produce: Use a brush to scrub produce with hard skins (such as potatoes, cucumbers, or carrots) to remove fallout particles. Peeling the skin adds an extra layer of safety.

Safe Food Storage and Handling Indoors

Clean Work Surfaces: Before handling food, clean all work surfaces with a disinfectant to prevent any fallout particles from contaminating your food preparation area.

Protective Clothing: Wear protective clothing when handling outdoor crops or harvesting from a greenhouse to avoid bringing contaminants into the shelter.

Testing Food for Radiation

Testing food for radioactive contamination can help you determine whether it is safe to eat, especially if it was grown or harvested outdoors.

Radiation Detectors: Use a handheld Geiger counter or dosimeter to test your food and water for radiation. If radiation levels are detected above safe thresholds, discard the contaminated food.

Soil Testing Kits: If you're growing food outdoors, regularly test the soil for radiation levels. If the soil is found to be contaminated, switch to raised beds or indoor farming until the contamination levels decrease.

Preventing Secondary Contamination in Storage

Once your food is properly stored, it's essential to prevent secondary contamination from fallout particles that may enter the shelter.

Decontaminate Before Entering the Storage Area: Before handling stored food, ensure that all individuals decontaminate by removing outer clothing and washing exposed skin with clean water. This helps prevent fallout particles from spreading in the food storage area.

Seal Storage Areas: Keep your food storage areas sealed and protected from airflow that could carry in contaminated dust or particles. Use airtight containers for all stored food.

Protecting your food supply from contamination is essential for ensuring survival in the aftermath of a nuclear event. By storing food in airtight containers, growing crops in controlled environments, filtering water, and practicing proper food handling techniques, you can significantly reduce the risk of radioactive contamination. Additionally, regularly testing food and water for radiation ensures that your food remains safe to consume.

Through careful planning and attention to detail, you can create a food system that provides clean, uncontaminated nourishment, even in the most challenging conditions. By implementing these strategies, you'll protect both your immediate and long-term food supply from the dangers of fallout contamination.

Surviving in a Shelter: What You Need to Know

Surviving in a fallout shelter during a nuclear event is a complex task that requires careful preparation, knowledge, and adaptability. The shelter will be your home and your lifeline for an extended period, potentially lasting weeks, months, or even years, depending on the severity of the event. This chapter explores what you need to know to survive inside a shelter, covering essential systems, supplies, and practical advice to ensure your safety and well-being in a confined environment.

The Basics of Shelter Survival

Surviving in a shelter goes beyond just waiting for radiation levels to drop. It involves managing your resources, maintaining your physical and mental health, and dealing with the everyday challenges of living in a confined space. Key areas to focus on include:

Essential Supplies

Managing Air Quality

Water and Food Management

Maintaining Sanitation and Hygiene

Power and Lighting

Physical and Mental Health

Communication and Information

Routine and Structure

Waste Management

Essential Supplies

To survive in a shelter for an extended period, you'll need a variety of essential supplies. It's important to stockpile enough resources to last for the duration of your stay, as leaving the shelter to replenish supplies may not be possible.

Food and Water

The most critical supplies are food and water. You should store enough non-perishable food and clean water to last for several months, with the ability to replenish supplies if needed.

Food: Store canned goods, dried foods, freeze-dried meals, and high-energy snacks. Ensure your diet includes a variety of nutrients to maintain health.

Water: Store at least 1 gallon (3.8 liters) of water per person per day. Consider ways to filter or treat water if your shelter has access to external sources.

First Aid and Medical Supplies

Injuries, illnesses, and other health issues may arise, so it's essential to have a well-stocked first aid kit. Include basic medical supplies like bandages, antiseptics, pain relievers, prescription medications, and any specialized treatments you may need, such as potassium iodide (for radiation exposure).

Clothing and Bedding

Prepare for different temperature conditions inside the shelter. Store warm clothing, blankets, sleeping bags, and thermal underwear for cold conditions, and lighter clothing for warmer days.

Tools and Maintenance Equipment

You'll need tools to maintain your shelter and make minor repairs. Include items such as:

Multi-tools

Wrenches, screwdrivers, and pliers

Duct tape

Spare parts for essential systems like air filters and water filtration units

Batteries and power banks

Managing Air Quality

Air quality is one of the most critical factors for survival in a shelter. Without proper ventilation and filtration, the air inside can become contaminated with carbon dioxide, radioactive particles, or other harmful substances.

Air Filtration Systems

Install an air filtration system to ensure the air inside your shelter is clean and free from radioactive fallout or other contaminants. Use HEPA or NBC (nuclear, biological, chemical) filters to remove dangerous particles from the air. Make sure to maintain and replace filters regularly.

Ventilation

Your shelter must have proper ventilation to circulate fresh air and prevent carbon dioxide buildup. Air intake and exhaust vents, equipped with blast valves and filters, are essential. In case of power loss, have a manual air circulation system, such as a hand-crank fan or pump.

Water and Food Management

Proper management of water and food supplies is vital for long-term survival. Mismanagement of these resources can lead to shortages or contamination.

Water Management

Conservation: Use water sparingly to make your supply last as long as possible. Limit water usage for hygiene and cooking.

Filtration and Purification: If you need to replenish your water supply from external sources, use a reliable water filtration system. Reverse osmosis or activated carbon filters are effective for removing radioactive particles and contaminants.

Food Rationing

Rationing: Develop a plan to ration your food supply to ensure it lasts for the full duration of your stay. Rotate through your stored food to avoid spoilage, and prioritize non-perishable items with long shelf lives.

Cooking: Use portable stoves, solar ovens, or non-electric methods to cook food. Ensure proper ventilation when cooking inside the shelter to avoid the buildup of harmful gases.

Maintaining Sanitation and Hygiene

In a confined space, maintaining sanitation and hygiene is critical to prevent the spread of illness. Poor hygiene can lead to infections and other health issues.

Personal Hygiene

Handwashing and Cleaning: Wash hands frequently with soap and clean water or use hand sanitizer to maintain hygiene. Use wet wipes for cleaning when water is limited.

Showers and Baths: If water supplies allow, take sponge baths to keep clean. In situations where water is scarce, focus on cleaning essential areas like hands, face, and armpits.

Toilets and Waste Disposal

Portable Toilets: Use portable or composting toilets to manage human waste. Make sure these systems are well-sealed to prevent odors and contamination.

Waste Storage: Store waste in sealed containers away from living areas to avoid the risk of contamination and disease.

Power and Lighting

Power is essential for running air filtration systems, lighting, and other critical devices. Plan for both short-term and long-term power solutions.

Generators and Solar Power

Generators: A fuel-powered generator can provide electricity for essential systems. Ensure you store enough fuel and have a well-ventilated area to run the generator safely.

Solar Power: Solar panels with a battery storage system can provide a renewable source of power, especially for long-term survival. Position solar panels outside the shelter, if safe to do so.

Lighting

LED Lights: Use energy-efficient LED lights to minimize power consumption. Have battery-powered lanterns and flashlights available in case of power outages.

Physical and Mental Health

Maintaining physical and mental health is crucial in a survival situation. The stress of confinement and isolation can lead to psychological issues, while lack of movement can affect physical health.

Exercise

Staying active is important for both physical and mental well-being. Even in a confined space, you can do bodyweight exercises, stretching, and resistance band workouts to stay fit.

Mental Health and Coping

Routine: Establish a daily routine to give structure to your time in the shelter. Include time for exercise, relaxation, and practical tasks.

Entertainment: To maintain morale, have entertainment options like books, games, music, or puzzles. Staying mentally engaged can help reduce stress and boredom.

Communication and Information

Staying informed about the situation outside the shelter is critical for assessing when it's safe to leave. Communication devices will help you receive emergency updates and stay in contact with others.

Radios and Communication Devices

Battery-Powered Radios: Use a battery-powered or hand-crank emergency radio to receive broadcasts from government or emergency services about radiation levels and rescue efforts.

Two-Way Communication: If possible, use a two-way radio or satellite phone to communicate with others or call for help.

Routine and Structure

Establishing a routine and structure in the shelter helps create a sense of normalcy and keeps your mind focused. Here are some ideas to implement:

Task Schedule: Assign daily tasks, such as cleaning, food preparation, and maintenance, to everyone in the shelter.

Rest and Relaxation: Schedule time for rest and relaxation, whether that's reading, meditating, or engaging in hobbies.

Waste Management

Efficient waste management is essential to prevent the spread of disease and maintain a clean living environment. All waste, including human waste, food scraps, and general trash, must be handled properly.

Waste Disposal

Sealed Containers: Store waste in sealed containers to prevent odors and contamination. Ensure the containers are located away from living areas.

Composting Toilets: Composting toilets are an efficient way to manage human waste in a confined environment, reducing the need for frequent waste disposal.

Surviving in a shelter for an extended period requires careful planning, resource management, and attention to both physical and mental health. By ensuring you have essential supplies, maintaining proper air quality, managing food and water resources, and keeping the shelter clean, you can significantly increase your chances of long-term survival.

Establishing a routine, maintaining communication, and focusing on mental well-being will help keep morale high and reduce stress in what can be a challenging and isolating experience. Proper preparation and ongoing attention to these key factors will make your time in the shelter safer and more bearable until it's safe to return to the outside world.

Nuclear Fallout: Dressing for Survival

In the aftermath of a nuclear event, one of the most critical steps for survival is protecting yourself from radioactive fallout. While seeking shelter is the top priority, there may be times when you need to go outside, whether to collect supplies, check on your surroundings, or take care of essential tasks. When venturing outside during or after a fallout, proper clothing and protective gear can help reduce exposure to harmful radioactive particles.

This chapter will guide you through how to dress for survival in a fallout environment, including choosing the right clothing materials, layering techniques, and the types of protective equipment that are necessary to keep you and your family safe from radioactive exposure.

Understanding Fallout and Its Risks

Nuclear fallout consists of radioactive dust and debris carried by the wind after a nuclear explosion. These particles can settle on the ground, water, buildings, and any exposed surface, including your skin and clothing. Exposure to fallout can result in both short-term and long-term health effects, such as radiation burns, acute radiation sickness, cancer, and other illnesses caused by radiation poisoning.

The primary way fallout affects the human body is through **gamma radiation**, which can penetrate clothing and skin, as well as **beta particles** and **alpha particles**, which pose a threat when inhaled, ingested, or absorbed through the skin. Proper clothing and gear can limit your exposure to these harmful particles and reduce your overall radiation dose.

Key Principles for Dressing in a Fallout Environment

Cover as Much Skin as Possible

Use Durable, Non-Permeable Materials

Layering for Protection

Protecting the Respiratory System

Eye Protection

Decontaminating Clothing and Equipment

Cover as Much Skin as Possible

The goal of fallout survival clothing is to minimize the amount of exposed skin that can come into direct contact with radioactive particles. Covering your entire body helps reduce the risk of radioactive contamination settling on your skin and being absorbed into your body.

Long-Sleeved Clothing

Full-Coverage Clothing: Wear long-sleeved shirts, pants, and jackets that fully cover your arms and legs. Ensure that there are no gaps where radioactive particles can come into contact with your skin.

Tight Seams and Closures: Choose clothing with tight seams and cuffs. Seal any gaps around the wrists, neck, and ankles using duct tape or elastic bands to prevent fallout from getting underneath your clothes.

Gloves and Boots

Gloves: Wear durable, non-permeable gloves to protect your hands from contact with radioactive particles. Rubber or nitrile gloves are ideal for this purpose, as they are resistant to chemicals and easy to decontaminate.

Boots: Wear sturdy, waterproof boots to protect your feet from radioactive dust and debris on the ground. Ensure that the boots can be easily cleaned or decontaminated after use. Seal the tops of your boots with tape or gaiters to prevent particles from entering through your pant legs.

Use Durable, Non-Permeable Materials

Choosing the right materials for your clothing is essential for minimizing radiation exposure. Certain fabrics and materials are better at preventing radioactive particles from penetrating your clothing and coming into contact with your skin.

Outer Layer Materials

Tyvek Suits: Tyvek is a non-woven, durable, and water-resistant material commonly used in hazmat suits and industrial protective gear. A disposable Tyvek suit is an excellent option for shielding your body from radioactive dust and particles. These suits are designed to prevent particulates from penetrating the fabric, offering a barrier between you and the fallout.

PVC or Rubber Suits: PVC (polyvinyl chloride) or rubber suits are also effective for blocking radioactive particles. These materials are non-porous, preventing dust and debris from sticking to or seeping through the fabric.

Waterproof or Water-Resistant Clothing

In the absence of a full-body suit, you can wear waterproof or water-resistant outerwear, such as raincoats or jackets. These garments provide an additional layer of protection by repelling fallout particles and making decontamination easier. Synthetic materials like nylon and polyester are preferable to cotton, as they are less likely to trap particles and are easier to clean.

Layering for Protection

Layering your clothing provides an added level of protection by increasing the distance between your skin and any radioactive particles that may land on your outer layers. Each layer acts as a barrier, trapping particles before they can reach your skin.

Base Layer

The base layer should be made of breathable, comfortable fabric, as this layer will be in direct contact with your skin. Choose moisture-wicking materials like synthetic fabrics (polyester or polypropylene) that will keep sweat and moisture away from your body, as moisture can cause particles to adhere more easily.

Mid-Layer

The mid-layer serves as insulation, helping regulate your body temperature while providing additional protection from radioactive particles. This layer can consist of fleece, wool, or insulated jackets. Choose materials that provide warmth but are not overly bulky, allowing you to move comfortably.

Outer Layer

The outermost layer is the most important for protection from fallout. It should be made of durable, non-permeable materials like Tyvek, PVC, or rubber. Ensure this layer is easy to clean or replace after exposure, as it will be the first to come into contact with radioactive dust and debris.

Protecting the Respiratory System

Your respiratory system is one of the most vulnerable parts of your body in a fallout environment, as inhaling radioactive particles can lead to internal contamination. Wearing a proper respirator or mask is critical for protecting your lungs from inhaling harmful particles.

Respirators and Masks

N95 Respirators: N95 respirators are designed to filter out at least 95% of airborne particles, making them effective for protecting against radioactive dust. Ensure the mask fits snugly to your face with no gaps around the edges.

Full-Face Respirators: For added protection, a full-face respirator with a P100 filter can block nearly all airborne particles and provide additional protection for your eyes. These respirators use advanced filters to prevent even small radioactive particles from entering your lungs.

Improvised Masks: If you do not have access to a respirator, you can create a makeshift mask using a damp cloth or bandana. While not as effective as a certified respirator, this can still provide some protection by filtering out larger particles.

Eye Protection

Radiation exposure can damage your eyes, especially if you are in a high-radiation area. Wearing protective eyewear helps shield your eyes from radioactive dust and fallout particles.

Goggles: Wear safety goggles or a full-face shield to protect your eyes from fallout particles. Ensure that the goggles fit securely to your face and leave no gaps for particles to enter.

Visors: In some cases, a visor or face shield can be worn over a full-face respirator to provide additional protection from debris and contamination.

Decontaminating Clothing and Equipment

After spending time outside in a fallout environment, it's essential to decontaminate your clothing and equipment to prevent radioactive particles from entering your shelter or coming into contact with your skin.

Removing and Cleaning Clothing

Doffing Procedure: Remove your outer layer of clothing carefully to avoid spreading fallout particles. Peel the clothing away from your body, starting with your gloves and outer garments. Do not shake or disturb the clothing, as this can release particles into the air.

Sealing Contaminated Clothing: Place contaminated clothing in a sealed plastic bag or container to prevent further exposure. Dispose of the clothing if it is not reusable, or wash it thoroughly in clean, filtered water.

Decontaminating Equipment

Rinsing: If you have worn reusable protective gear, such as boots, goggles, or gloves, rinse them with clean, filtered water to remove radioactive particles. Ensure the rinse water does not pool in areas where it could contaminate surfaces.

Wiping Down Surfaces: Use wet wipes or a damp cloth to wipe down any surfaces or objects that may have been exposed to fallout. Dispose of the wipes safely in sealed bags.

Dressing for survival in a fallout environment requires careful attention to detail and the right combination of clothing, protective gear, and equipment. By covering your skin, wearing non-permeable materials, using proper respiratory protection, and layering your clothing, you can reduce your exposure to harmful radioactive particles. Decontaminating yourself and your gear after exposure is just as important as wearing protective clothing. Follow proper procedures for removing and cleaning clothing to prevent bringing radioactive particles into your shelter.

Essential Protective Equipment for You and Your Family

In a nuclear fallout survival scenario, having the right protective equipment can be the difference between life and death. Protective equipment shields you from the harmful effects of radiation, radioactive particles, and other potential dangers in the post-nuclear environment. This chapter will outline the essential protective equipment you and your family will need, covering both personal protective gear and essential survival tools to keep you safe during and after a nuclear event.

Why Protective Equipment is Necessary

After a nuclear event, radioactive fallout can contaminate the air, water, soil, and surfaces around you. Exposure to this fallout can result in radiation sickness, burns, and long-term health risks such as cancer. The right protective equipment can reduce your exposure by shielding your body, respiratory system, and eyes from radioactive particles. It can also help you safely handle contaminated materials, water, and food, ensuring you avoid unnecessary exposure.

The protective equipment you'll need falls into two categories:

Personal Protective Gear: Items worn to shield your body from radioactive particles and contaminants.

Essential Survival Tools: Equipment used to maintain clean air, water, food, and hygiene during fallout conditions.

Personal Protective Gear

Respirators and Masks

One of the most critical pieces of protective equipment is a mask or respirator that can filter out radioactive particles and dust from the air. Breathing in radioactive fallout can cause severe internal damage, so protecting your respiratory system is essential.

N95 Respirators: These masks filter out 95% of airborne particles, including radioactive dust, making them suitable for short-term use in lower-risk environments. Ensure a tight fit for maximum protection.

P100 Respirators: These offer even higher protection, filtering out nearly 100% of harmful particles. A half-face or full-face respirator with a P100 filter provides better protection in areas with higher concentrations of fallout.

Full-Face Respirators: A full-face respirator offers both respiratory and eye protection, making it ideal for areas with heavy fallout. This type of respirator covers the entire face and filters out harmful particles while protecting the eyes from contamination.

Protective Clothing

Clothing that fully covers your body can protect you from coming into direct contact with radioactive dust and particles. The right materials can block particles and are easy to decontaminate after exposure.

Tyvek Suits: These are lightweight, disposable coveralls that protect against radioactive particles, dust, and chemicals. They are non-porous and prevent fallout from seeping through the material, making them an ideal choice for full-body protection.

PVC or Rubber Suits: If Tyvek suits are unavailable, heavy-duty PVC or rubber rain gear can also provide protection. These suits are non-permeable, making them resistant to fallout particles.

Gloves: Wear durable, non-permeable gloves to protect your hands from contamination. Rubber or nitrile gloves are best for handling contaminated materials, and they should be taped at the wrists to prevent particles from entering through gaps.

Boots: Waterproof, high-quality boots made of rubber or leather are essential to protect your feet from contaminated surfaces. Use tape or gaiters to seal the tops of the boots where they meet your pants to keep fallout out.

Eye Protection

Your eyes are vulnerable to radioactive particles, which can cause irritation, burns, or long-term damage. Proper eye protection is essential, especially in areas with heavy fallout or high winds that can spread particles.

Safety Goggles: Wear safety goggles that seal around your eyes to protect them from radioactive dust. Goggles should fit snugly, leaving no gaps for particles to enter.

Full-Face Shields: If using a half-face respirator, pair it with a full-face shield to protect both your eyes and face from debris and fallout particles.

Dosimeter or Geiger Counter

Monitoring radiation exposure is crucial to avoid dangerous levels of radiation. Personal dosimeters and Geiger counters help measure the radiation levels around you, allowing you to take appropriate action to protect yourself.

Personal Dosimeters: These small, wearable devices measure your cumulative exposure to radiation over time. They alert you if radiation levels reach dangerous thresholds, helping you avoid areas with high exposure.

Geiger Counters: These handheld devices measure radiation levels in your immediate surroundings. Use them to check areas for safety before entering or to assess contamination levels on objects, water, or food.

Face Shields or Balaclavas

In extreme fallout conditions, covering all skin surfaces is crucial. Use face shields, scarves, or balaclavas to cover any gaps between your respirator, goggles, and clothing. This extra layer ensures no radioactive particles come into direct contact with your skin.

Essential Survival Tools

Beyond personal protective gear, there are critical survival tools you'll need to maintain clean air, water, food, and hygiene during a nuclear fallout. These tools help reduce the risks associated with exposure to radioactive contamination.

Air Filtration Systems

Your shelter needs a reliable air filtration system to remove radioactive particles from the air and ensure safe breathing conditions. Without this system, the air inside the shelter can become contaminated with fallout, even if you're sealed off from the outside.

HEPA Filters: A HEPA (High-Efficiency Particulate Air) filter removes up to 99.97% of particles from the air, including radioactive dust. A HEPA filter system is essential for purifying air inside the shelter.

NBC Filters: These are specialized filters that protect against nuclear, biological, and chemical contaminants. If you expect higher levels of fallout or chemical exposure, an NBC filtration system offers maximum protection.

Water Filtration Systems

Water sources may become contaminated with fallout particles, so having a reliable water filtration system is critical for survival. You will need a system capable of removing both radioactive particles and chemical contaminants.

Reverse Osmosis Systems: Reverse osmosis is one of the most effective filtration methods for removing radioactive particles from water. Install a reverse osmosis system in your shelter to ensure a safe water supply.

Portable Water Filters: If you need to collect water from external sources, use portable water filters such as a gravity-fed system or a pump filter. Look for filters that use activated carbon or ceramic elements to remove contaminants.

Water Purification Tablets: Keep water purification tablets (iodine or chlorine) in your supplies as a backup method for killing pathogens. While these won't remove radioactive particles, they will eliminate harmful bacteria and viruses.

Radiation Suits

In high-radiation areas, a full-body radiation suit offers the highest level of protection. These suits are designed to prevent exposure to radioactive particles and radiation, keeping you safe during critical tasks.

Hazmat Suits: Full-body hazmat suits, made of materials like Tychem or ChemMAX, provide protection from nuclear, biological, and chemical threats. Paired with a respirator and gloves, these suits shield you from direct contact with fallout particles.

Lead-Lined Suits: For extreme radiation environments, consider lead-lined suits, which block gamma radiation. These suits are heavier and bulkier but offer enhanced protection.

Decontamination Supplies

After exposure to fallout, decontaminating yourself and your equipment is vital to avoid spreading radioactive particles inside the shelter. Have decontamination supplies readily available to clean surfaces and clothing.

Decontamination Showers: Set up a decontamination area or shower outside your shelter. This space allows you to rinse off fallout particles before entering the shelter, reducing the risk of contamination inside.

Wet Wipes and Soap: Use soap, water, and wet wipes to clean exposed skin, equipment, and surfaces. Wash thoroughly to remove any particles that may have settled on your body or belongings.

Disposable Clothing: Keep disposable coveralls or Tyvek suits on hand for tasks that require going outside. After use, discard the clothing in a sealed plastic bag to prevent further contamination.

Additional Essential Gear

Emergency Radios: Keep a battery-powered or hand-crank emergency radio to receive updates on radiation levels, weather, and rescue efforts from authorities.

Emergency Lighting: Battery-powered lanterns, flashlights, and glow sticks are essential for navigating in the dark, especially if power is lost. Ensure you have extra batteries or a solar-powered charger.

Multi-Tools and Basic Equipment: A multi-tool, duct tape, and basic hand tools are essential for repairs, sealing gaps, or handling emergencies in the shelter.

Protecting yourself and your family from radiation exposure requires a combination of personal protective gear and essential survival tools. Respirators, protective clothing, and eye protection are critical for reducing direct exposure to radioactive particles, while air and water filtration systems ensure you have access to clean, uncontaminated resources.

By assembling the right protective equipment and learning how to use it effectively, you significantly improve your chances of surviving in a fallout environment. Decontamination procedures, dosimeters for radiation monitoring, and basic survival tools further enhance your preparedness, helping you create a safe environment in the midst of extreme danger. Properly equipping yourself and your family will allow you to face the challenges of nuclear fallout with greater confidence and resilience.

Guarding Against Desperate People in a Post-Nuclear World

In the wake of a nuclear event, societal structures may collapse, leaving survivors in a state of chaos and desperation. As resources like food, water, and medical supplies become scarce, some individuals may resort to violence and theft to survive. In such a volatile environment, protecting yourself and your family from desperate people becomes a key component of survival. This chapter explores strategies for guarding against threats posed by others in a post-nuclear world, focusing on situational awareness, home fortification, defensive tactics, and conflict de-escalation.

Understanding the Threat of Desperation

In a post-nuclear environment, scarcity drives desperation. Food, water, shelter, and medicine may be hard to come by, leading people to act out of fear, hunger, or hopelessness. As law enforcement breaks down and infrastructure deteriorates, you may encounter individuals or groups willing to do anything to secure resources.

These desperate people may not be hardened criminals but rather regular individuals pushed to extreme behavior by the need to survive. Understanding this mindset is crucial for protecting yourself while also preventing unnecessary confrontations. In such a scenario, you need to prepare for the possibility of theft, attacks, or home invasions by taking measures to secure your family and your resources.

Key Strategies for Guarding Against Desperate People

Situational Awareness

Fortifying Your Shelter or Home

Establishing a Defense Plan

Dealing with Intruders

Conflict De-escalation and Negotiation

Forming Alliances and Community Networks

Situational Awareness

The first line of defense against potential threats is situational awareness. Being aware of your surroundings, the people in your vicinity, and changes in behavior or activity levels can help you identify potential threats before they escalate.

Observation

Monitor Your Environment: Regularly assess the area surrounding your shelter or home. Use binoculars or other optical devices to watch for unusual activity, such as strangers loitering, groups of people moving toward your location, or signs of desperation like looting or shouting.

Stay Informed: Listen to emergency radio broadcasts to stay updated on local conditions. Knowing about movements of displaced people, food shortages, or civil unrest can give you an early warning to heighten your defenses.

Blend In

Maintain a Low Profile: Don't advertise your resources or preparedness. Avoid overt signs of stockpiling or well-being that could attract attention, such as visible stockpiles of food or water or outward signs that your shelter is fully powered while others struggle. Avoid wearing flashy or clean clothing that signals wealth or readiness.

Noise Discipline: Keep noise levels to a minimum. Loud noises such as generators, music, or even conversation can draw attention to your location, signaling that you have resources others may covet.

Fortifying Your Shelter or Home

A well-fortified shelter or home can be a deterrent to would-be intruders. Making it difficult for desperate people to breach your space can give you the time needed to defend yourself or avoid conflict altogether.

Reinforce Entry Points

Doors and Windows: Secure doors with heavy-duty locks, reinforced steel, or solid wood. Install deadbolts and consider adding crossbars for added protection. Windows should be reinforced with metal grates, bars, or polycarbonate panels to prevent easy entry. Use blackout curtains to prevent light from escaping and revealing your presence at night.

Secondary Barriers: If possible, create secondary barriers inside your shelter or home, such as additional locked doors or hidden rooms. These barriers can slow down intruders, giving you time to respond.

Perimeter Security

Fencing and Barriers: Erect fencing or natural barriers like thorny bushes around your property. These barriers can discourage intruders or slow them down as they attempt to approach your shelter. Even simple obstacles like debris piles or tangled wire can deter would-be intruders.

Tripwires and Alarms: Set up simple tripwire systems connected to noise-making devices like cans, bells, or even homemade alarms. This can provide an early warning of approaching intruders and give you time to prepare.

Lighting: While light discipline is important to avoid attracting attention, motion-activated lighting or solar-powered floodlights can startle intruders and make them think twice about approaching your home.

Escape Routes

Designate Safe Exits: Always have a plan for retreat. Identify safe exits from your shelter or home that allow you to escape in the event of a breach. Ensure these routes are not obvious to outsiders and can be accessed quickly in an emergency.

Establishing a Defense Plan

Defending yourself and your family requires preparation and coordination. Develop a defense plan that outlines how you will respond to potential threats, both in terms of active defense and evasion.

Non-Lethal Defense Options

Pepper Spray and Tasers: Non-lethal defense tools like pepper spray, bear spray, or tasers can incapacitate an attacker without resorting to deadly force. These tools can help protect your family while avoiding escalation to lethal violence.

Blunt Instruments: Simple items like baseball bats, crowbars, or large sticks can serve as effective weapons if you are forced to defend yourself. Keep such tools easily accessible in case of an emergency.

Firearms and Lethal Defense

Firearms: If you are trained and legally permitted to carry a firearm, this can be a critical part of your defense plan. Ensure you have sufficient ammunition and are familiar with using and maintaining your firearm. However, use firearms only as a last resort, as lethal force can escalate a situation quickly.

Strategic Positioning: Identify areas within your home or shelter where you can take cover while defending yourself, such as corners with clear views of entry points or elevated positions. These spots should provide protection from incoming threats while allowing you to defend your position.

Plan for Family Safety

Roles and Responsibilities: Assign roles to each family member so that everyone knows what to do in the event of an intruder. For example, one person may be responsible for alerting the group while another takes cover or guards a specific entry point.

Safe Zones: Designate safe zones within your home or shelter where family members can retreat if danger is imminent. These zones should be as secure as possible and difficult for intruders to access.

Dealing with Intruders

If someone breaches your perimeter or approaches your home, knowing how to deal with intruders effectively can prevent conflict from escalating.

Initial Response

Verbal Warnings: If it is safe to do so, issue verbal warnings from inside your home or from a secure position. Let potential intruders know that you are armed and prepared to defend your property. Sometimes, a clear and firm warning can deter people from attempting an invasion.

Non-Violent Deterrents: Use non-violent methods to scare off intruders, such as making loud noises, activating bright lights, or using decoys that give the impression of a larger, well-defended group.

Avoiding Confrontation

Stay Hidden: If you believe intruders are only passing by or are unlikely to break in, it may be safer to stay hidden and allow them to move on. Drawing attention to yourself unnecessarily can invite further problems.

Offer Minimal Resources: In some cases, desperate people may be willing to negotiate. Offering minimal resources like food or water, while keeping your larger stockpile hidden, might defuse the situation without conflict. Only do this if you are confident it won't make you a target for future raids.

Conflict De-escalation and Negotiation

Not all confrontations need to end in violence. Knowing how to de-escalate a situation or negotiate with desperate people can sometimes prevent a dangerous encounter.

Stay Calm

Keep Your Composure: If you are faced with desperate individuals, staying calm and speaking firmly but non-threateningly can help defuse tensions. Avoid shouting or aggressive behavior, as this may escalate the situation.

Listen: Sometimes, simply listening to the needs of desperate people can create an opportunity for peaceful resolution. Offering them something small or guiding them to another location may help them move on without conflict.

Negotiation

Bartering: In a resource-scarce environment, you may be able to barter for peace. Offering small amounts of non-essential items like food, water, or basic supplies could convince desperate individuals to leave without violence. However, always be cautious when revealing any resources, as this could make you a target for future thefts.

Forming Alliances and Community Networks

In a post-nuclear world, forming alliances with trustworthy neighbors or nearby survivors can enhance your security and reduce the likelihood of attacks. A strong community can provide mutual protection and a greater sense of safety.

Create a Defense Group

Group Protection: If you live near other survivors or have family nearby, consider forming a small defense group. Working together to share resources and monitor your surroundings can deter threats and provide a stronger line of defense.

Watch Rotations: Establish a rotating watch system where group members take turns monitoring the perimeter of your shelter or neighborhood. This ensures that someone is always on alert to spot potential threats.

Build a Support Network

Trade and Barter: Establish relationships with others for trade and bartering. A small but trusted network can help you acquire necessary items without the risks of venturing into unknown areas.

Shared Resources: By pooling resources with a trusted group, you can increase the overall supply of essential items like food, water, and medical supplies, while reducing individual vulnerability. Shared knowledge and skills within the group can also strengthen your ability to survive and protect against potential threats.

Guarding Against Organized Groups

While some desperate individuals may act alone, others may band together to form groups that seek to take control of resources. These groups may be more dangerous than individuals and can pose a greater threat to your safety and supplies.

Recognizing Organized Threats

Watch for Patterns: Pay attention to groups of people who may be scouting or patrolling your area. Organized groups often exhibit behaviors like coordinated movements, posted lookouts, or deliberate reconnaissance of potential targets.

Signs of Authority: Some groups may attempt to establish control over areas by claiming authority, either through force or intimidation. Be cautious of groups that present themselves as law enforcement or military without proper credentials, as they may be imposters looking to exploit others.

Avoid Drawing Attention

Remain Low-Profile: Continue to practice low-profile behavior, avoiding visible signs of wealth or resources. Do not engage with larger groups unless absolutely necessary, as this can increase the risk of being targeted.

Defensive Positioning: If you suspect that an organized group is planning an attack or raid, position yourself and your family defensively, and prepare escape routes. Have weapons ready if you are trained to use them, but avoid unnecessary confrontation with groups that have superior numbers or firepower.

Plan for Long-Term Defense

Rotate Defenses: For long-term survival, rotate your defenses, such as different hiding spots for resources or alternate patrol routes, to avoid becoming predictable. This can help prevent organized groups from identifying weaknesses in your defenses.

Relocation: If your location becomes too dangerous, plan a potential relocation strategy. Ensure you have alternative safe houses or hideouts in case your current location is compromised by a larger, organized group.

Trust and Security

Surviving in a post-nuclear world often requires interaction with others, but trusting the wrong people can put you and your family at risk. It's important to balance building alliances with maintaining your security.

Be Selective About Who You Trust

Assess Behavior: Watch for signs of desperation, manipulation, or deceit in people who approach you. While some may genuinely need help, others may be seeking to take advantage of your resources. Trust should be earned gradually, based on consistent, trustworthy behavior.

Limit Information Sharing: Keep your stockpiles, defense plans, and supplies confidential. Even among allies, revealing too much about your resources can make you a target if circumstances change.

Internal Group Security

Monitor Internal Dynamics: If you are part of a survival group, pay attention to changes in behavior, morale, or conflicts. Desperation can cause even close friends or family to act unpredictably. Ensure that everyone understands the rules and consequences of breaking group trust or jeopardizing the group's safety.

Guarding against desperate people in a post-nuclear world requires a combination of awareness, preparation, and decisive action. By fortifying your shelter, staying vigilant, and creating a well-thought-out defense plan, you can protect yourself and your family from potential threats.

While it's important to avoid unnecessary confrontations, you must be ready to defend your resources and your loved ones if the situation calls for it. Balancing self-defense with conflict de-escalation, forming alliances, and maintaining strong situational awareness will help you navigate the dangers of a world where desperation drives people to extremes. By planning ahead and staying prepared, you improve your chances of surviving the fallout and the human dangers that follow.

Defending Your Shelter and Supplies from Outside Threats

In a post-nuclear world, defending your shelter and supplies from outside threats becomes essential for your survival. As society crumbles and resources become scarce, your shelter may attract the attention of desperate individuals or organized groups seeking to take what you have. Successfully defending your shelter involves more than just physical fortification—it requires strategic planning, vigilance, and the ability to adapt to evolving threats.

In this chapter, we will discuss how to effectively defend your shelter and supplies, focusing on establishing a secure perimeter, deterrence strategies, defensive tactics, and handling potential confrontations.

Key Elements of Shelter Defense

Fortifying Your Shelter

Perimeter Security

Deterrence Tactics

Arming Yourself for Defense

Handling Group Attacks

Dealing with Confrontations

Psychological Defense

Fortifying Your Shelter

The first step in defending your shelter is making it as difficult as possible for intruders to break in. A well-fortified shelter acts as both a physical barrier and a psychological deterrent to those who may be looking to steal your supplies.

Reinforce Doors and Windows

Doors: Install solid, reinforced doors made from steel or heavy wood. Add deadbolt locks, door bars, and cross braces to make it harder for intruders to force entry. If possible, use a steel security door with reinforced hinges.

Windows: Windows are typically weak points in a shelter's defense. Cover them with metal grates or bars to prevent break-ins. You can also reinforce windows with polycarbonate panels, which are shatter-resistant. Keep all windows covered with blackout curtains to avoid attracting attention with light.

Seal Vulnerabilities

Basements and Attics: Check for potential entry points through your basement or attic. Reinforce basement doors or windows, and ensure attic spaces cannot be accessed from the outside.

Roof Access: If your shelter is part of a larger building, such as a house, secure access to the roof to prevent intruders from entering through skylights or vents. Install security measures like locks, bars, or alarms on roof hatches.

Create Internal Barriers

In case intruders do manage to breach the outer defenses, having internal barriers can buy you time to react or retreat to a safer location within your shelter.

Secondary Doors: Install secondary doors or internal barriers inside your shelter, such as reinforced doors leading to key storage areas or safe rooms. These barriers can slow down intruders and give you time to prepare a defense.

Hidden Rooms: If possible, create hidden rooms or storage spaces that are difficult for intruders to find. These areas can be used to hide family members or supplies during an attack.

Perimeter Security

Securing the area around your shelter is critical for defending against outside threats. A well-protected perimeter can give you early warning of approaching danger and prevent intruders from getting too close.

Fencing and Physical Barriers

Fences: Erect sturdy fences around your property, preferably made of chain-link, wood, or metal. While a fence won't stop determined attackers, it can slow them down and act as a first line of defense.

Natural Barriers: Use natural barriers such as thorny bushes, brambles, or rock piles to make it more difficult for people to approach your shelter. Even small obstacles can delay intruders and give you time to react.

Surveillance and Early Warning Systems

Cameras and Alarms: If you have power, install surveillance cameras around your property. Battery-operated or solar-powered cameras are ideal for maintaining constant monitoring. Motion-activated alarms can alert you to movement near your shelter and scare off potential intruders.

Tripwires: Set up tripwires around your shelter, connected to noise-making devices like bells, cans, or even improvised alarms. These systems can alert you to anyone attempting to approach undetected.

Guard Dogs

Dogs as Deterrents: Guard dogs can be an excellent early warning system and deterrent for intruders. Well-trained dogs can alert you to nearby movement, defend your perimeter, and discourage would-be attackers with their presence alone. Make sure you have enough food and water for your dogs to keep them healthy and alert.

Deterrence Tactics

Deterrence is a key aspect of defense. Making your shelter look unappealing or too difficult to attack can discourage outsiders from even attempting to breach it.

Appear Unremarkable

Avoid Standing Out: Try to make your shelter appear as unremarkable as possible. The less attention you draw, the less likely you are to be targeted. Avoid outward signs of wealth or preparedness, such as visible solar panels, large stockpiles, or well-maintained exteriors.

Disguising Supplies: If possible, disguise your supplies and resources by keeping them out of view. Hide food, water, and medical supplies in areas where intruders would not immediately think to look, or in hidden compartments.

Signs of Occupation

Signs and Warnings: Sometimes, visible deterrents such as warning signs (e.g., "Private Property," "Armed Homeowners," or "Keep Out") can dissuade intruders. Pair these signs with visible signs of security measures, like fencing or cameras, to reinforce the message.

Visible Defenders: When safe to do so, make it known that your shelter is defended. If potential attackers see people patrolling or hear barking dogs, they may think twice before attempting a breach.

Arming Yourself for Defense

While avoidance and deterrence are preferable, sometimes the situation will require direct defense of your shelter and supplies. Having the right weapons and knowing how to use them can significantly increase your ability to protect your family and resources.

Non-Lethal Weapons

Pepper Spray and Tasers: Non-lethal weapons like pepper spray, bear spray, and tasers can incapacitate attackers without the need for deadly force. These are effective for close-range defense when lethal force is not warranted or desired.

Blunt Weapons: Tools like baseball bats, crowbars, or sturdy sticks can be used for defense if firearms are unavailable or not an option.

Lethal Weapons

Firearms: If you are trained in the use of firearms and legally permitted to own them, a firearm can be a powerful tool for defense. Shotguns, rifles, and handguns are all effective in close-quarters defense. Make sure you have enough ammunition and are familiar with the weapon's operation and maintenance.

Knives and Bladed Weapons: Knives or machetes can serve as both tools and weapons. Keep a sharp, sturdy blade on hand in case of close combat situations.

Handling Group Attacks

In some cases, you may face not just individuals but organized groups attempting to raid your shelter for supplies. These situations require special consideration and tactics.

Defensive Positioning

Strategic Positions: Place yourself and your family in defensible positions within the shelter where you can observe entry points without exposing yourselves. Look for high ground or positions with good visibility.

Bottleneck Entry Points: Use furniture or other objects to create bottlenecks at key entry points. This can slow down intruders and force them to enter one at a time, making it easier to defend.

Divide and Delay Tactics

Divide the Group: If a group attacks, try to divide them by isolating individuals. Use barriers, locked doors, or distractions to separate attackers, making them easier to deal with one-on-one or in smaller numbers.

Delay the Attack: If possible, use delay tactics like barricades, smoke bombs, or other distractions to buy time or force attackers to retreat.

Dealing with Confrontations

In a post-nuclear world, conflicts may arise unexpectedly. Knowing how to handle confrontations—whether through negotiation or self-defense—is crucial.

Conflict De-escalation

Stay Calm and Composed: In some cases, talking can defuse a situation before it becomes violent. If you have the opportunity, calmly engage with intruders or groups. Offer them minimal resources, such as water or small amounts of food, to avoid escalation.

Appeal to Logic: Sometimes appealing to reason—explaining that attacking your shelter would be too costly or dangerous for them—can convince would-be attackers to move on.

Last Resort: Fighting Back

Self-Defense: If negotiation fails and your shelter is under attack, you may have no choice but to fight back. Use your weapons, defensive positions, and barriers to protect your family and your supplies.

Know When to Retreat: In some cases, defending your shelter may not be feasible. If the threat is overwhelming, you may need to evacuate using a pre-planned escape route and retreat to a secondary safe location. Your life and the safety of your family are always the top priority.

Psychological Defense

Defending your shelter is not just about physical barriers or weapons—it's also about maintaining your psychological resilience in the face of danger.

Confidence and Control

Confidence: Show confidence in your ability to defend your shelter. Desperate people may try to intimidate or threaten you, but a calm and confident demeanor can make them reconsider. They may assume that your readiness indicates strength, which could deter an attack.

Maintain Control: Establish and maintain a routine within your shelter to give your family a sense of control and normalcy. In the face of external threats, maintaining a calm, organized environment inside the shelter helps reduce stress and allows everyone to focus on survival. A routine that includes tasks such as checking supplies, fortifying defenses, and rotating watch shifts will also keep everyone engaged and alert.

Managing Fear and Stress

Mental Health: Prolonged exposure to stress, fear, and the uncertainty of a post-nuclear world can take a toll on mental health. Acknowledge these feelings and create strategies for coping, such as deep breathing exercises, meditation, or even group discussions where family members can express their concerns and support one another.

Breaks and Distraction: Constant vigilance is important, but so is rest. Schedule breaks where you and your family can engage in simple activities like reading, playing games, or other forms of entertainment to take your mind off the dangers outside. Keeping morale high is critical for long-term survival.

Dealing with Trauma

Post-Conflict Trauma: Defending your shelter or experiencing an attack can be traumatic. After a confrontation, take time to process the event with your family. Encourage open communication about what happened and how everyone is feeling. Address any lingering fears and plan for improving security so everyone feels safer moving forward.

Encourage Resilience: Promote a mindset of resilience by focusing on what has been done well—whether it's reinforcing the shelter, successfully warding off an attack, or simply surviving another day. This helps build a positive outlook and reinforces the idea that, as a family or group, you can get through these challenges together.

Defending your shelter and supplies from outside threats in a post-nuclear world requires a combination of physical fortifications, tactical planning, and psychological resilience. By reinforcing entry points, establishing a secure perimeter, and using deterrence tactics, you can significantly reduce the likelihood of an attack. If conflict arises, having the right defensive tools and strategies will allow you to protect your family and resources.

However, defense isn't just about physical confrontation. Understanding how to de-escalate situations, strategically retreat, and manage the emotional toll of living under constant threat is just as important. By preparing both mentally and physically, you can improve your chances of surviving in an unstable and hostile environment. Ultimately, staying aware, staying calm, and being ready to act when necessary will ensure you can protect those you care about and preserve your vital supplies for the long-term challenges ahead.

Medical Issues during Long-Term Shelter Living

Long-term living in a fallout shelter brings unique medical challenges that must be anticipated and addressed to ensure survival. Without access to hospitals or professional medical care, managing health issues on your own becomes critical. This chapter focuses on common medical problems that can arise during extended periods of shelter living and offers practical solutions to prevent and manage these issues. We will cover physical health concerns, mental health management, hygiene, and first aid.

Key Medical Issues during Long-Term Shelter Living

Radiation Sickness and Exposure

Infectious Diseases

Malnutrition and Dehydration

Chronic Illness Management

Mental Health and Psychological Well-being

Injuries and Wound Care

Hygiene-Related Illnesses

Maintaining a First Aid and Medical Supply Kit

Radiation Sickness and Exposure

After a nuclear event, radiation exposure becomes one of the primary medical concerns. Prolonged exposure to high levels of radiation can result in acute radiation sickness (ARS), while low-level, long-term exposure may lead to cancer or other serious health problems.

Symptoms of Radiation Sickness

Early Symptoms: Nausea, vomiting, diarrhea, fatigue, and headache are common early signs of radiation sickness. These symptoms can appear within hours of exposure and may subside before worsening again in more severe cases.

Severe Symptoms: As radiation sickness progresses, it can lead to internal bleeding, infections, and damage to the bone marrow, which severely weakens the immune system. Symptoms include hair loss, bloody stools, high fever, and extreme fatigue.

Treatment and Prevention

Potassium Iodide (KI): Potassium iodide tablets can block the thyroid from absorbing radioactive iodine, helping to prevent thyroid cancer and other related health issues. KI should be taken as soon as possible after exposure to fallout.

Decontamination: If you suspect radiation exposure, immediately decontaminate by removing clothing and washing the skin with clean, filtered water to remove radioactive particles. Proper sheltering and limiting exposure time are the best ways to avoid radiation sickness.

Hydration and Rest: Ensure that the affected person drinks plenty of clean water and rests. In severe cases, medical care may be needed, but in a shelter scenario, focus on hydration, nutrition, and preventing secondary infections.

Infectious Diseases

Living in close quarters for an extended period increases the risk of infectious diseases spreading within your shelter. Respiratory infections, gastrointestinal illnesses, and skin infections are particularly common in confined spaces.

Common Illnesses

Respiratory Infections: Colds, flu, and pneumonia can spread quickly in close quarters. Symptoms include coughing, fever, sore throat, and difficulty breathing.

Gastrointestinal Infections: Contaminated food or water can cause illnesses such as diarrhea, vomiting, and stomach cramps. Common causes include bacteria like E. coli and Salmonella.

Skin Infections: Poor hygiene can lead to skin infections like boils, fungal infections, and rashes.

Prevention and Treatment

Vaccinations: If possible, ensure that all family members are up-to-date on their vaccines, particularly for diseases like influenza, pneumonia, and tetanus.

Hygiene Practices: Encourage frequent handwashing with soap and clean water or hand sanitizer. Disinfect commonly touched surfaces regularly to prevent the spread of germs.

Quarantine Sick Individuals: Isolate anyone who shows symptoms of illness to prevent the spread of infections to others. Provide them with their own bedding, eating utensils, and hygiene supplies.

Treatments: Stock up on over-the-counter medications like cough syrup, decongestants, pain relievers, and antidiarrheal medications. Use antibiotics responsibly, only when bacterial infections are confirmed.

Malnutrition and Dehydration

Long-term shelter living may limit access to fresh, nutritious food and clean water, increasing the risk of malnutrition and dehydration.

Signs of Malnutrition

Early Signs: Fatigue, dizziness, irritability, and weakness can indicate nutrient deficiencies. Lack of vitamins like Vitamin C (scurvy) and Vitamin D can cause long-term damage, including weakened bones and impaired immune function.

Severe Signs: In severe cases, malnutrition leads to weight loss, muscle atrophy, weakened immune system, and cognitive decline.

Prevention and Treatment

Balanced Diet: Focus on maintaining a balanced diet, even with limited food supplies. Rotate your stored food and supplement it with preserved fruits, vegetables, and protein sources like beans, freeze-dried meat, or protein powders. Multivitamins can help fill nutritional gaps.

Water Management: Store enough water for drinking and hygiene, and use filtration systems to ensure the water is safe. Symptoms of dehydration include dry mouth, dark urine, dizziness, and confusion. Treat dehydration by encouraging small, frequent sips of water or oral rehydration solutions.

Chronic Illness Management

Individuals with chronic conditions such as diabetes, asthma, hypertension, or heart disease face additional challenges during long-term shelter living. Without access to routine medical care, managing these conditions becomes a priority.

Medication Supply

Stockpile Medications: Ensure you have enough medication to manage chronic conditions for at least several months. This includes insulin, inhalers, blood pressure medication, and other essential prescriptions.

Monitoring Health: If someone has a condition like diabetes, stock up on glucose testing supplies. Blood pressure cuffs and pulse oximeters can help monitor conditions like hypertension and respiratory diseases.

Alternative Therapies

Managing Without Medications: In case medications run out, know alternative ways to manage chronic conditions. This may include diet modifications, exercise, and relaxation techniques to control hypertension or respiratory exercises for asthma management.

Mental Health and Psychological Well-being

Long-term shelter living can lead to psychological challenges like stress, anxiety, depression, and post-traumatic stress disorder (PTSD). The confined space, isolation, and ongoing fear of the outside world can take a toll on mental health.

Common Psychological Issues

Stress and Anxiety: Fear of the unknown, lack of control, and daily survival pressures can result in high levels of anxiety and chronic stress.

Depression: The loss of normal life, social isolation, and prolonged confinement may lead to feelings of hopelessness and depression.

PTSD: Surviving a nuclear event or violent encounters can cause trauma, leading to flashbacks, hypervigilance, and difficulty sleeping.

Coping Strategies

Maintain a Routine: Establish a daily routine that includes time for work, rest, exercise, and social interaction within the shelter. Structure helps create a sense of normalcy.

Exercise and Physical Activity: Physical activity, even in a confined space, helps reduce stress and anxiety. Simple exercises like stretching, yoga, and bodyweight movements can improve mental well-being.

Mental Health Support: Encourage open communication about fears, worries, and feelings of isolation. Create a support system within the family or group, and engage in activities that foster social interaction, like playing games or group discussions.

Relaxation Techniques: Practice mindfulness, deep breathing exercises, and meditation to reduce anxiety and promote calmness.

Injuries and Wound Care

Without access to professional medical care, injuries must be treated immediately to prevent infection and complications. Common injuries include cuts, burns, sprains, and fractures.

First Aid for Common Injuries

Cuts and Scrapes: Clean wounds with clean water and apply antiseptic ointment. Cover with sterile bandages or gauze to prevent infection.

Burns: For minor burns, cool the area with clean water and apply burn ointment or aloe vera. Cover with a sterile dressing. For severe burns, keep the affected area clean and seek medical help as soon as possible.

Sprains and Strains: Immobilize the injured area and apply ice or a cold pack to reduce swelling. Elevate the injured limb and use a compression bandage if available.

Fractures: Immobilize the fractured area with a splint or makeshift brace. Keep the injured person as comfortable as possible and seek help if available.

Preventing Infection

Sterile Dressings: Keep a supply of sterile bandages, gauze, and adhesive tape to cover wounds. Change dressings regularly to prevent infection.

Antibiotics: If an infection develops, use antibiotics if available. Be mindful of antibiotic resistance and only use them when absolutely necessary.

Hygiene-Related Illnesses

Poor hygiene in a confined shelter can lead to the spread of illnesses like fungal infections, rashes, and gastrointestinal diseases.

Preventing Hygiene-Related Illnesses

Personal Hygiene: Even with limited water, maintain basic hygiene practices. Use wet wipes, hand sanitizer, and sponge baths to stay clean.

Sanitation Systems: Properly manage human waste with portable toilets or composting toilets. Ensure waste is stored or disposed of safely to prevent contamination.

Lice and Scabies: Infestations like lice or scabies can spread in close quarters. Keep bedding clean, wash clothes regularly, and treat infestations immediately with over-the-counter treatments if possible.

Maintaining a First Aid and Medical Supply Kit

A well-stocked first aid kit is essential for addressing medical issues during long-term shelter living. Your medical supply kit should be comprehensive, covering everything from basic first aid to chronic illness management, infection prevention, and mental health support. Maintaining this kit ensures that you can respond to emergencies, manage minor injuries, and address common health issues without needing to rely on outside assistance.

Essentials for a First Aid Kit

Bandages and Dressings: Sterile bandages, gauze pads, adhesive bandages (Band-Aids), medical tape, and elastic bandages for securing dressings or supporting sprains.

Antiseptics and Ointments: Antiseptic wipes, hydrogen peroxide, alcohol pads, and antibiotic ointments to clean and disinfect wounds.

Burn Cream and Aloe Vera: To treat minor burns and skin irritations.

Cold Packs and Heating Pads: For treating sprains, strains, or muscle aches.

Splints and Slings: For immobilizing injured limbs and providing support to sprains or fractures.

Gloves and Masks: Disposable medical gloves for treating injuries, and face masks to reduce the risk of spreading respiratory infections.

Tweezers and Scissors: For removing splinters or debris from wounds and cutting bandages.

Thermometer: To monitor body temperature and check for fever, a common sign of infection.

Medications to Stockpile

Pain Relievers: Over-the-counter painkillers such as ibuprofen, acetaminophen, and aspirin for managing pain, headaches, and fever.

Antibiotics: If possible, stock antibiotics for treating bacterial infections. Be aware of expiration dates and proper usage to avoid resistance.

Antihistamines: For allergic reactions, stock over-the-counter medications like Benadryl or Claritin.

Antidiarrheals: Medications like loperamide (Imodium) to treat diarrhea, which can quickly lead to dehydration in a survival situation.

Laxatives: Constipation can be a problem in confined living situations, so stock a supply of mild laxatives.

Antacids: To manage indigestion, heartburn, or stomach discomfort.

Vitamins: Multivitamins to supplement nutrition and prevent deficiencies, particularly when fresh food sources are limited.

Electrolyte Solutions: Oral rehydration salts or electrolyte powders to treat dehydration from illness or lack of water.

Cough Syrup and Decongestants: For treating respiratory infections and relieving symptoms like coughing and congestion.

Supplies for Chronic Illness Management

If anyone in your family has a chronic condition, make sure your kit includes necessary medications and supplies for managing their health:

Blood Pressure Monitor: For individuals with hypertension, a blood pressure cuff is essential for monitoring.

Diabetes Supplies: For diabetics, stock blood glucose monitors, testing strips, insulin, and syringes. Also, stock sugary snacks or glucose tablets for managing hypoglycemia.

Asthma Inhalers: Ensure you have a sufficient supply of inhalers or other medications for asthma management.

d. Mental Health Support Items

Maintaining mental health during long-term shelter living is vital, so consider including items that help reduce stress and provide emotional support:

Books and Puzzles: Keeping the mind occupied with reading materials or puzzles can help reduce anxiety and stress.

Comfort Items: Items like familiar blankets, stuffed animals for children, or family photos can provide comfort in difficult times.

Journals and Writing Materials: Writing or journaling can be a healthy outlet for processing emotions and thoughts during long periods of confinement.

Long-term shelter living presents numerous medical challenges that must be managed effectively to ensure the health and well-being of everyone in the shelter. By preparing for common medical issues, such as radiation sickness, infections, injuries, and chronic illness management, you can mitigate the risks associated with living in a confined space for an extended period.

Maintaining hygiene, proper nutrition, and mental health is just as important as having a well-stocked first aid kit. With the right medical supplies, knowledge, and preventative measures in place, you can address health concerns as they arise and ensure your family is well-prepared for the long-term survival challenges of a post-nuclear world.

Your ability to adapt, stay vigilant, and respond to medical issues efficiently will be critical to overcoming the physical and emotional toll of prolonged shelter living, ensuring your family remains strong, healthy, and ready to face whatever challenges lie ahead.

Psychological Strain in a Fallout Shelter: Coping Mechanisms

Living in a fallout shelter for an extended period can take a severe toll on mental health. The isolation, fear, and uncertainty, coupled with the claustrophobic environment, can lead to intense psychological strain. Addressing this strain is as vital to survival as food and water, as maintaining mental health ensures that you and your family can function, make decisions, and cope with the daily stresses of long-term shelter living.

In this chapter, we will explore common psychological challenges faced during extended confinement in a fallout shelter and introduce coping mechanisms that can help you and your family manage the emotional and mental burden.

Psychological Challenges in a Fallout Shelter

Isolation and Loneliness

Fear and Anxiety

Depression and Hopelessness

Claustrophobia and Cabin Fever

Post-Traumatic Stress Disorder (PTSD)

Interpersonal Conflicts

Boredom and Lack of Stimulation

Isolation and Loneliness

Extended periods of isolation in a fallout shelter can lead to feelings of loneliness, even if you are with family or a group. The lack of contact with the outside world, friends, or extended family may intensify these feelings.

Coping Mechanisms:

Routine Communication: If possible, maintain communication with the outside world via radio, phone, or other devices. Hearing the voices of others, even through broadcasts, can reduce feelings of isolation.

Strengthen Relationships Within the Shelter: Engage in regular conversations and activities with the people around you. Create time for shared experiences like playing games, storytelling, or group discussions to foster a sense of community.

Create a Social Routine: Schedule daily or weekly gatherings where everyone can check in with each other emotionally. Even structured conversations, like sharing what each person is grateful for, can help reduce the feeling of being disconnected.

Fear and Anxiety

The constant threat of radiation, the uncertainty of what lies ahead, and the fear of the unknown can lead to high levels of anxiety. Fear can affect your ability to make rational decisions and increase stress levels among those in the shelter.

Coping Mechanisms:

Focus on What You Can Control: Reduce anxiety by focusing on tasks within your control. Whether it's managing food supplies, maintaining hygiene, or keeping the shelter clean, staying productive helps ground your mind in the present.

Breathing Exercises: Practice deep breathing techniques to calm the body and reduce stress. Try the "4-7-8" breathing method: breathe in for four counts, hold for seven counts, and exhale for eight counts. This helps activate the body's relaxation response.

Limit Exposure to Negative Information: If communication channels like radios or news reports are available, avoid overloading on potentially distressing information. Keeping informed is important, but obsessing over bad news can worsen anxiety.

Mindfulness and Meditation: Incorporate daily mindfulness or meditation exercises. These practices help clear the mind and reduce racing thoughts. You can also engage in guided meditations, if available, or simply sit quietly and focus on your breath.

Depression and Hopelessness

Living through a catastrophic event, coupled with a lack of normalcy and no clear end in sight, can lead to depression and feelings of hopelessness. You may find it difficult to stay motivated or see the purpose in daily routines.

Coping Mechanisms:

Set Small, Achievable Goals: Break tasks down into manageable steps. Accomplishing even small tasks, such as organizing supplies or preparing a meal, can provide a sense of achievement and boost your mood.

Stay Active: Physical activity is a powerful mood booster. Regular exercise releases endorphins, which improve mood and reduce feelings of depression. Incorporate simple exercises like stretching, bodyweight workouts, or yoga into your daily routine.

Keep a Journal: Writing down your thoughts and feelings can help process emotions. Journaling also allows you to track your mental state and reflect on small victories or progress made over time, which can improve your outlook.

Create a Sense of Purpose: Remind yourself and others of the importance of survival and the hope for a better future. Having a reason to endure hardship, whether it's for family, future generations, or a return to normal life, can provide the mental strength needed to overcome difficult moments.

Claustrophobia and Cabin Fever

Being confined in a small space for long periods can cause feelings of claustrophobia and stir up cabin fever. This restlessness can lead to frustration, irritability, and even panic attacks.

Coping Mechanisms:

Structure Your Space: Create designated areas in the shelter for different activities—sleeping, eating, exercise, and relaxation. This separation, even within a small space, can create a sense of order and reduce feelings of being trapped.

Take Breaks from the Group: While shelter life encourages close interaction, taking personal time for solitude can help reduce tensions. Find a quiet corner or area to decompress when needed, allowing for some mental space away from others.

Visualization Exercises: Practice visualization to mentally escape the confines of the shelter. Imagine yourself in open, peaceful environments like a forest, beach, or mountain. This mental exercise can help reduce feelings of confinement.

Physical Movement: Engage in stretching or walking around the shelter space to alleviate physical restlessness. Moving your body regularly helps reduce pent-up energy and alleviates the sensation of being trapped.

Post-Traumatic Stress Disorder (PTSD)

Surviving a nuclear event, witnessing death or destruction, or dealing with violent encounters can cause PTSD. Symptoms may include flashbacks, nightmares, heightened anxiety, and difficulty sleeping.

Coping Mechanisms:

Grounding Techniques: When feeling overwhelmed by flashbacks or intrusive thoughts, use grounding techniques to bring yourself back to the present. Focus on your surroundings, such as what you can hear, see, or feel, to anchor yourself in the moment.

Seek Support: Talk openly with trusted family members or companions about your experiences. Simply sharing your thoughts can provide emotional relief, especially if others are experiencing similar trauma.

Create a Calming Environment: If PTSD symptoms disrupt sleep, create a soothing nighttime routine. Use soft lighting, calming sounds, or relaxing activities like reading to help ease into sleep.

Interpersonal Conflicts

Living in close quarters for long periods can lead to conflict between individuals. Differences in opinions, stress, and lack of privacy can increase tensions, leading to arguments and strained relationships.

Coping Mechanisms:

Communication Skills: Encourage open communication and foster an environment where everyone feels heard. Use active listening techniques, where each person gets a chance to speak without interruption.

Conflict Resolution Strategies: When conflicts arise, address them early before they escalate. Use calm, respectful language and focus on resolving the issue rather than assigning blame. Having clear rules and agreed-upon responsibilities can also reduce disagreements.

Time Apart: Take breaks from one another if tensions become too high. A short time alone can help diffuse emotions and allow everyone to regroup mentally before re-engaging with others.

Boredom and Lack of Stimulation

Without access to the outside world, and with limited resources for entertainment, boredom can set in. A lack of stimulation can lead to frustration, restlessness, and feelings of purposelessness.

Coping Mechanisms:

Creative Projects: Engage in creative activities like drawing, writing, storytelling, or even crafting with available materials. These activities provide mental stimulation and can serve as an emotional outlet.

Learning and Education: Use time in the shelter to learn new skills or teach one another practical survival techniques, first aid, or even academic subjects for children. Learning something new can instill a sense of accomplishment and purpose.

Games and Puzzles: Keep a supply of board games, puzzles, card games, or even simple paper and pencil games like Sudoku or crossword puzzles. These activities can pass the time, engage the mind, and offer light-hearted breaks from the stress of shelter life.

Conclusion

The psychological strain of long-term fallout shelter living is one of the most significant challenges you and your family will face. It's important to recognize that mental health is just as crucial as physical health during a survival scenario. By addressing issues like isolation, fear, depression, and interpersonal conflict early on, you can reduce the toll of prolonged confinement.

Coping mechanisms like establishing routines, engaging in physical activity, practicing mindfulness, and fostering open communication can go a long way in maintaining emotional well-being. With the right mental health strategies, you'll be better equipped to face the challenges of shelter living and emerge stronger on the other side.

Caring for Pets and Livestock During a Nuclear Winter

During a nuclear winter, the survival of your pets and livestock can become an essential part of your overall survival strategy. Pets provide companionship and emotional support, while livestock can serve as a vital food source. However, the extreme conditions of a nuclear winter—including radiation, scarcity of resources, and harsh living environments—make caring for animals particularly challenging.

This chapter will discuss how to care for pets and livestock in such an environment, focusing on their protection from radiation, maintaining their health, ensuring their food and water supply, and creating a sustainable living arrangement for long-term survival.

Key Considerations for Caring for Pets and Livestock

Radiation Protection for Pets and Livestock

Sheltering Animals During a Nuclear Winter

Food and Water Supply for Pets and Livestock

Maintaining Health and Hygiene

Managing Stress and Behavioral Changes

Sustainability of Livestock for Long-Term Survival

Waste Management and Hygiene for Animals

Radiation Protection for Pets and Livestock

After a nuclear event, radioactive fallout can contaminate the environment, putting your pets and livestock at risk. Radiation exposure can lead to serious health problems in animals, just as it does in humans, so protecting them from radioactive particles is a top priority.

Limit Outdoor Exposure

Indoor Sheltering: Whenever possible, keep pets and livestock indoors or in covered, sheltered areas to protect them from radioactive fallout. If your animals must go outside, limit their exposure and ensure they are not grazing or ingesting contaminated plants or soil.

Protective Gear: For pets, you can create makeshift protective clothing using plastic sheeting or raincoats to limit contact with radioactive particles. For livestock, keeping them in covered pens or barns will provide better protection.

Decontamination

Cleaning Pets: If your pets are exposed to fallout, clean them immediately. Brush their fur thoroughly to remove any dust, and wash them with soap and clean water. Pay close attention to their paws, as radioactive particles can cling to the fur and cause contamination.

Decontaminating Livestock: Livestock may be harder to clean, but the same principles apply. Use water and brushes to clean their coats and hooves, and try to avoid exposing them to contaminated areas.

Radiation Symptoms in Animals

Signs of Radiation Sickness: Pets and livestock exposed to high levels of radiation may show symptoms like vomiting, diarrhoea, lethargy, hair loss, and unusual behavior. If radiation sickness is suspected, isolate the affected animal to prevent further contamination and provide supportive care, including hydration and nutrition.

Sheltering Animals during a Nuclear Winter

Proper sheltering is essential to protect animals from the cold, radiation, and other harsh conditions of a nuclear winter. The shelter you provide will depend on the type and size of the animal, but the key is to create a space that is insulated, secure, and stocked with necessary supplies.

For Pets

Indoor Space: Pets like dogs, cats, or small animals should be kept indoors with you. Set aside a space in the shelter where they can stay warm and comfortable. Keep bedding, toys, and food in their designated area to maintain a sense of normalcy for them.

Exercise Areas: Pets need some form of physical activity, even in confined spaces. Create a small exercise area where they can stretch, move, and engage in play.

For Livestock

Barn or Stable Shelter: Livestock, such as chickens, goats, or cattle, should be housed in a barn, stable, or shed with proper insulation. Ensure the shelter is ventilated but protected from wind and fallout. Keeping livestock close to your primary shelter can make care easier.

Temperature Control: During a nuclear winter, the cold can be a significant threat to livestock. Add extra insulation to barns or pens using straw, blankets, or other materials. Provide heat sources such as wood stoves or solar heaters if available, but ensure proper ventilation to prevent carbon monoxide buildup.

Food and Water Supply for Pets and Livestock

Ensuring a consistent supply of clean food and water for your pets and livestock is crucial for their survival. Contaminated food or water can expose animals to radiation and other hazards, so taking precautions is necessary.

Food Supplies

For Pets: Stock up on long-term pet food, such as dry kibble, canned food, or freeze-dried food. Make sure you have enough to last for several months, and store it in sealed, airtight containers to prevent contamination. In an emergency, pets may need to consume human food, so keep that option in mind.

For Livestock: Store hay, grain, and feed in secure, covered areas to protect them from fallout. If you are growing crops for livestock, use hydroponic or greenhouse methods to avoid contamination from radioactive particles. You may also need to ration feed carefully to ensure it lasts over the long term.

Water Supply

Filtered Water: Pets and livestock need access to clean, uncontaminated water. Use filtered or purified water to ensure their hydration needs are met. If you rely on rainwater or surface water, filter it through reverse osmosis or other effective filtration methods to remove radioactive particles.

Water Storage: Store water in large, sealed containers for both pets and livestock. Monitor water intake to prevent dehydration, and regularly check water supplies for signs of contamination or spoilage.

Maintaining Health and Hygiene

During a nuclear winter, maintaining the health and hygiene of your pets and livestock can prevent disease outbreaks and ensure their well-being.

Veterinary Supplies

Stock Veterinary Medications: Keep a supply of basic veterinary medications, including flea and tick treatments, de-wormers, and any specific medications your pets or livestock require. First aid kits for animals should also include antiseptic solutions, bandages, and wound care items.

Regular Check-ups: Even in a shelter environment, regularly check your animals for signs of illness, injury, or unusual behavior. Early detection of problems can help you address issues before they become severe.

Hygiene

Waste Management: For pets, establish a designated area for waste elimination. Regularly clean litter boxes or outdoor pens to prevent unsanitary conditions. For livestock, manage manure by removing it from their shelter and storing it away from their living space.

Grooming: Regular grooming is important to maintain the health of pets and livestock. Brush pets' fur to prevent matting, and clean hooves or claws in livestock to avoid infections.

Managing Stress and Behavioral Changes

Animals, like humans, can experience stress in unfamiliar and confined environments. Prolonged shelter living, reduced activity, and changes in routine can lead to behavioral changes in pets and livestock.

For Pets

Maintain Routine: Keep a consistent feeding, exercise, and play routine for your pets. This helps reduce anxiety and gives them a sense of stability.

Provide Comfort: Give your pets familiar items like their favorite toys, blankets, or beds. Familiar smells and objects can help soothe them in stressful times.

Behavioral Issues: If your pets exhibit anxiety, restlessness, or aggression, try to provide calming stimuli, such as soft music or extra attention. Training and patience will be required to manage any unwanted behaviors.

For Livestock

Reduce Stressors: Livestock may become agitated in confined spaces or with limited access to grazing. Try to provide space for movement and maintain a calm, quiet environment to minimize stress.

Group Dynamics: Monitor group-living livestock for signs of aggressive behavior toward one another, especially if food and space are limited. Separate animals if necessary to prevent fights or injuries.

Sustainability of Livestock for Long-Term Survival

If you are raising livestock, their role in your long-term survival plan will likely involve providing food such as eggs, milk, or meat. Sustainability becomes key when caring for livestock over a prolonged period in a nuclear winter.

Breeding Livestock

Repopulation: If you rely on livestock for food, breeding animals becomes essential for long-term sustainability. Ensure you have a breeding pair or a small group of animals capable of reproducing to maintain a steady supply of food.

Animal Care during Breeding: Pregnant livestock require special care and additional nutrition. Monitor the health of pregnant animals closely and provide them with extra warmth and space as needed.

Utilizing Livestock Products

Eggs and Dairy: Chickens can provide a regular source of eggs, while goats or cows offer milk. Make sure their living conditions support healthy production, and monitor their food intake to sustain their output.

Butchering Livestock: If you are raising livestock for meat, make sure you have the tools and skills to butcher animals humanely and efficiently. Store the meat in cool, dry conditions, or use preservation techniques like drying, salting, or freezing.

Waste Management and Hygiene for Animals

Proper waste management is essential to prevent unsanitary conditions in confined spaces, which can lead to disease outbreaks.

Waste Disposal for Pets

Litter and Waste Bags: For pets like dogs and cats, maintain a supply of waste bags and litter. Dispose of waste outside the shelter or in designated waste areas to keep living conditions sanitary.

Livestock Manure Management

Composting Manure: Livestock waste can be managed through composting. This reduces the risk of contamination and creates fertilizer for future gardening efforts. Keep the composting area far from your shelter and livestock housing to minimize odors and the potential spread of bacteria or contaminants. Regularly turn the compost to aid decomposition and reduce harmful pathogens.

Waste Removal Systems

Livestock Pens: For livestock kept in pens or stables, ensure you have a waste removal system in place. Regularly clean out manure and bedding to maintain hygiene and prevent the buildup of harmful gases like ammonia, which can affect both animals and humans in confined spaces.

Bedding Management: Replace bedding, such as straw or hay, regularly to keep the shelter clean. Bedding should be kept dry and free of contamination, as wet or dirty bedding can lead to infections and respiratory issues in livestock.

Caring for pets and livestock during a nuclear winter presents unique challenges, but with proper planning, it is possible to maintain their health and well-being even in the most extreme conditions. Protecting animals from radiation exposure, ensuring their food and water supplies, and maintaining hygiene are crucial for their survival. Pets offer emotional support, and their well-being can help improve morale within the shelter, while livestock provides essential food sources like eggs, milk, and meat for long-term survival. By establishing proper shelters, managing waste, and addressing stress and health issues, you can ensure that your animals continue to thrive during the difficult conditions of a nuclear winter. With careful attention to both their physical and emotional needs, pets and livestock can become a valuable part of your survival strategy, contributing to your food supply, emotional well-being, and overall resilience in the face of a nuclear winter.

Protecting Livestock from Radiation Exposure

Protecting livestock from radiation exposure during a nuclear event is crucial for ensuring their health and long-term viability as a food source. Livestock, like humans, can suffer severe health consequences from radiation, such as radiation sickness, genetic mutations, and organ failure. Prolonged exposure to radioactive fallout can also contaminate their food and water, leading to further health risks. In this chapter, we'll explore strategies to shield your livestock from radiation and minimize the impact of fallout on their living environment.

Key Strategies for Protecting Livestock from Radiation Exposure

Understanding the Risks of Radiation for Livestock

Sheltering Livestock from Fallout

Feeding Livestock Safely

Water Supply Protection

Decontaminating Livestock

Monitoring Livestock Health for Radiation Exposure

Creating a Sustainable Environment for Long-Term Protection

Understanding the Risks of Radiation for Livestock

Livestock, like humans, are vulnerable to the effects of radiation exposure, which can have both short-term and long-term consequences. The degree of harm depends on the type of radiation (alpha, beta, or gamma) and the amount of exposure.

Short-Term Effects of Radiation Exposure

Radiation Sickness: Livestock exposed to high levels of radiation may show symptoms of radiation sickness, including vomiting, diarrhoea, fatigue, and loss of appetite. In severe cases, radiation can damage internal organs, leading to death.

Skin and Fur Damage: Livestock can suffer radiation burns and hair loss, particularly if exposed to radioactive particles on their skin or fur.

Long-Term Effects of Radiation Exposure

Reproductive Issues: Exposure to radiation can cause infertility or birth defects in livestock, which may affect the ability to breed healthy offspring.

Cancer and Organ Damage: Prolonged exposure increases the risk of cancer and other long-term health problems, such as weakened immune systems and organ damage, which can compromise the animal's ability to survive.

Sheltering Livestock from Fallout

Sheltering your livestock from radioactive fallout is one of the most effective ways to reduce their radiation exposure. The goal is to keep them indoors or in areas where radioactive particles cannot reach them.

Building or Reinforcing Livestock Shelters

Barns and Stables: Ensure that your barns and stables are well-constructed and can provide sufficient protection from fallout. Reinforce roofs and walls with materials like concrete, steel, or heavy timber to block radiation. Consider adding insulation and lining the walls with lead sheets or concrete blocks for added protection from gamma radiation.

Enclosed Pens: For smaller livestock like chickens, goats, or pigs, enclosed pens can offer protection. Use metal or plastic covers to shield pens from falling radioactive dust. Ensure these enclosures are tightly sealed to prevent contamination.

Underground or Semi-Underground Shelters

Below-Ground Shelters: If possible, construct below-ground shelters for your livestock. Underground shelters offer excellent protection from radiation, as the soil and earth act as natural barriers. These shelters can be as simple as trenches covered with heavy-duty plastic and soil or more permanent structures made from concrete.

Semi-Underground Structures: Another option is to build semi-underground shelters, where part of the structure is below ground and part above. These structures can be reinforced with thick walls and roofs to shield livestock from radiation while still allowing ventilation.

Ventilation and Air Filtration

Filtered Air: If your livestock are in an enclosed or underground shelter, proper ventilation is critical. Install air filtration systems, such as HEPA or NBC (nuclear, biological, chemical) filters, to prevent radioactive particles from entering the shelter.

Natural Ventilation: Ensure that air circulation is adequate to maintain the health of the livestock, but take precautions to filter any air entering from outside. Screened windows or vents covered with plastic sheeting and filters can allow airflow without letting in fallout.

Feeding Livestock Safely

Radioactive fallout can contaminate soil, plants, and animal feed, making it essential to protect your livestock's food supply. Preventing your animals from consuming contaminated food is critical to their long-term survival.

Stockpiling Clean Feed

Dry Feed: Stockpile a large amount of uncontaminated dry feed, such as hay, grain, and pellets, in a secure storage area away from exposure to fallout. Keep feed in airtight containers, covered bins, or sealed sheds to prevent radioactive particles from settling on it.

Supplemental Feeding: If grazing is not possible due to contamination, you will need to rely on stored feed for an extended period. Be prepared to ration feed appropriately to ensure it lasts for several months or longer.

Hydroponic or Indoor Feed Production

Growing Fodder Indoors: Consider setting up hydroponic or soil-based systems indoors to grow fodder for your livestock. Growing barley, wheatgrass, or other grains indoors under grow lights allows you to produce fresh feed without exposing the crops to fallout contamination.

Greenhouse Cultivation: If indoor space is limited, you can use greenhouses with filtered air systems to grow livestock feed. Cover the greenhouse with plastic sheeting or polycarbonate panels to prevent radioactive particles from entering.

Water Supply Protection

Contaminated water poses a significant risk to livestock health, as radioactive particles can settle in open water sources. Ensuring a safe, clean water supply is essential.

Filtered Water

Use of Water Filters: Provide filtered water to your livestock using systems like reverse osmosis or activated carbon filters. These systems can remove radioactive particles and other contaminants from water, making it safe for consumption.

Rainwater Collection: If you collect rainwater, use tarps, roofs, or other materials to channel water into closed containers. Ensure these containers are sealed and protected from fallout before filtering the water for livestock use.

Avoiding Surface Water

Prevent Access to Contaminated Water: Keep livestock away from surface water sources, such as ponds, rivers, or lakes, which may be contaminated with fallout particles. Ensure that drinking troughs or water containers are covered and protected from exposure to radioactive dust.

Decontaminating Livestock

If your livestock are exposed to radioactive particles, decontamination is necessary to prevent further exposure and contamination of their food, water, or shelter.

Brushing and Washing

Brushing: If livestock are exposed to fallout, brush their coats thoroughly to remove radioactive dust. Use gloves and protective clothing while handling contaminated animals to avoid direct exposure.

Washing: For more significant contamination, wash livestock with clean, filtered water and soap. Pay particular attention to areas that may have trapped particles, such as hooves or fur. Avoid using contaminated water for washing.

Clipping Fur

Fur Clipping: In extreme cases, where animals have long fur, consider clipping their fur to remove any particles trapped in their coats. This helps reduce the risk of further contamination and makes decontamination easier.

Monitoring Livestock Health for Radiation Exposure

Keeping a close eye on the health of your livestock is crucial in a fallout situation. Early detection of radiation sickness or contamination can help you take swift action.

Watch for Symptoms of Radiation Exposure

Common Symptoms: Monitor your livestock for symptoms of radiation sickness, such as lethargy, loss of appetite, vomiting, diarrhoea, hair loss, and skin irritation. If these symptoms appear, isolate the affected animals to prevent contamination of the rest of the herd or flock.

Testing for Radiation

Geiger Counters: If you have access to a Geiger counter, you can use it to measure the radiation levels in your livestock, feed, and water. Regular testing helps ensure that you are not exposing your animals to harmful levels of radiation.

Soil Testing: Test the soil in grazing areas or around shelters to ensure it is safe for your animals to come into contact with. If contamination is detected, move your livestock to a safer, enclosed area.

Creating a Sustainable Environment for Long-Term Protection

Protecting your livestock from radiation is not just about short-term solutions—it requires creating a sustainable, long-term environment that can support their health and productivity.

Breeding and Reproduction

Protecting Breeding Animals: Focus on protecting key breeding animals to ensure the long-term sustainability of your livestock. Maintaining the health of these animals is essential for reproducing and sustaining your food supply in the future.

Rotating Safe Grazing Areas

Soil Recovery: If possible, allow areas to recover from radiation exposure by rotating your livestock between different enclosures or areas. This helps reduce the buildup of contamination in any one location.

Fodder Management: Balance the use of stored feed, hydroponic production, and outdoor grazing to extend the longevity of your resources while protecting your animals from contaminated environments.

Protecting livestock from radiation exposure during a nuclear event requires careful planning and diligent management of their living conditions, food, and water. By sheltering your livestock, providing them with safe food and water, and regularly monitoring their health, you can reduce the risk of radiation sickness and ensure their survival.

Long-term sustainability involves protecting breeding animals, growing safe feed, and maintaining clean environments. With the right strategies in place, your livestock can continue to provide essential food resources, supporting your overall survival in the aftermath of a nuclear event.

Cleaning Contaminated Soil for Sustainable Agriculture

In the aftermath of a nuclear event, radioactive fallout can settle on soil, making it unsafe for agricultural use. Contaminated soil poses a significant risk to both human health and livestock, as crops grown in such soil can absorb radioactive particles. Cleaning and rehabilitating soil for sustainable agriculture becomes critical for long-term survival, as a self-sufficient food supply is essential when external resources are scarce.

This chapter explores methods for cleaning contaminated soil, managing radiation risks, and implementing sustainable agricultural practices that allow for the growth of safe, uncontaminated crops.

Key Steps for Cleaning Contaminated Soil

Understanding Soil Contamination

Assessing the Level of Soil Contamination

Phytoremediation: Using Plants to Remove Radiation

Soil Replacement and Dilution

Soil Immobilization Techniques

Using Organic Amendments to Bind Radioactive Particles

Testing Soil for Safety

Establishing a Sustainable Agricultural System

Understanding Soil Contamination

When radioactive fallout settles on the ground, it can become embedded in the soil, particularly in the top layer where plants draw their nutrients. The contamination may include harmful isotopes such as cesium-137, strontium-90, and iodine-131, all of which can be absorbed by plants and enter the food chain. These radioactive elements can persist in the environment for many years, creating a long-term hazard for agriculture.

Types of Radioactive Contamination

Cesium-137: This isotope is water-soluble and can be absorbed by plants and animals. It has a half-life of about 30 years, meaning it can persist in the soil for decades.

Strontium-90: Mimicking calcium, strontium-90 can be absorbed by plants and animals, leading to potential contamination of milk, meat, and vegetables.

Iodine-131: Though it has a shorter half-life of eight days, iodine-131 can contaminate crops in the immediate aftermath of a nuclear event, making early planting hazardous.

Assessing the Level of Soil Contamination

Before beginning any soil remediation efforts, it is essential to assess the level of contamination to determine the appropriate methods for cleaning the soil.

Soil Testing

Geiger Counters: A Geiger counter can help measure the radiation levels in your soil, indicating whether the contamination is at dangerous levels. This tool will give you an overall understanding of how widespread the contamination is and help you prioritize which areas need attention.

Soil Samples: Collect soil samples from different depths and areas of your land for more detailed analysis. These samples can be sent to specialized labs, if available, to determine which radioactive isotopes are present and at what concentration.

Phytoremediation: Using Plants to Remove Radiation

Phytoremediation is a natural method of cleaning contaminated soil using plants that absorb radioactive isotopes from the soil through their roots. This process can take time, but it offers a sustainable, low-cost solution for decontaminating soil over the long term.

Plants for Phytoremediation

Sunflowers: Sunflowers are one of the most effective plants for absorbing cesium-137 and strontium-90 from contaminated soil. They were successfully used to clean up radiation around Chernobyl and Fukushima.

Mustard Greens: These plants can also absorb radioactive elements and heavy metals from the soil, making them an ideal choice for soil remediation.

Amaranth: Another excellent plant for drawing contaminants out of the soil, amaranth is hardy and grows well in many climates.

b. Planting and Harvesting for Phytoremediation

Planting: Grow these plants in contaminated areas, focusing on regions with the highest radiation levels. Make sure the soil remains moist to encourage plant growth and absorption.

Harvesting: Once the plants have matured and absorbed radioactive particles, they must be harvested and disposed of properly. The plants themselves will now be radioactive, so do not compost them or use them for livestock feed. Instead, they must be safely disposed of in sealed containers, or in an isolated area where they can decay without causing harm.

Soil Replacement and Dilution

In areas with extremely high levels of contamination, replacing the top layer of soil may be necessary to create a safer environment for agriculture.

Topsoil Removal

Remove Contaminated Soil: Carefully remove the top 6 to 12 inches of contaminated soil where radioactive particles are most concentrated. This soil should be moved to an isolated location where it can be stored and monitored until it decays to safe levels.

Replace with Clean Soil: Bring in uncontaminated topsoil from a safe area to replace the removed soil. This process can be labor-intensive but provides immediate benefits by reducing the level of radiation in your agricultural land.

Soil Dilution

Mixing Contaminated and Clean Soil: If topsoil removal is not feasible, you can reduce radiation levels through soil dilution. Mix contaminated soil with clean soil from deeper layers or from areas with lower contamination to reduce the concentration of radioactive particles in any given area.

Soil Immobilization Techniques

Immobilization techniques involve adding materials to the soil that bind radioactive particles, preventing them from being absorbed by plants.

Clay and Zeolite Amendments

Clay: Adding clay to the soil can help bind cesium and strontium, preventing these isotopes from being absorbed by plants. The fine particles in clay act as a barrier, holding onto the contaminants.

Zeolites: Zeolites are natural minerals with a high capacity for trapping radioactive isotopes. Adding zeolite to contaminated soil can immobilize radioactive particles and prevent them from entering the food chain.

Using Organic Amendments to Bind Radioactive Particles

Organic amendments, such as compost or biochar, can be used to improve soil structure and reduce the bioavailability of radioactive particles.

Composting

Compost: Adding compost improves the organic matter in the soil, which can help trap radioactive particles and reduce their mobility. This makes it harder for plants to absorb radiation from the soil.

Biochar

Biochar: Biochar, a form of charcoal made from organic material, can be added to contaminated soil to bind radioactive particles. Biochar has a high surface area and can effectively trap heavy metals and radioactive elements, reducing their uptake by crops.

Testing Soil for Safety

After soil remediation efforts, regular testing is necessary to ensure that radiation levels have decreased to safe levels for agriculture.

Soil Monitoring

Regular Testing: Use Geiger counters or soil sample analysis to monitor radiation levels over time. Testing should be done before each planting season to ensure that the soil remains safe for growing crops.

Safe Levels for Agriculture

Government Guidelines: Follow local or international guidelines for radiation levels in agricultural soil. Safe radiation thresholds vary depending on the type of isotope present and local regulations, so it's important to stay informed about what constitutes a safe growing environment.

Establishing a Sustainable Agricultural System

Once you have reduced radiation levels in the soil, it is important to implement agricultural practices that will maintain soil health and continue to minimize radiation risks.

Crop Rotation

Rotating Crops: Practice crop rotation to prevent overuse of nutrients in the soil and to continue to deplete radioactive particles over time. Alternating between crops that absorb radiation and those that don't as much can help reduce contamination.

Greenhouses and Raised Beds

Greenhouses: Growing crops in greenhouses allows you to control the environment and prevent radioactive particles from contaminating the soil or plants. Use filtered air systems and cover crops with plastic or polycarbonate sheeting to protect them from fallout.

Raised Beds: Raised beds filled with uncontaminated soil provide a safer alternative to planting directly in contaminated ground. Raised beds can be covered during nuclear fallout events and are easier to monitor and manage.

Cleaning contaminated soil for sustainable agriculture is a vital step toward long-term survival in a post-nuclear environment. Using methods like phytoremediation, soil replacement, and the application of organic and mineral amendments, you can reduce the levels of radioactive contamination in your soil and make it safe for growing food again.While it may take time to fully rehabilitate contaminated land, persistence and regular testing will allow you to eventually create a sustainable agricultural system.

Combining these techniques with safe water management, greenhouse cultivation, and crop rotation ensures that you can grow nutritious, uncontaminated food even in the challenging conditions of a nuclear winter. By working steadily to clean and protect your soil, you'll establish a foundation for long-term food security and self-sufficiency.

The Best Vegetables and Crops to Grow in a Post-Nuclear World

Growing vegetables and crops in a post-nuclear world presents unique challenges due to radiation exposure, limited resources, and harsh environmental conditions. However, selecting the right crops can significantly increase your chances of survival by providing essential nutrients, caloric intake, and long-term sustainability. In this chapter, we'll explore the best vegetables and crops to grow in a post-nuclear environment, focusing on those that are hardy, nutritious, and adaptable to controlled or protected environments like greenhouses or raised beds.

Key Factors in Choosing Crops

Radiation Resistance and Absorption

Nutritional Value

Quick Growth and High Yield

Storage and Preservation

Adaptability to Protected Environments

Water and Soil Requirements

Radiation Resistance and Absorption

Some crops are better suited to growing in environments where soil and water may be exposed to low levels of radiation. Crops that are less prone to absorbing radioactive isotopes, such as cesium-137 or strontium-90, are ideal for cultivation in a post-nuclear world.

Crops Less Likely to Absorb Radiation

Potatoes: Potatoes have relatively low uptake of radioactive particles and can grow well in a variety of soils. Their ability to produce high yields in small spaces makes them an excellent choice for sustenance.

Carrots: Root vegetables like carrots tend to absorb fewer radioactive particles compared to leafy greens. They provide essential nutrients such as vitamin A and are hardy in different environments.

Sweet Potatoes: Similar to regular potatoes, sweet potatoes are a great crop for long-term storage and have a lower risk of absorbing radiation from the soil.

Crops to Avoid or Grow with Caution

Leafy Greens: While leafy greens like lettuce and spinach are nutritious, they are more likely to absorb radiation due to their exposed surface area and rapid growth. If grown, they should be cultivated in protected environments, such as greenhouses or raised beds with clean soil.

Mushrooms: Although mushrooms can grow in limited light and nutrient conditions, they are known to absorb higher levels of radioactive particles from the soil. Exercise caution when growing them in contaminated areas.

Nutritional Value

In a survival scenario, growing crops that provide the most essential nutrients is crucial for maintaining health and energy levels. Focus on crops that offer a balance of carbohydrates, protein, vitamins, and minerals to sustain you and your family.

Crops High in Carbohydrates

Corn: Corn is a staple crop that provides a high level of carbohydrates and can be used in a variety of meals. It also stores well and can be dried for long-term use.

Potatoes and Sweet Potatoes: These tubers are rich in carbohydrates, making them excellent energy sources. They are versatile and can be cooked in various ways.

Winter Squash: Squash varieties such as butternut or acorn squash are nutrient-dense and store well, providing both calories and essential vitamins like vitamin A and C.

Protein-Rich Crops

Beans: Legumes such as beans, peas, and lentils are an excellent source of plant-based protein. They can be dried and stored for long periods and are easy to grow in most environments.

Quinoa: Quinoa is a complete protein, meaning it contains all nine essential amino acids. It is relatively easy to grow and has a short growing season, making it ideal for survival gardens.

Soybeans: Rich in protein and fat, soybeans are another excellent crop for long-term sustenance. They can be used to make tofu, soy milk, and other products that provide essential nutrients.

Vitamin-Rich Crops

Carrots: High in beta-carotene, which converts to vitamin A, carrots are essential for eye health and immune function.

Peppers: Peppers, particularly bell peppers, are rich in vitamin C, which helps boost the immune system and aids in healing.

Kale and Spinach: Although these leafy greens should be grown in protected environments to avoid radiation exposure, they are packed with vitamins A, C, and K, as well as iron and calcium.

Quick Growth and High Yield

In a survival situation, it's important to grow crops that mature quickly and produce high yields. This ensures a steady food supply and allows you to rotate crops efficiently.

Fast-Growing Vegetables

Radishes: Radishes are one of the fastest-growing vegetables, often maturing in just 20-30 days. They can be grown in small spaces and provide a quick source of food.

Spinach: While spinach should be grown in clean soil or controlled environments, it grows rapidly and can be harvested in as little as four to six weeks.

Lettuce: Similar to spinach, lettuce is a quick-growing crop that can be grown indoors or in greenhouses. It provides essential vitamins and can be harvested multiple times during the growing season.

High-Yield Crops

Zucchini: Zucchini and other summer squashes are prolific producers, often yielding large quantities of food in a short time. They are relatively easy to grow and can be eaten fresh or preserved.

Beans: Beans, especially bush and pole varieties, can produce a large amount of food in a small area. They can be harvested fresh or allowed to dry for long-term storage.

Tomatoes: Tomatoes produce a high yield and can be grown in raised beds, greenhouses, or pots. They provide essential vitamins and can be preserved through canning or drying for long-term use.

Storage and Preservation

Growing crops that store well is essential for maintaining a food supply throughout the year, especially when conditions prevent planting or harvesting.

Root Vegetables for Long-Term Storage

Potatoes and Sweet Potatoes: These tubers store exceptionally well in cool, dark conditions and can last for several months without refrigeration.

Carrots and Beets: Root vegetables like carrots, beets, and parsnips store well in root cellars or cool areas, making them ideal for long-term food supplies.

Onions and Garlic: Both onions and garlic store well and provide flavor and nutritional benefits. They can also be grown indoors in pots or greenhouses.

Crops for Preservation

Corn: Corn can be dried and stored for long-term use, either ground into cornmeal or used in soups and stews.

Beans: Dry beans are one of the best crops for preservation, as they can last for years when properly stored.

Winter Squash: Winter squash varieties, such as butternut or spaghetti squash, can be stored for months in cool conditions without spoiling.

Adaptability to Protected Environments

In a post-nuclear world, growing crops in controlled environments like greenhouses, hydroponic systems, or raised beds with clean soil can protect them from radiation and extreme weather conditions.

Greenhouse-Friendly Crops

Tomatoes: Tomatoes thrive in greenhouse environments, where they can be protected from contaminated soil and unpredictable weather.

Lettuce and Spinach: Both of these leafy greens can be grown year-round in greenhouses or hydroponic systems, providing a steady source of fresh vegetables.

Cucumbers: Cucumbers grow well in greenhouses and produce high yields, making them a valuable addition to a post-nuclear garden.

Raised Bed Crops

Carrots and Radishes: These root vegetables grow well in raised beds with clean, uncontaminated soil. Raised beds can be covered during nuclear fallout events to prevent exposure to radioactive particles.

Beans and Peas: Both beans and peas grow well in raised beds and provide an excellent source of protein and fiber.

Water and Soil Requirements

Selecting crops with low water and soil quality requirements can be crucial when resources are limited or contaminated.

Drought-Tolerant Crops

Amaranth: Amaranth is a drought-tolerant grain that provides both edible leaves and seeds. It is easy to grow in dry conditions and has high nutritional value.

Quinoa: Quinoa is another crop that can tolerate dry conditions, making it ideal for environments where water is scarce.

Crops for Poor Soil

Beans: Beans are nitrogen-fixing plants, meaning they improve soil fertility by adding nitrogen back into the soil. This makes them excellent for growing in areas with poor soil quality.

Sweet Potatoes: Sweet potatoes can thrive in less-than-ideal soil conditions, making them a versatile and resilient crop for difficult growing environments.

Growing vegetables and crops in a post-nuclear world requires careful planning and consideration of factors like radiation exposure, nutritional value, and environmental adaptability. By choosing crops that are hardy, nutrient-dense, and capable of growing in protected environments, you can establish a sustainable food supply that will support your long-term survival.

Crops like potatoes, beans, carrots, and squash provide essential nutrients and can be stored or preserved for future use, while fast-growing crops like radishes, spinach, and lettuce ensure a steady supply of fresh produce. With careful selection, proper protection, and efficient use of resources, your post-nuclear garden can become a critical lifeline in maintaining health and self-sufficiency.

Cooking and Preparing Food Safely After Fallout

In a post-nuclear world, preparing food safely becomes a top priority for survival. Radioactive fallout can contaminate crops, water, and livestock, making it essential to handle food with caution to avoid radiation exposure. Proper cooking and food preparation techniques will help reduce the risks associated with fallout contamination while ensuring you and your family can still get the necessary nutrients for survival.

This chapter will cover key practices for preparing food safely after fallout, including strategies for decontaminating food, safe cooking methods, and handling contaminated water and food supplies.

Key Considerations for Cooking and Preparing Food After Fallout

Understanding Fallout Contamination in Food

Decontaminating Fresh Vegetables and Fruits

Cooking and Preparing Meat Safely

Water Safety and Purification Techniques

Preserving and Storing Food Safely

Choosing Safe Cooking Methods

Monitoring Radiation Levels in Food

Avoiding Cross-Contamination

Understanding Fallout Contamination in Food

After a nuclear event, radioactive fallout can settle on crops, water, and livestock, contaminating food sources with dangerous isotopes like cesium-137, iodine-131, and strontium-90. These radioactive particles can be ingested if proper precautions are not taken, leading to internal radiation exposure, which can cause severe health problems.

Surface Contamination vs. Internal Contamination

Surface Contamination: Fallout particles can land on the surface of fruits, vegetables, and meat, making them dangerous to consume without washing or peeling. This type of contamination can often be reduced or eliminated through proper cleaning techniques.

Internal Contamination: Some radioactive isotopes may be absorbed into plants and animals through the soil and water, leading to internal contamination. In these cases, cooking or cleaning alone may not completely remove the radiation, and caution must be taken when consuming such food.

Decontaminating Fresh Vegetables and Fruits

To safely consume vegetables and fruits that may have been exposed to fallout, it is crucial to decontaminate them thoroughly.

Washing Vegetables and Fruits

Rinse with Clean Water: Thoroughly rinse all vegetables and fruits with clean, uncontaminated water. Use a scrubbing brush to remove any dirt or fallout particles from the surface. If possible, use filtered or purified water to avoid adding contaminants during washing.

Soaking in Salt Water or Baking Soda Solution: After rinsing, soak vegetables and fruits in a solution of water and salt or baking soda for 10 to 15 minutes. This helps dislodge any remaining radioactive particles. Rinse them again with clean water afterward.

Peeling Vegetables and Fruits

Peel the Outer Layer: For root vegetables like carrots, potatoes, and beets, or fruits with thicker skins like apples and cucumbers, peel away the outer layer to remove any remaining surface contamination. Discard the peel in a safe, isolated area to avoid re-contaminating other food.

Boiling and Blanching

Boiling: For additional safety, boil vegetables before consuming them. Boiling can help reduce the concentration of radioactive particles by drawing them out of the food and into the water, which should then be discarded. However, boiling does not eliminate internal contamination.

Blanching: For leafy greens like spinach or lettuce, blanching them briefly in boiling water before consumption can help remove surface contaminants. Be cautious with leafy greens, as they are more likely to absorb radiation.

Cooking and Preparing Meat Safely

Meat from livestock or wild animals may also be contaminated by radioactive fallout, especially if the animals have been exposed to contaminated water or feed.

Removing Contaminated Fat and Organs

Trim Fat: Radioactive particles tend to concentrate in the fatty tissues of animals. Before cooking, trim off as much fat as possible from meat cuts to reduce the risk of contamination.

Avoid Organ Meat: Organs like the liver and kidneys may contain higher levels of radioactive isotopes, as these organs filter contaminants. Avoid consuming organ meats in the aftermath of a nuclear event.

Cooking Meat Thoroughly

High-Heat Cooking: Cooking meat at high temperatures can reduce the risk of contamination by breaking down surface-level radioactive particles. Use methods like grilling, boiling, or baking at high temperatures to ensure the meat is cooked thoroughly.

Avoid Smoking or Curing: Smoking or curing meat does not effectively reduce radioactive contamination. Instead, focus on cooking meat in ways that allow for the safe removal of contaminated fat and juices.

Water Safety and Purification Techniques

Water contaminated by radioactive fallout can be one of the most dangerous sources of exposure. Drinking or cooking with contaminated water can lead to serious health risks, so it's essential to purify your water before use.

Filtering Water

Water Filters: Use high-quality water filters, such as reverse osmosis systems, to remove radioactive particles from your water supply. Standard household filters like carbon filters may not be effective at removing all types of radioactive isotopes.

Clay or Ceramic Filters: Some ceramic or clay water filters can help reduce the presence of radioactive particles in water. Be sure to use filters specifically rated for removing heavy metals or radioactive contaminants.

Boiling Water

Boiling: Boiling water kills biological contaminants like bacteria and viruses but does not remove radioactive particles. It is important to filter water first before boiling, especially if fallout has contaminated the source.

Rainwater Collection

Collect Rainwater Safely: If you are relying on rainwater for drinking or cooking, use tarps, roofs, or other clean surfaces to collect it. Ensure that collected rainwater is filtered and purified before use, as fallout particles can settle in the water.

Preserving and Storing Food Safely

Proper storage and preservation of food are critical to preventing further contamination and ensuring you have enough supplies to last throughout the nuclear winter.

Sealing Food in Airtight Containers

Airtight Storage: Store all food in airtight containers to protect it from airborne fallout particles. This is especially important for dry goods like grains, flour, and beans, as well as fresh produce.

Vacuum Sealing: If possible, use vacuum sealing to preserve food and create a protective barrier against contamination.

Drying and Dehydrating Food

Dehydration: Drying or dehydrating food is an effective way to preserve it for long-term use. Make sure that food is dried in a clean, protected environment to avoid exposure to fallout during the process.

Sun Drying Precautions: Avoid drying food outdoors in open air, as it can become contaminated by radioactive particles in the environment. Use solar dehydrators or indoor methods instead.

Canning and Bottling

Pressure Canning: Pressure canning is a reliable method for preserving meats, vegetables, and fruits. It also helps sterilize the food, reducing the risk of contamination. Ensure that all cans are properly sealed and stored in a cool, dark area.

Bottling Liquids: For liquids like water, soups, or broths, use sterilized glass bottles with airtight seals to prevent contamination.

Choosing Safe Cooking Methods

Certain cooking methods are better suited for reducing the risk of contamination and ensuring food safety in a post-nuclear environment.

Boiling

Effective for Vegetables and Meat: Boiling is one of the best methods for cooking vegetables and meats, as it can help reduce surface-level contaminants. Be sure to discard the water after boiling, as it may contain radioactive particles.

Baking and Roasting

High-Heat Cooking: Baking or roasting food in a covered oven at high temperatures can reduce contamination risks by ensuring that the food is cooked thoroughly and any juices or fats are safely discarded.

Grilling

Removing Contaminated Fat: Grilling meats allows fat to drip off during cooking, which helps remove some contaminants that may be concentrated in fatty tissues. Ensure the meat is cooked evenly and to a high internal temperature.

Monitoring Radiation Levels in Food

If you have access to a radiation detection device like a Geiger counter, you can monitor the radiation levels in your food before consuming it.

Using a Geiger Counter

Scan Food Items: Use the Geiger counter to scan food items before preparing or consuming them. This can help detect any significant levels of radiation present on the surface or in the food.

Threshold for Consumption: Follow guidelines for safe levels of radiation in food. If the radiation levels exceed safe thresholds, avoid consuming the food.

Avoiding Cross-Contamination

Cross-contamination occurs when radioactive particles from contaminated food or surfaces are transferred to clean food, utensils, or preparation areas. Proper food handling and sanitation practices can prevent this from happening.

Sanitize Utensils and Surfaces

Clean Utensils Thoroughly: Wash all cooking utensils, cutting boards, and countertops with clean, filtered water and soap to prevent fallout particles from contaminating your food.

Separate Contaminated Items: Keep contaminated food, packaging, or containers away from clean items. Use separate utensils for handling raw food that may have been exposed to fallout.

Use Protective Gloves and Clothing

Wear Protective Gear: When handling contaminated food, wear protective gloves and clothing to avoid transferring radioactive particles to your skin or other surfaces. Dispose of or clean the gear after use.

Cooking and preparing food safely in a post-nuclear world requires diligence and knowledge about how radioactive contamination can affect food sources. From decontaminating fruits and vegetables to ensuring meat is cooked thoroughly, the techniques outlined in this chapter will help reduce the risk of radiation exposure and ensure your food supply remains as safe as possible. Proper handling, cleaning, and cooking methods can significantly minimize the presence of radioactive particles, helping you and your family maintain your health during long-term fallout scenarios.

Using safe water, avoiding cross-contamination, and preserving food correctly are critical aspects of food preparation in such extreme conditions. Additionally, regularly monitoring radiation levels in your food and environment, using tools like a Geiger counter, will give you greater control over what is safe to consume.

While some contamination is inevitable, especially in the immediate aftermath of a nuclear event, the steps you take to prepare and handle your food can greatly reduce your exposure to radiation. With proper knowledge, tools, and precautions, you can continue to provide nutritious, safe meals that support long-term survival in a nuclear winter environment.

How to Regenerate the Local Environment After Nuclear Contamination

Regenerating the local environment after nuclear contamination is a long and challenging process, but it is essential for restoring ecosystems, agriculture, and human habitability. While radioactive fallout can have severe and lasting impacts on soil, water, plants, and wildlife, it is possible to rehabilitate affected areas through a combination of natural processes and human intervention. This chapter will focus on practical strategies for regenerating the local environment after nuclear contamination, addressing both immediate remediation efforts and long-term recovery plans.

Key Strategies for Environmental Regeneration

Understanding the Effects of Nuclear Contamination on the Environment

Assessing the Extent of Contamination

Soil Remediation Techniques

Water Purification and Management

Revitalizing Plant Life through Phytoremediation

Restoring Wildlife and Biodiversity

Long-Term Monitoring and Sustainability

Building Community Involvement in Environmental Recovery

Understanding the Effects of Nuclear Contamination on the Environment

Nuclear contamination affects the environment in multiple ways. Radioactive fallout can contaminate the soil, water, and air, disrupting ecosystems and rendering areas unsafe for human and animal life. The severity of the contamination depends on factors like the type of radioactive isotopes released, the duration of exposure, and the weather patterns that spread fallout across a region.

Radiation's Impact on Soil and Water

Soil Contamination: Fallout particles can bind to soil, especially in the top layers, affecting its fertility and making it hazardous for plant growth. Radioactive isotopes like cesium-137 and strontium-90 can persist for decades, continuing to pose risks to agriculture and ecosystems.

Water Contamination: Radioactive particles can settle in water bodies such as rivers, lakes, and groundwater. Contaminated water sources not only harm aquatic life but also spread radiation to other parts of the environment through the food chain.

Impact on Plant and Animal Life

Plant Life: Plants can absorb radioactive particles from the soil, which can stunt growth, cause mutations, or lead to long-term damage in species that are critical for the health of ecosystems.

Wildlife: Animals may ingest radioactive particles through contaminated food and water sources, leading to sickness, reproductive issues, or death. Additionally, radioactive contamination can decimate biodiversity, leading to ecological imbalances.

Assessing the Extent of Contamination

Before beginning regeneration efforts, it is essential to assess the extent of contamination in the affected area. This involves measuring radiation levels in soil, water, and plants to determine where remediation is most needed and which methods are appropriate.

Soil Testing

Geiger Counters: Use Geiger counters to test surface-level radiation in the soil. This will give you a general idea of which areas are most contaminated and require immediate attention.

Soil Sample Analysis: Collect soil samples from different depths and locations for more detailed analysis. This can help determine which radioactive isotopes are present and at what concentrations.

Water Testing

Testing Water Sources: Use specialized water testing kits to detect radioactive particles in rivers, lakes, and groundwater. Test for isotopes like caesium, strontium, and iodine, which are particularly harmful in water sources.

Soil Remediation Techniques

Rehabilitating contaminated soil is one of the most crucial steps in environmental recovery. There are several techniques to remove or neutralize radioactive particles from soil, each suited to different levels of contamination.

Topsoil Removal

Remove Contaminated Soil: In heavily contaminated areas, the most effective solution may be to remove the top layer of soil. This contaminated soil should be safely stored or buried in an isolated area until the radiation decays to safe levels.

Replace with Clean Soil: After removing the contaminated soil, bring in clean soil from uncontaminated areas to restore the fertility of the land. This is particularly useful for agricultural areas.

Soil Washing

Washing the Soil: Soil washing involves flushing contaminated soil with water or chemical solutions to extract radioactive particles. This process is resource-intensive but effective for certain types of contamination.

Water Filtration: The water used to wash the soil must be carefully filtered to prevent further contamination of the environment.

Phytoremediation

Using Plants to Clean Soil: Phytoremediation involves planting specific types of plants, like sunflowers or mustard greens, that can absorb radioactive particles through their roots. Over time, these plants help extract contaminants from the soil, which can then be harvested and safely disposed of.

Rotating Crops: In less contaminated areas, rotating phytoremediative crops with other vegetation can gradually reduce radiation levels while still allowing for some agricultural activity.

Water Purification and Management

Cleaning contaminated water sources is vital to the regeneration of ecosystems and safe human habitation. Water purification efforts should focus on removing radioactive particles and preventing further contamination.

Water Filtration Systems

Activated Carbon Filters: Activated carbon filters can remove some radioactive particles from water, though they are more effective at removing organic contaminants. More advanced filtration systems, such as reverse osmosis, may be required for heavily contaminated water.

Clay or Ceramic Filters: Certain ceramic or clay filters can remove heavy metals and radioactive particles. These are especially useful for filtering drinking water in smaller quantities.

Natural Water Filtration

Wetland Filtration: Wetlands naturally filter water by allowing plants and microorganisms to absorb contaminants. Creating or restoring wetlands near contaminated water sources can help clean the water over time.

Sedimentation Ponds: Sedimentation ponds allow radioactive particles to settle out of the water, making it safer for further treatment or use.

Revitalizing Plant Life through Phytoremediation

Restoring plant life is critical to environmental regeneration. Plants not only absorb radioactive particles to help clean the soil but also provide habitat and food for wildlife, helping ecosystems recover.

Planting Radiation-Resistant Species

Radiation-Resistant Plants: Start by planting species known to tolerate or thrive in slightly radioactive environments. These plants will help stabilize the soil and begin the process of natural regeneration.

Using Phytoremediation Plants

Sunflowers and Mustard Greens: As mentioned earlier, sunflowers and mustard greens are excellent choices for absorbing radioactive particles. These plants should be grown in contaminated areas, harvested carefully, and safely disposed of to remove radiation from the ecosystem.

Restoring Native Vegetation

Reintroduce Native Plants: Once radiation levels are sufficiently reduced, begin replanting native vegetation to restore the natural ecosystem. Native plants are better suited to the local climate and conditions, helping to promote biodiversity and stability.

Restoring Wildlife and Biodiversity

As the environment begins to recover, it is important to reintroduce and protect wildlife populations to restore ecological balance.

Creating Safe Habitats

Protected Areas: Establish wildlife sanctuaries or protected areas where animals can live and breed without fear of contamination. These areas should be monitored to ensure they remain safe and free of radiation.

Supporting Biodiversity

Encourage Pollinator Species: Pollinators like bees and butterflies play a key role in ecosystem health. Encourage the growth of pollinator-friendly plants to support their populations and promote biodiversity.

Wildlife Corridors: Create wildlife corridors that connect different habitats, allowing animals to move safely between areas and encouraging the return of a balanced ecosystem.

Long-Term Monitoring and Sustainability

Environmental regeneration is a long-term process that requires ongoing monitoring and management. Regular testing and adjustment of remediation efforts will ensure that contamination levels continue to decline and that the environment remains safe for future generations.

Ongoing Radiation Testing

Regular Soil and Water Tests: Continue to test the soil and water for radiation levels over the coming years. This will allow you to track progress and adjust your remediation strategies as needed.

Sustainable Land Management Practices

Conservation Farming: Use sustainable farming practices such as crop rotation, cover cropping, and organic farming to maintain soil health and prevent further degradation of the land.

Sustainable Forestry: If forests are part of the affected area, practice sustainable logging and reforestation efforts to ensure that the ecosystem can regenerate while still providing resources for human use.

Building Community Involvement in Environmental Recovery

Environmental regeneration after a nuclear event is a large-scale effort that requires community involvement. Encourage collaboration and collective action to ensure long-term success.

Community Education

Teach Remediation Techniques: Educate community members about soil remediation, water purification, and sustainable agriculture. This will empower individuals to contribute to the regeneration process and foster a sense of collective responsibility.

Collective Reforestation and Planting

Community Replanting Efforts: Organize replanting drives to encourage the growth of native species, phytoremediative plants, and food crops. Involving the entire community in these efforts accelerates environmental recovery and strengthens social bonds.

Regenerating the local environment after nuclear contamination is a complex, long-term effort, but it is essential for restoring ecosystems, ensuring food security, and making the area safe for human and animal life. By using a combination of soil remediation, water purification, and the reintroduction of plants and wildlife, you can help bring the environment back to health. Through sustained effort and community involvement, you can transform contaminated land into fertile ground, restore biodiversity, and create a sustainable, thriving environment for future generations.

Establishing a Safe and Functional Society for Survivors

Establishing a safe and functional society for survivors in the aftermath of a nuclear event is one of the most important aspects of long-term survival. As the initial threats of radiation exposure, fallout, and contaminated resources diminish, the focus must shift to rebuilding a stable social structure that ensures safety, access to essential resources, and collaboration among survivors. This chapter will cover key elements of rebuilding society, addressing leadership, security, governance, resource management, and the establishment of social norms and laws.

Key Elements for Establishing a Safe and Functional Society

Leadership and Governance

Security and Protection

Resource Management and Distribution

Healthcare and Medical Infrastructure

Education and Skill Development

Social Norms and Laws

Conflict Resolution and Justice Systems

Communication and Information Sharing

Leadership and Governance

Effective leadership is essential for establishing order and guiding the rebuilding process. Without a clear structure for governance, chaos, power struggles, and disorganization can threaten the stability of the group.

Types of Leadership

Democratic Leadership: In a post-nuclear society, a democratic or participatory leadership structure can provide a sense of fairness and shared responsibility. This can include the formation of councils or assemblies where decisions are made collectively by representatives of different survivor groups.

Authoritative Leadership: In some cases, a more authoritative leadership may be necessary to make quick decisions during times of crisis. Leaders with strong survival skills, knowledge of crisis management, or expertise in areas like agriculture or medicine may take on leadership roles temporarily.

Delegation of Responsibilities

Task Assignments: Assign roles based on each individual's skills and expertise. For example, those with medical knowledge can oversee healthcare, while those with experience in agriculture or engineering can focus on food production and infrastructure rebuilding.

Forming Teams: Create teams or committees focused on specific tasks such as security, resource allocation, medical care, and education. These teams will report back to leadership to ensure that decisions are made efficiently and effectively.

Security and Protection

In a post-nuclear society, security is a primary concern. Protecting survivors from external threats, such as desperate individuals, hostile groups, or environmental hazards, is essential for maintaining order and safety.

Establishing a Defense System

Perimeter Security: Secure the boundaries of your community by setting up a perimeter with barriers, fences, or natural obstacles to deter intruders. Maintain watchposts or assign guards to monitor for potential threats.

Forming a Defense Group: Establish a group responsible for defense, which could be made up of volunteers or individuals with military or law enforcement experience. They should be trained in handling weapons, maintaining security, and de-escalating conflicts when necessary.

Internal Safety Protocols

Safety Rules: Establish clear rules for behavior within the community, including restrictions on weapon use, curfews, or procedures for reporting suspicious activities. Internal security is as important as external protection, and everyone should know how to respond to emergencies.

Conflict Prevention: Regular community meetings and open channels of communication can prevent misunderstandings and conflicts from escalating into violence.

Resource Management and Distribution

Managing limited resources—such as food, water, shelter, and medical supplies—is one of the most critical aspects of rebuilding society. A fair and efficient system for distributing these resources helps prevent conflict and ensures everyone has access to essentials.

Inventory and Rationing

Resource Inventory: Keep an accurate inventory of all available resources, from food supplies to medical equipment. This will help you determine how long the supplies will last and identify any gaps in resources.

Rationing System: Establish a fair rationing system that ensures all community members receive an equitable share of resources. Create specific guidelines for distributing food, water, and other necessities based on the size of each family or group.

Sustainable Practices

Food Production: As the environment recovers, focus on sustainable agricultural practices, including growing crops, raising livestock, and hunting or fishing as appropriate. Set up community gardens or farms to ensure a steady food supply.

Water Management: Prioritize the purification and conservation of clean water. Establish systems for collecting rainwater, filtering groundwater, or desalinating water from nearby sources if available.

Healthcare and Medical Infrastructure

The physical and mental well-being of survivors is paramount for building a functional society. Establishing a healthcare system, even on a small scale, ensures that medical needs are addressed, and health risks are minimized.

Medical Facilities

Setting up Clinics: Designate a central location for healthcare, where medical supplies, first aid, and professional assistance are available. Ideally, this would be staffed by individuals with medical training, but basic care and first aid should be available to all.

Stockpiling Medical Supplies: Maintain a stockpile of essential medical supplies, such as antibiotics, painkillers, bandages, and basic surgical tools. These supplies should be rationed and used carefully to avoid running out.

Mental Health Support

Psychological Care: Prolonged survival situations take a toll on mental health. Offer psychological support by establishing regular counseling sessions or group discussions where survivors can express their feelings, anxieties, and trauma.

Education and Skill Development

Rebuilding society involves more than just physical survival. Educating survivors and passing on knowledge and skills is essential for creating a self-sustaining community capable of long-term survival.

Teaching Essential Skills

Survival Skills: Teach essential survival skills such as first aid, food cultivation, foraging, building shelters, and basic engineering. Everyone should contribute to the survival effort by learning practical skills that benefit the community.

Technical Training: Provide training in technical skills, such as water purification, renewable energy systems, and mechanical repairs. These skills will become invaluable as infrastructure is rebuilt and technology is restored.

Education for Children

Basic Education: Ensure that children receive basic education in reading, writing, and arithmetic. This is essential for the long-term development of society, as these children will become the future leaders, scientists, and builders of the community.

Social Norms and Laws

Rebuilding society requires the establishment of clear social norms and laws to guide behavior, resolve disputes, and create a sense of order. These laws should be agreed upon by the community and enforced fairly.

Creating a Legal System

Community Agreement on Laws: Gather community members to agree on basic laws and social norms that everyone will follow. These could include rules about theft, violence, resource sharing, and personal conduct.

Enforcement: Designate a group of trusted individuals to enforce these laws fairly and without bias. They should be responsible for ensuring that everyone abides by the agreed-upon norms and that justice is applied consistently.

Rewarding Positive Behavior

Community Recognition: Recognize and reward positive behavior, such as acts of kindness, selflessness, and contributions to the community's well-being. This helps reinforce social cohesion and encourages others to contribute positively.

Conflict Resolution and Justice Systems

Conflict is inevitable in any society, but how it is resolved will determine the community's long-term stability. Establishing systems for fair conflict resolution ensures that grievances are addressed and prevents violence.

Mediation and Arbitration

Mediators: Designate neutral mediators within the community to help resolve conflicts before they escalate. These individuals should be respected by all parties and work to find fair solutions to disputes.

Arbitration Process: In more serious conflicts, set up an arbitration process where community leaders or a council can make binding decisions on disputes. Ensure this process is transparent and fair to all involved.

Punishment and Rehabilitation

Fair Punishment: For serious offenses, establish a clear system of consequences that fits the crime. Avoid excessively harsh punishments, but ensure that the community understands that certain behaviors will not be tolerated.

Rehabilitation: Where possible, offer offenders a chance for rehabilitation, allowing them to make amends through community service or by contributing to rebuilding efforts.

Communication and Information Sharing

Clear and effective communication is essential for organizing efforts, making decisions, and fostering collaboration among survivors. Establish reliable methods for sharing information within the community.

Community Meetings

Regular Meetings: Hold regular community meetings where leaders can share updates, discuss important issues, and hear concerns from all members. These meetings help keep everyone informed and involved in decision-making processes.

Transparency: Ensure that all decisions regarding resources, laws, and security are made transparently and that the reasoning behind decisions is explained to the community.

Communication Systems

Radio and Bulletin Systems: If technology permits, set up communication systems like radios or bulletin boards where important messages can be relayed. This can be especially useful for coordinating security efforts or responding to emergencies.

Establishing a safe and functional society after a nuclear event is no small task, but with effective leadership, security, resource management, and a shared commitment to rebuilding, survivors can create a stable and thriving community. By focusing on fair governance, sustainable practices, healthcare, education, and conflict resolution, a post-nuclear society can not only survive but eventually flourish. The road to recovery will be long, but with collaboration and resilience, a new and functional society can emerge from the ashes of disaster.

What a Two-Year Survival Plan Looks Like for a Family

Creating a two-year survival plan for a family after a nuclear event is essential for ensuring safety, health, and long-term survival. This plan covers all critical aspects of survival, including shelter, food, water, medical care, security, and mental well-being. In this chapter, we'll outline what a comprehensive two-year survival plan looks like, focusing on the practical strategies a family would need to thrive in a post-nuclear world.

Key Elements of a Two-Year Survival Plan

Shelter: Safe Housing and Living Conditions

Food Supply: Growing, Storing, and Preserving Food

Water Supply: Clean and Sustainable Water Sources

Health and Medical Care: Preventative and Emergency Care

Security: Protecting Your Family and Resources

Energy and Power: Ensuring Sustainable Energy Sources

Mental and Emotional Well-Being: Maintaining Stability and Routine

Skills and Education: Developing Knowledge for Long-Term Survival

Contingency Planning: Adapting to Change and Unforeseen Challenges

Shelter: Safe Housing and Living Conditions

The first priority after a nuclear event is establishing a safe and secure shelter that can protect your family from radiation, weather, and other external threats. A well-built shelter ensures long-term safety and comfort.

Choosing the Right Shelter Type

Underground Shelter: Ideally, your family's primary shelter should be underground or semi-underground, as the earth provides natural radiation shielding. Constructing an underground shelter with at least 3 feet of earth cover is recommended to protect from fallout.

Above-Ground Reinforced Shelter: If an underground shelter is not possible, reinforce your home with thick materials like concrete or steel. Seal windows and doors to prevent radioactive dust from entering, and establish an air filtration system.

Living Conditions in the Shelter

Ventilation: Ensure proper ventilation with filtered air systems. Use HEPA or NBC (nuclear, biological, chemical) filters to reduce the risk of radiation exposure.

Temperature Control: Maintain a stable, livable temperature, especially during colder months. Insulation, passive solar heating, or wood stoves can be effective methods for heating your shelter.

Space and Comfort: Create a designated space for each family member to sleep, store belongings, and carry out daily tasks. Even in tight quarters, personal space helps maintain mental well-being.

Food Supply: Growing, Storing, and Preserving Food

A sustainable food plan is critical for survival. The first step is ensuring that your family has enough stored food to last through the initial months, followed by growing your own food to maintain a long-term supply.

Stored Food Supplies

Stockpile of Non-Perishables: Stock up on non-perishable foods like dried beans, rice, pasta, canned vegetables, and freeze-dried meals. Ensure you have enough food for at least six months to a year.

Rotating Food Supplies: Rotate stored food to prevent spoilage. Use the oldest items first, and replace them as you go.

Growing Your Own Food

Garden Setup: After the initial phase of survival, establish a garden using raised beds or a greenhouse to protect plants from contamination. Grow a variety of nutrient-dense crops such as potatoes, carrots, beans, and leafy greens.

Livestock: If space permits, keep small livestock like chickens, rabbits, or goats for a sustainable source of eggs, meat, and milk.

Food Preservation

Canning and Drying: Learn food preservation techniques such as canning, drying, and fermenting to extend the life of your harvest. These methods help ensure a steady food supply throughout the year.

Water Supply: Clean and Sustainable Water Sources

Access to clean water is critical for both drinking and food preparation. A two-year survival plan must include multiple sources of water and purification methods.

Water Storage

Stored Water: Store at least 100 gallons of water per person for the first few months of survival. Water should be stored in BPA-free containers and kept in a cool, dark place to avoid contamination.

Rainwater Collection: Set up a rainwater collection system using tarps, gutters, or tanks to supplement your stored water supply. Rainwater is a renewable source, but it must be purified before use.

Water Purification

Filtration Systems: Use water filters, such as reverse osmosis or ceramic filters, to remove radioactive particles and other contaminants from collected or surface water.

Boiling and Chemical Treatment: Boil water for at least one minute to kill bacteria and viruses, but this method does not remove radioactive particles. Use purification tablets or bleach for added safety if filters are not available.

Health and Medical Care: Preventative and Emergency Care

Ensuring the health of your family in the long term involves having a well-stocked medical kit, knowledge of basic first aid, and the ability to manage chronic conditions or injuries.

Medical Supplies

First Aid Kit: Your survival plan should include a comprehensive first aid kit with bandages, antiseptics, pain relievers, and basic medical tools.

Stockpiling Medication: Ensure you have a supply of essential medications, such as antibiotics, insulin (if needed), inhalers, and prescription drugs for any chronic conditions.

Preventative Health

Hygiene: Maintain strict hygiene standards to prevent illness. Use soap, water, and hand sanitizer to reduce the spread of disease. Regularly clean living spaces to minimize the risk of infections.

Mental Health: Encourage communication and support within the family to manage stress, anxiety, and depression. Mental health is just as important as physical health for long-term survival.

Security: Protecting Your Family and Resources

Maintaining security over the long term involves protecting your family and resources from both external threats and internal disputes.

Perimeter Defense

Perimeter Surveillance: Set up a perimeter defense around your shelter using fences, natural barriers, or improvised obstacles. Keep a watch on the surroundings and be ready to detect potential intruders.

Defense Plan: Train family members in basic self-defense and the safe handling of firearms, if applicable. Have a plan for responding to potential threats without resorting to unnecessary violence.

Resource Protection

Securing Supplies: Keep essential supplies, such as food, water, and fuel, secured in locked areas to prevent theft or contamination.

Barter and Trade Security: If trading with other survivors, ensure that exchanges happen in neutral, secure areas to reduce the risk of conflict or deception.

Energy and Power: Ensuring Sustainable Energy Sources

Power is essential for cooking, heating, and communication. Your survival plan should include renewable energy sources and backup systems.

Renewable Energy

Solar Power: Install solar panels to provide electricity for basic needs like lighting, communication devices, and small appliances.

Wind or Water Power: If you live near a river or in a windy area, consider using small wind turbines or water-powered generators for additional energy.

Backup Energy Sources

Generators: Keep a gas or diesel-powered generator as a backup for emergencies, but use it sparingly to conserve fuel.

Wood Stove: A wood stove is a versatile tool for both cooking and heating, especially in colder climates. Stockpile firewood or learn to safely gather it.

Mental and Emotional Well-Being: Maintaining Stability and Routine

Mental resilience is key to enduring a long-term survival situation. Creating routines, engaging in activities, and maintaining relationships are all important for mental health.

Daily Routines

Structure and Tasks: Establish a daily routine that includes chores, meal preparation, exercise, and family activities. This structure helps reduce stress and gives everyone a sense of purpose.

Recreational Activities: Set aside time for relaxation and entertainment, such as reading, playing games, or engaging in creative projects. These activities can help maintain morale.

Community and Communication

Staying Connected: If possible, maintain communication with other survivors via radio or other methods. Establishing connections with nearby groups can provide emotional support and exchange of useful information.

Skills and Education: Developing Knowledge for Long-Term Survival

In a long-term survival situation, continuously learning and improving skills is essential for adapting to new challenges and becoming more self-sufficient.

Survival Skills

Food and Water Skills: Teach all family members essential skills like purifying water, foraging for food, gardening, and preserving food.

Mechanical and Technical Skills: Learn basic mechanical skills, such as repairing tools, equipment, or shelter structures, to ensure long-term functionality.

Educational Plans for Children

Home-schooling: Continue educating children using available resources, such as books and educational materials. Focus on practical skills like math, science, and reading, as well as survival-related knowledge.

Contingency Planning: Adapting to Change and Unforeseen Challenges

A two-year survival plan must account for the unexpected. Flexibility and adaptability are crucial for dealing with unforeseen challenges.

Backup Plans for Relocation

Secondary Shelter: In case your primary shelter becomes compromised, have a backup shelter or evacuation plan in place. Ensure everyone knows the location and the route to the secondary shelter.

Mobile Supplies: Keep a bag packed with essential supplies, often referred to as a "bug-out bag," ready in case your family needs to relocate quickly. This bag should contain food, water, first aid supplies, clothing, and key survival tools for each family member.

Adapting to Resource Shortages

Alternate Food Sources: If food supplies begin to dwindle, be prepared to adapt by foraging, hunting, or fishing. Learn how to identify edible plants in your local area and use traps or fishing techniques to supplement your diet.

Fuel and Energy Alternatives: In case fuel for generators or wood for heating becomes scarce, explore alternative energy sources, such as solar power or using animal fat or oil for lamps. Consider energy conservation methods to make your current supply last longer.

Responding to New Threats

Environmental Hazards: The post-nuclear world may present additional environmental hazards, such as extreme weather or changes in the local ecosystem. Monitor the environment closely and adjust your survival plan accordingly.

Interpersonal Conflicts: As the situation evolves, conflicts may arise within your family or community. Have conflict resolution strategies in place and encourage open communication to manage tensions before they escalate.

A two-year survival plan for a family after a nuclear event involves careful preparation, flexibility, and adaptability. By focusing on shelter, food, water, health, security, and maintaining a sense of normalcy, you can create a sustainable way of living that ensures your family's safety and well-being. Continuously learning and developing skills, while adapting to unforeseen challenges, will help your family survive and thrive in the post-nuclear environment.

With the right mindset and resources, your family can not only endure the immediate aftermath but also lay the foundation for a stable, self-sufficient life over the next two years and beyond.

Food Supplies: What You'll Need for Two Years of Shelter Living

Preparing a two-year supply of food for shelter living in the aftermath of a nuclear event requires careful planning and a focus on nutrition, longevity, and ease of storage. Your goal is to ensure that your family has access to a well-balanced diet that can support long-term health, while also making the best use of limited space and resources. This chapter will outline what types of food to store, how to preserve them, and how to create a stockpile that will sustain a family for two years.

Key Considerations for a Two-Year Food Supply

Nutritional Balance: Meeting Daily Needs

Food Storage and Preservation

Types of Food to Stockpile

Cooking and Preparation Considerations

Water and Food Preparation

Managing Food Supplies: Rationing and Rotation

Bartering and Supplementing Your Food Supply

Nutritional Balance: Meeting Daily Needs

A well-balanced diet is essential for long-term survival, as it ensures your family gets the necessary nutrients to maintain health and energy. Your food supply should cover three main categories: carbohydrates, proteins, and fats, with additional vitamins and minerals to prevent deficiencies.

Daily Caloric Intake

Adults: Each adult will need approximately 2,000 to 2,500 calories per day, depending on activity levels. In cold environments or during physically demanding activities, caloric needs may increase.

Children: Adjust calorie intake based on the age and size of children. Generally, children may require between 1,200 to 1,800 calories per day.

Macronutrients

Carbohydrates: Provide the bulk of calories and are essential for energy. Sources like rice, pasta, oats, and flour are long-lasting staples.

Protein: Necessary for muscle repair and immune function. Stock up on beans, lentils, canned meat, and freeze-dried meat or fish.

Fats: Fats are important for energy and brain function. Store cooking oils, nuts, seeds, and canned butter or ghee to ensure adequate fat intake.

Micronutrients

Vitamins and Minerals: Without access to fresh food, it's easy to develop nutrient deficiencies. Stock multivitamins, dried fruits, and fortified cereals to help meet these needs.

Food Storage and Preservation

Storing food properly is crucial to prevent spoilage and contamination. You'll need to plan for a long-term storage solution that keeps food safe, fresh, and ready to use when needed.

Storage Conditions

Temperature: Keep stored food in a cool, dry place. Temperature fluctuations can reduce shelf life, so maintaining a stable environment (ideally between 50°F and 70°F) is important.

Humidity: Low humidity is key to preventing mold growth and spoilage. Consider using moisture absorbers in food storage areas to maintain dryness.

Darkness: Keep food away from direct sunlight, as exposure to light can degrade the nutritional content of food, especially items like oils and grains.

Packaging

Airtight Containers: Store bulk foods like rice, beans, and flour in airtight containers or Mylar bags with oxygen absorbers to extend their shelf life.

Canned Foods: Canned goods are ideal for long-term storage because they are shelf-stable, resistant to spoilage, and do not require refrigeration.

Vacuum-Sealed Bags: For items like dried fruits, jerky, or grains, vacuum-sealed bags can help preserve freshness and prevent pests.

Types of Food to Stockpile

To build a two-year food supply, focus on foods that are high in calories, nutritionally dense, and have a long shelf life. A combination of dried, canned, freeze-dried, and preserved foods will give you flexibility and variety in your diet.

Carbohydrates

Rice: White rice can last 30+ years if stored properly. It's a versatile staple that provides a steady source of calories.

Pasta: Dried pasta is easy to store and has a long shelf life, offering another source of carbohydrates.

Oats: Rolled oats or steel-cut oats are a great breakfast option and are high in fiber, with a shelf life of up to 25 years if stored in Mylar bags.

Flour and Grains: Store all-purpose flour, cornmeal, and wheat berries (which can be ground into flour) for making bread and other baked goods.

Proteins

Beans and Lentils: Dried beans and lentils are high in protein, fiber, and essential nutrients. They can last up to 30 years when stored properly.

Canned Meat and Fish: Canned tuna, chicken, beef, and salmon are excellent sources of protein and can last several years.

Peanut Butter: Peanut butter (or powdered peanut butter) is calorie-dense and a good source of both protein and fat, with a shelf life of up to two years.

Freeze-Dried Meat: Freeze-dried meat, while expensive, offers an excellent long-term source of protein with a shelf life of up to 25 years.

Fats

Cooking Oils: Stock up on oils like olive oil, coconut oil, and vegetable oil. Rotate them frequently, as oils have a shorter shelf life compared to other stored foods (up to two years).

Nuts and Seeds: Store vacuum-sealed nuts and seeds for an additional source of healthy fats. Opt for vacuum-sealed packages to extend their shelf life.

Powdered Butter or Ghee: Powdered butter and ghee (clarified butter) are shelf-stable and provide fat for cooking and baking.

Fruits and Vegetables

Canned Vegetables: Stock canned vegetables like green beans, corn, carrots, and tomatoes for essential vitamins and minerals.

Freeze-Dried Vegetables: Freeze-dried vegetables retain most of their nutrients and can last up to 25 years if stored properly.

Dried Fruits: Store dried fruits like raisins, apricots, and apples. These provide essential vitamins and can satisfy sugar cravings.

Baking Essentials and Condiments

Sugar and Honey: Both sugar and honey are excellent for baking and sweetening and have long shelf lives. Honey, in particular, never spoils.

Salt: Stockpile salt for both cooking and food preservation. It's essential for flavor and long-term food storage.

Spices and Seasonings: Store a variety of spices to keep meals flavorful. Spices like pepper, garlic powder, and dried herbs can last for years when stored in airtight containers.

Cooking and Preparation Considerations

When planning your two-year food supply, it's important to consider how you will prepare meals, especially if access to power or cooking fuel is limited.

Cooking Equipment

Wood Stove or Camp Stove: In case of power loss, a wood stove or camp stove can be used for cooking. Ensure you have adequate fuel (wood, propane, etc.) stored safely.

Solar Ovens: A solar oven is an energy-efficient way to cook food without fuel, especially useful in sunny climates.

Manual Tools: Have manual kitchen tools like can openers, grinders, and manual mixers to prepare food without electricity.

Emergency Cooking Methods

Boiling Water: Boiling is one of the simplest ways to prepare many dried or canned foods. Ensure you have enough clean water stored or a way to purify water for cooking.

Dehydrated Meals: If you have access to freeze-dried or dehydrated meals, they can be rehydrated quickly with boiling water, making meal preparation easy.

Water and Food Preparation

Cooking and preparing your stored food requires a reliable source of clean water. Plan for at least one gallon of water per person per day, accounting for drinking, cooking, and cleaning.

Water Filtration Systems

Water Filters: Use a high-quality water filtration system, such as a Berkey filter or portable reverse osmosis system, to ensure clean water for drinking and cooking.

Water Storage

Stored Water: Store large quantities of water in BPA-free barrels or containers. Rotate stored water every six months to keep it fresh, or treat it with water purifying tablets.

Managing Food Supplies: Rationing and Rotation

Rationing your food supply and keeping track of expiration dates is crucial for ensuring that your stockpile lasts the full two years.

Rationing

Portion Control: Calculate the necessary portions to provide enough calories for each family member without waste. Stick to pre-planned meal sizes and avoid overeating.

Meal Planning: Create a meal plan that outlines daily and weekly food usage. This ensures variety and helps manage your supplies over the long term.

Rotation System

First In, First Out: Always use the oldest food items first. This prevents spoilage and ensures your food stays as fresh as possible.

Labeling: Label all food containers with purchase dates and expiration dates for easy tracking.

Bartering and Supplementing Your Food Supply

In a post-nuclear scenario, bartering with other survivors may be necessary to supplement your food supply, especially if certain items run low or become scarce.

Bartering Essentials

High-Value Items: Stockpile items that are highly valued for bartering, such as salt, sugar, coffee, tobacco, and alcohol. These commodities tend to be in high demand during extended survival situations and can be used to trade for food, medicine, or other essentials.

Bartering Food Items

Dried Goods: Excess supplies of dried foods like rice, beans, or flour can be useful for trading. These items are easy to store, transport, and offer in trade for other goods or services.

Canned Goods: If you have an abundance of canned goods, they can also be valuable in a barter system. However, ensure you always maintain enough for your family's needs before considering trade.

Skills as Bartering Currency

Offering Services: Beyond physical goods, your skills can become a form of currency in a bartering economy. If you are skilled in areas like food preservation, medical care, or mechanical repair, you can trade your knowledge or labor in exchange for food or other necessities.

Agricultural Knowledge: If you're able to grow more food than your family needs, trading seeds, plants, or knowledge about gardening and crop management can help build stronger relationships within a survivor community.

A well-thought-out two-year food supply for shelter living is essential for long-term survival in a post-nuclear world. By focusing on nutrient-rich, long-lasting foods, maintaining a strict storage and rotation system, and ensuring a reliable source of clean water, you can secure your family's sustenance through the most challenging conditions. The key is to balance between stored supplies, growing your own food, and having the flexibility to trade or adapt if necessary.

Planning for food is more than just calories; it's about ensuring your family has enough variety to maintain health, morale, and energy for the duration of the shelter living period. With careful preparation, a robust food storage system, and the ability to adapt to unforeseen circumstances, your family can confidently navigate the challenges of extended survival.

Water: Ensuring a Long-Term Clean Supply

Ensuring a long-term supply of clean water is critical for survival in a post-nuclear environment. Water is not only essential for drinking but also for cooking, cleaning, and hygiene. After a nuclear event, water sources can become contaminated with radioactive fallout, requiring careful planning and management to secure a safe and reliable supply for your family. This chapter will guide you through the methods for collecting, purifying, and storing water for long-term survival.

Key Elements for Securing a Long-Term Water Supply

Understanding Water Contamination

Water Storage for Long-Term Use

Collecting Water Safely

Purification Methods: Removing Contaminants

Managing Water Consumption

Backup and Alternative Water Sources

Tools and Equipment for Water Management

Understanding Water Contamination

After a nuclear event, radioactive fallout can contaminate surface water, groundwater, and even rainwater. Radioactive particles like iodine-131, cesium-137, and strontium-90 can enter water sources, making them dangerous to consume. Understanding the risks and identifying potential contamination is the first step in ensuring a clean water supply.

Types of Contamination

Radioactive Particles: Fallout can introduce harmful isotopes into water, which can cause radiation poisoning if ingested.

Chemical Contaminants: In addition to radiation, industrial and agricultural chemicals may enter water sources after infrastructure breaks down, adding to the contamination.

Biological Contaminants: Microorganisms like bacteria, viruses, and parasites can also become more prevalent as water systems fail, especially in untreated sources like rivers or ponds.

Water Storage for Long-Term Use

Storing a sufficient amount of clean water is essential to provide your family with a buffer in case your water collection or purification systems are temporarily disrupted. A basic rule is to store at least one gallon of water per person per day for drinking, cooking, and hygiene.

How Much Water to Store

Initial Supply: For two years of shelter living, calculate your family's total water needs. For a family of four, you would need at least 1,460 gallons of water (365 days x 4 people = 1,460 gallons) just for drinking and basic hygiene.

Additional Storage: If space allows, store more water to account for emergencies, such as illness, extra cleaning needs, or hotter climates where people will drink more.

Water Storage Containers

Food-Grade Containers: Use BPA-free, food-grade plastic containers or barrels specifically designed for water storage. These should be sealed tightly to prevent contamination.

Water Barrels: Large water barrels (55 gallons or more) are ideal for long-term storage. Keep these barrels in a cool, dark place to prevent algae growth and deterioration of the container.

Water Bottles and Jugs: Smaller, portable water containers can be useful for day-to-day use and in case you need to leave your shelter. Keep these stored with your emergency supplies.

Rotating Stored Water

Rotation Schedule: Even properly stored water needs to be rotated periodically. Replace stored water every 6 to 12 months to ensure it remains fresh and uncontaminated.

Collecting Water Safely

If stored water runs low or you need to supplement your supply, collecting water from natural sources can be essential. However, all collected water must be treated before use to remove contaminants.

Rainwater Collection

Gutters and Barrels: Set up a rainwater collection system using gutters, pipes, and sealed barrels. Collect water from your roof, but make sure it's filtered and purified before use, as fallout particles can contaminate rainwater.

Covering Collection Systems: During periods of fallout, cover your collection systems with plastic or tarps to prevent radioactive particles from contaminating the collected water. Once the immediate threat has passed, you can resume collecting rainwater, but all water must still be purified.

Surface Water Sources

Rivers, Lakes, and Ponds: Surface water from lakes, rivers, and ponds may be used as a supplemental source, but it is highly susceptible to contamination from fallout, chemicals, and microorganisms.

Groundwater Sources: Wells can provide a more secure water source, but they can also be contaminated if fallout particles reach the groundwater. Testing and treating well water is critical in such cases.

Purification Methods: Removing Contaminants

Water from any source, including stored water, may need purification before consumption. Different purification methods target various types of contamination, such as biological pathogens, radioactive particles, and chemicals.

Boiling

Effectiveness: Boiling water for at least one minute kills bacteria, viruses, and parasites. However, boiling does not remove radioactive particles or chemical contaminants, so it should be used in conjunction with filtration.

Filtration Systems

Activated Carbon Filters: Activated carbon filters help remove chemical contaminants and improve taste but may not remove all radioactive particles or microorganisms. These filters work best when combined with other methods.

Ceramic Filters: Ceramic filters are effective at removing bacteria, protozoa, and some radioactive particles from water. Use these as part of a multi-stage purification process.

Reverse Osmosis Systems: Reverse osmosis systems are among the most effective methods for removing radioactive particles and other contaminants from water. While expensive and complex, they provide a high level of protection.

Chemical Purification

Water Purification Tablets: These tablets, typically iodine or chlorine-based, can disinfect water by killing harmful microorganisms. However, they do not remove radioactive or chemical contaminants.

Chlorine Bleach: Unscented household bleach can be used to purify water. Use 8 drops per gallon of water, stir, and let it sit for 30 minutes. This method is effective against bacteria and viruses but does not eliminate chemical or radioactive contaminants.

Distillation

Solar Distillation: Solar distillation involves evaporating water using sunlight and collecting the clean condensation. This method removes radioactive particles, salts, and many other contaminants, making it one of the most reliable long-term purification options.

Distillation Units: If possible, invest in a distillation unit that can purify water from any source. Distillation is highly effective for removing radioactive particles, heavy metals, and other contaminants.

Managing Water Consumption

Efficient use of water is critical to ensuring your supply lasts as long as possible. Conservation measures should be in place from the start.

Daily Rationing

Drinking Water: Set strict guidelines for water use. Each person should drink at least one gallon per day, but limit water use for non-essential activities like bathing or cleaning.

Prioritizing Hygiene: Prioritize water use for hygiene in small amounts to prevent the spread of illness. Use hand sanitizer or alcohol-based wipes when possible to conserve water.

Recycling Water

Greywater Systems: Consider implementing a greywater system to reuse water from bathing, dishwashing, or laundry for purposes like flushing toilets or irrigation.

Rainwater for Cleaning: Rainwater, after being purified, can be used for tasks such as washing clothes or cleaning dishes.

Backup and Alternative Water Sources

Planning for alternative water sources is essential if your primary supply is compromised.

Portable Water Filters

Lifestraw or Sawyer Filters: Portable water filters like the Lifestraw or Sawyer can be lifesaving in emergencies. These filters allow you to drink directly from contaminated water sources, removing bacteria and protozoa.

Desalination Units

Saltwater Desalination: If you live near the ocean, a desalination unit can turn seawater into drinkable water. Portable desalination units are available, but they require significant energy, so they are more suited to emergency use.

Condensation and Dew Collection

Dew Traps: In dry environments, condensation traps can collect moisture from the air overnight. These systems are low-maintenance and provide a small but steady water supply.

Solar Water Stills: Solar stills use the sun to evaporate contaminated water or moisture and collect clean water through condensation. This method is slow but effective for emergency water collection.

Tools and Equipment for Water Management

Having the right tools and equipment on hand will make collecting, purifying, and managing water much easier.

Water Testing Kits

Radiation Detection: Use water testing kits that include radiation detection tools to check for radioactive contamination in collected water.

Chemical Testing: Keep chemical water testing kits to ensure that water is safe from industrial or agricultural pollutants.

Water Pumps and Filters

Hand Pumps for Wells: If you rely on a well for water, a hand pump can be a reliable backup in case of power outages or equipment failure.

Gravity-Fed Water Filters: These filters use gravity to push water through purification systems. They are effective for long-term use when power is unavailable.

Ensuring a long-term supply of clean water is one of the most important aspects of surviving in a post-nuclear environment. By combining proper water storage, safe collection methods, and effective purification techniques, your family can have reliable access to water for drinking, cooking, and hygiene. Careful management of water resources, along with backup plans for emergencies, will help your family survive and thrive over the two-year period in a shelter. Proper preparation, including regular testing and purification of water, ensures that your family remains healthy and hydrated, even in the most challenging conditions. Having the right tools, knowledge, and strategies in place is critical to maintaining a safe water supply for the long term. Below, we will cover a few final points to ensure a robust water management plan.

Regular Testing and Purification

Consistent Testing

Regular Water Testing: Even if you believe your water sources are secure, test the water regularly for contamination, especially if using surface water or groundwater. This ensures that radioactive particles, chemicals, and pathogens do not enter your drinking supply undetected.

Routine Maintenance of Filters: Keep your water filtration and purification systems in good condition. Clean and replace filters as recommended by the manufacturer to maintain efficiency.

Purification as a Habit

Treat All Water: Make it a rule to purify all collected water, whether from rain, surface water, or even stored sources. You never know when water may become compromised, so it's essential to make purification a routine.

Sustainable Water Usage and Community Collaboration

Conserving Water over Time

Gradual Adjustments: Over time, teach family members to use water more conservatively. Adjust habits as needed, using the minimum amount of water necessary for drinking, cooking, and hygiene.

Greywater Recycling: As mentioned earlier, develop systems to reuse greywater for non-drinking purposes like irrigation or cleaning. This helps stretch your clean water supply.

Community Water Sharing

Building a Water Network: If your shelter is part of a larger survivor community, consider sharing water resources, skills, and equipment. For instance, communities can invest in larger-scale filtration systems or desalination units, which may be too costly or resource-intensive for individual families.

Water Management Agreements: Establish agreements and protocols within your community for fair water use, including sharing access to wells, collection systems, and purification units.

Preparing for Unforeseen Challenges

Emergency Water Plans

Secondary Water Sources: Always have alternative water sources lined up. Whether it's rainwater collection, surface water, or even desalination, having multiple backup options reduces the risk of running out.

Emergency Water Kits: Pack portable water purification systems in your emergency evacuation kits. Ensure these kits include compact filters, purification tablets, and collapsible water containers.

Adapting to Environmental Changes

Changing Water Needs: In different seasons, your water needs may fluctuate. Hotter months may require more drinking water, while colder months might see higher water use for heating and hygiene. Be prepared to adjust your storage and collection efforts accordingly.

Weather Monitoring: Pay attention to weather patterns to optimize rainwater collection and adjust purification systems. Regularly monitor for contamination from natural events like floods, which can spread pollutants.

A long-term water plan ensures the survival of your family by providing reliable access to clean, safe water throughout a two-year shelter living period. By properly storing water, setting up sustainable collection systems, and using thorough purification methods, you can mitigate the risks posed by nuclear fallout and other environmental hazards. Regular testing, conservation efforts, and backup strategies will ensure that your water supply remains steady and safe even under difficult circumstances.

With a solid water management plan in place, your family can confidently face the challenges of living in a post-nuclear environment, secure in the knowledge that one of the most critical survival needs—clean water—is accounted for.

Clothing and Gear for the Long Haul

Clothing and gear are vital for long-term survival in a post-nuclear environment, as they offer protection from radiation, harsh weather, and other environmental hazards. The right clothing and gear can help your family stay warm, dry, and safe, while also providing practical advantages for daily tasks like cooking, cleaning, and defending your shelter. This chapter will explore how to select, store, and care for the clothing and gear you'll need to endure two years of shelter living.

Key Considerations for Long-Term Survival Clothing and Gear

Protecting Against Radiation

Layering for Temperature Control

Choosing Durable, Functional Clothing

Protective Gear for Outdoor Ventures

Footwear for Comfort and Safety

Maintaining and Repairing Clothing

Other Essential Gear for Survival

Protecting Against Radiation

In a post-nuclear world, clothing and gear serve as a first line of defense against radioactive fallout. While no clothing can completely block radiation, certain materials and gear can reduce exposure, particularly if you need to venture outdoors in the days and weeks following a nuclear event.

Protective Clothing

NBC (Nuclear, Biological, Chemical) Suits: These specialized suits are designed to protect against radioactive particles and chemical agents. If available, equip each family member with a suit, along with gloves and boots designed for radiation protection.

Tyvek Suits: If you can't access an NBC suit, Tyvek suits can offer some protection against radioactive dust and particles. These are lightweight, disposable coveralls that can be worn over regular clothing when venturing outdoors or handling contaminated materials.

Masks and Respirators: A high-quality mask, such as an N95 respirator or, ideally, a full-face respirator, is critical for filtering out radioactive particles in the air. Make sure each family member has a properly fitting mask, and keep extra filters or cartridges on hand.

Head and Face Protection

Goggles: Use safety goggles or a full-face shield to protect your eyes from dust, debris, and airborne radioactive particles.

Head Coverings: Wear hats or hoods to keep radioactive dust from settling in your hair, as particles can cling to hair and skin.

Layering for Temperature Control

Since you may experience a range of weather conditions over two years, layering clothing is essential for regulating body temperature, from cold winter months to warmer seasons. Layering also helps you adapt to temperature fluctuations within your shelter.

Base Layers

Moisture-Wicking Fabrics: Choose moisture-wicking base layers made from merino wool or synthetic materials to keep sweat away from your skin. This helps regulate body temperature and reduces the risk of hypothermia or heat rash.

Thermal Underwear: In colder climates, thermal underwear is a crucial base layer for staying warm.

Insulating Layers

Fleece and Wool: Fleece jackets and wool sweaters are excellent insulating layers that trap heat without adding excessive bulk. Wool, in particular, is durable, moisture-resistant, and retains warmth even when wet.

Down or Synthetic Insulation: For extreme cold, a down or synthetic insulated jacket provides lightweight warmth. Consider jackets with removable liners for versatility in various temperatures.

Outer Layers

Waterproof and Windproof Jackets: A durable, waterproof jacket or parka is essential for protecting against rain, snow, and wind. Look for jackets with sealed seams and adjustable hoods for maximum protection.

Ventilated Layers for Warm Climates: In warmer weather, opt for lightweight, breathable outer layers made from materials like cotton or nylon to keep you cool while still protecting against environmental elements.

Choosing Durable, Functional Clothing

The clothing you wear during long-term survival must be durable, functional, and easy to maintain. Prioritize fabrics that can withstand wear and tear, resist moisture, and dry quickly.

Durable Fabrics

Canvas and Ripstop Nylon: Pants, jackets, and gloves made from canvas or ripstop nylon are durable and resistant to tearing. These fabrics are ideal for outdoor work, such as gathering firewood or constructing shelter components.

Wool: Wool clothing is naturally insulating, flame-resistant, and moisture-wicking, making it an excellent choice for long-term survival situations. Wool socks, gloves, and sweaters are essential items.

Pockets and Storage

Multi-Pocket Pants and Jackets: Choose clothing with multiple pockets for carrying small tools, gear, and personal items. Cargo pants and utility vests with large, reinforced pockets allow you to keep essentials within easy reach.

Tool Belts or Pouches: Consider wearing a tool belt or tactical pouch for hands-free storage of essential survival tools, such as knives, multi-tools, or radios.

Protective Gear for Outdoor Ventures

When you need to venture outside your shelter to gather supplies, tend to crops, or perform other tasks, wearing protective gear is essential for safety.

Gloves

Heavy-Duty Work Gloves: For tasks like chopping wood, moving debris, or building structures, wear heavy-duty work gloves made from leather or reinforced synthetic materials. These gloves protect your hands from cuts, abrasions, and splinters.

Disposable Gloves: Keep a supply of disposable gloves on hand for handling potentially contaminated materials. These can be worn under heavier gloves for added protection during cleanup tasks.

Headlamps and Flashlights

Hands-Free Lighting: A high-quality headlamp is invaluable for working in low-light conditions or at night. Ensure it's waterproof and has adjustable brightness settings. Keep extra batteries or a rechargeable model to ensure ongoing use.

Flashlights: In addition to headlamps, store several durable flashlights with a long battery life, and make sure they are easily accessible.

Footwear for Comfort and Safety

Proper footwear is critical for both indoor and outdoor activities. Over the course of two years, you'll need shoes that provide comfort, durability, and protection in various environments.

Sturdy Boots

Work Boots: Invest in high-quality work boots that provide ankle support, grip, and protection. Waterproof boots are ideal for wet conditions, while insulated boots will keep your feet warm in cold climates.

Rubber Boots: For wet or contaminated environments, rubber boots offer protection from water, mud, and hazardous materials.

Socks

Wool Socks: Wool socks are excellent for long-term use because they wick moisture, provide warmth, and resist odors. Stock up on several pairs, and rotate them regularly to extend their lifespan.

Liners for Extra Warmth: In very cold conditions, consider wearing sock liners underneath your wool socks for added warmth and moisture-wicking.

Maintaining and Repairing Clothing

To ensure your clothing and gear last for the full two years, you'll need to care for them properly and have the tools to make repairs as necessary.

Cleaning Clothing

Handwashing: If water is limited, you may need to handwash clothing in small batches. Use biodegradable soap, and dry clothes in direct sunlight to kill bacteria and keep garments fresh.

Rotating Clothing: Rotate your clothing frequently to reduce wear and tear. Having multiple sets of essential items (socks, underwear, shirts) ensures that no single item gets worn out too quickly.

Repair Kits

Sewing Kit: Keep a basic sewing kit with heavy-duty thread, needles, patches, and buttons. Learn basic sewing techniques to mend tears, replace buttons, and patch holes.

Fabric Glue and Tape: Fabric glue and duct tape can be useful for temporary repairs on clothing, tents, and gear in situations where sewing is impractical.

Other Essential Gear for Survival

In addition to clothing, you'll need a variety of survival gear to aid in daily tasks and protect your family in a long-term shelter situation.

Multi-Tools and Knives

Multi-Tools: A multi-tool, such as a Swiss Army knife or Leatherman, is invaluable for survival. It can be used for cutting, prying, screwing, and repairing items in various situations.

Fixed-Blade Knife: A sturdy fixed-blade knife is essential for outdoor tasks like cutting rope, preparing food, or building shelters. Make sure it is durable and easy to sharpen.

Blankets and Sleeping Bags

Wool Blankets: Wool blankets provide warmth, even in damp conditions, and are ideal for cold-weather shelter living.

Sleeping Bags: Invest in high-quality, four-season sleeping bags that can withstand low temperatures. These are essential for maintaining body heat during colder months or in unheated environments.

Tactical Backpacks

Durable Backpacks: A rugged, tactical backpack is essential for carrying supplies during outdoor excursions or evacuations. Look for packs with padded straps, multiple compartments, and MOLLE (Modular Lightweight Load-carrying Equipment) webbing for attaching additional gear.

Clothing and gear are essential components of long-term survival in a post-nuclear environment. By selecting durable, functional clothing, protective gear, and essential survival tools, you can ensure your family stays safe, warm, and prepared for any challenges that arise over the next two years. Proper maintenance, repair, and adaptation of your clothing and gear will extend their lifespan and allow you to make the most of limited resources. Thoughtful preparation ensures that even in difficult conditions, your family will be equipped to handle the physical challenges of long-term survival, stay protected from environmental hazards, and maintain comfort and functionality throughout their time in the shelter.

Final Thoughts on Clothing and Gear

As you prepare for the long haul, remember that versatility is key when selecting clothing and gear. Every item you choose should serve multiple purposes if possible, ensuring that you make the most of limited storage space and resources. Always prioritize comfort, protection, and durability, as these will have the most significant impact on your family's well-being over an extended period.

Regular maintenance of both your clothing and gear is essential to their longevity. Rotate items as often as possible, make repairs as soon as they're needed, and ensure that your tools and equipment are cleaned and properly stored when not in use.

Finally, as your family adapts to shelter living and outdoor survival, remain flexible and be prepared to adjust your clothing and gear strategies to meet new challenges. Whether it's layering for extreme temperatures or using creative methods for gear repair, your ability to adapt will be one of the most important factors in your long-term survival. By carefully planning, selecting, and maintaining the right clothing and gear, you can ensure that your family remains protected, comfortable, and well-prepared to face the challenges of surviving in a post-nuclear world.

Tools and Equipment for Shelter Construction and Maintenance

In a post-nuclear survival scenario, building and maintaining a shelter that can protect your family for an extended period is essential. The tools and equipment you choose for shelter construction and upkeep play a critical role in ensuring your shelter is secure, durable, and capable of withstanding the environmental challenges that may arise. This chapter will focus on the essential tools and equipment needed for constructing, fortifying, and maintaining your shelter over a two-year period.

Key Considerations for Shelter Construction and Maintenance

Basic Hand Tools for Construction

Power Tools and Their Alternatives

Specialized Tools for Shelter Fortification

Materials for Repair and Maintenance

Essential Equipment for Energy and Heating

Tools for Water Management

Personal Safety and Protective Gear

Storage and Care of Tools and Equipment

Basic Hand Tools for Construction

Whether you are constructing a new shelter or fortifying an existing structure, a reliable set of basic hand tools is indispensable. These tools should be durable, multi-functional, and capable of handling various tasks such as cutting, fastening, and assembling.

Hammers

Framing Hammer: A heavy-duty hammer is essential for driving nails, securing beams, and performing basic construction tasks.

Claw Hammer: A claw hammer, with its curved side for removing nails, is useful for smaller jobs and repairs.

Saws

Hand Saw: A versatile hand saw is ideal for cutting wood or other materials if you don't have access to power tools.

Coping Saw: This smaller saw is helpful for cutting precise shapes and intricate curves in wood or metal.

Measuring and Levelling Tools

Tape Measure: A sturdy tape measure is essential for accurate measurements when constructing or repairing your shelter.

Level: Ensuring that walls, floors, and ceilings are level is crucial for structural stability. A spirit level (bubble level) can help you achieve precision.

Screwdrivers and Wrenches

Multi-Bit Screwdriver: A multi-bit screwdriver with interchangeable heads (Phillips, flathead, etc.) is useful for tightening and loosening screws of various types.

Adjustable Wrench: An adjustable wrench is versatile and can be used on bolts and nuts of different sizes, making it essential for assembly and maintenance tasks.

Pry Bars

Crowbar or Pry Bar: For demolition or removing obstacles, a crowbar is invaluable. It can also be used to pry apart nailed boards or leverage heavy objects.

Power Tools and Their Alternatives

While power tools can speed up the construction process, you may not have reliable access to electricity in a long-term shelter situation. Therefore, it's important to have both power tools and non-electric alternatives.

Cordless Power Drill

Battery-Powered Drill: A cordless power drill is essential for drilling holes, driving screws, and building structures quickly. Ensure you have a set of drill bits and extra batteries that can be recharged using solar or alternative energy sources.

Hand Drill: A manual hand drill can serve as a backup when power is unavailable. It is slower but effective for smaller drilling tasks.

Power Saw

Circular Saw: A cordless circular saw can handle larger cutting jobs and is especially useful when building or cutting large beams and boards. Extra batteries and blades are necessary.

Bow Saw: A bow saw is a suitable non-electric alternative for cutting wood, and its simplicity makes it easy to maintain in low-tech environments.

Specialized Tools for Shelter Fortification

Fortifying your shelter to withstand environmental hazards, radiation, and potential intruders requires additional specialized tools that can reinforce structural integrity.

Concrete and Masonry Tools

Trowels: If you are using concrete or mortar for construction, a masonry trowel is necessary for spreading and smoothing these materials.

Masonry Drill Bits: If you need to drill into concrete or stone, use masonry drill bits that are designed to handle tougher materials.

Rebar Cutter: Rebar can be used to reinforce walls and ceilings, especially in underground shelters. A rebar cutter is essential for cutting this tough material to the correct size.

Axes and Hatchets

Axe: An axe is essential for chopping wood, whether for building materials or firewood. Choose a sturdy, sharp axe that can handle heavy-duty work.

Hatchet: A hatchet is useful for smaller chopping jobs, such as clearing brush or splitting kindling for fires.

Shovels and Digging Tools

Shovel: A strong, durable shovel is indispensable for digging trenches, creating drainage systems, and moving dirt or debris.

Pickaxe: A pickaxe is helpful for breaking through hard ground, rock, or tough materials when building underground or reinforcing your shelter.

Materials for Repair and Maintenance

Over the course of two years, your shelter will likely require ongoing maintenance and repairs. Having the right materials on hand ensures that you can address these issues as they arise.

Nails, Screws, and Fasteners

Variety of Sizes: Keep an assortment of nails, screws, bolts, and washers in various sizes. These fasteners are essential for repairs and construction.

Heavy-Duty Anchors: For anchoring structures into concrete or brick, use heavy-duty wall anchors and screws that can bear significant weight.

Sealants and Adhesives

Silicone Sealant: Silicone sealant is ideal for sealing gaps and cracks around windows, doors, and other openings to prevent water and air leaks.

Epoxy and Glue: Epoxy adhesives are useful for bonding materials like metal, wood, and plastic. Stock up on heavy-duty glue for smaller repair jobs.

Plywood and Lumber

Plywood Sheets: Plywood is a versatile material that can be used for patching walls, reinforcing doors, or creating temporary barriers. Keep a supply of plywood sheets stored in a dry area.

Lumber for Framing: 2x4s or similar-sized lumber are useful for framing, structural repairs, and fortifying doors or windows.

Plastic Sheeting and Tarps

Plastic Sheeting: Thick plastic sheeting can be used to seal windows, create temporary barriers, or cover holes in walls or roofs.

Tarps: Heavy-duty tarps are essential for covering supplies, sealing off areas of the shelter, or providing temporary protection from rain or fallout.

Essential Equipment for Energy and Heating

Maintaining power and heat in your shelter is crucial, particularly in colder climates. The right tools and equipment ensure that you can generate energy and stay warm.

Solar Panels and Chargers

Portable Solar Panels: Solar panels are a reliable source of renewable energy for powering small electronics, tools, and lighting. Pair solar panels with rechargeable batteries for maximum efficiency.

Solar Chargers: Solar-powered chargers can be used to charge small devices like radios, lanterns, and emergency communication equipment.

Wood Stove and Chimney Maintenance Tools

Wood Stove: If your shelter includes a wood stove for heating and cooking, make sure to have a supply of firewood and tools for chopping and splitting wood.

Chimney Brush: To prevent chimney fires, regularly clean your chimney with a chimney brush to remove soot and debris.

Generators and Fuel

Portable Generator: A gas or propane-powered generator can provide backup power for critical tools and appliances. Keep enough fuel stored safely, and use the generator sparingly to conserve resources.

Wind-Up Generator: A hand-crank or wind-up generator can be used to charge small devices without relying on fuel.

Tools for Water Management

Water management is critical for shelter maintenance, particularly if you are collecting rainwater or dealing with potential flooding.

Water Filtration and Purification Tools

Water Filters: Portable water filtration systems, such as gravity filters or pump filters, are necessary for purifying water for drinking and washing.

Water Purification Tablets: Keep a supply of purification tablets for emergencies when filtration isn't possible.

Buckets and Water Storage

Heavy-Duty Buckets: Buckets are invaluable for carrying water, mixing concrete, and other tasks. Make sure to have durable, multi-purpose buckets on hand.

Rainwater Collection Barrels: Set up rainwater collection systems with barrels to ensure a steady water supply. Pair with filters to purify the collected water.

Personal Safety and Protective Gear

To safely carry out construction and maintenance tasks, it's important to have protective gear that keeps you safe from injury or exposure to hazardous materials.

Protective Eyewear and Masks

Safety Goggles: Protect your eyes from dust, debris, and chemicals while working on construction or repairs.

Respirators and Masks: When dealing with dust, mold, or chemical fumes, use an N95 mask or respirator to protect your lungs.

Work Gloves

Heavy-Duty Gloves: Use leather or reinforced gloves to protect your hands from cuts, abrasions, and burns during construction and maintenance tasks.

Hard Hats

Hard Hat: A hard hat is crucial for protecting your head while working on construction tasks, especially if you are moving heavy materials or working in confined spaces.

Storage and Care of Tools and Equipment

Maintaining your tools and equipment is essential for ensuring their longevity and functionality throughout the two-year period. Proper storage, regular maintenance, and preventive care will ensure that your tools are always ready for use when needed.

Tool Storage Solutions

Toolbox or Tool Chest: Invest in a durable, waterproof toolbox or tool chest to keep your hand tools organized, secure, and protected from the elements. Keep similar tools grouped together for easy access.

Wall Racks and Pegboards: If space allows, use wall-mounted racks or pegboards to hang tools like hammers, saws, and wrenches. This keeps them off the ground and easily accessible when working on projects.

Shelving Units: Store larger equipment like power tools, buckets, and fuel containers on sturdy shelving units. Keep heavy items on lower shelves to prevent accidents.

Preventive Maintenance

Cleaning Tools: After each use, clean your tools to remove dirt, grease, or other debris that could cause rust or damage. Wipe down metal parts with a cloth, and dry them thoroughly before storing.

Lubrication: Keep moving parts of tools like hinges, drills, and saw blades lubricated with oil to prevent rust and ensure smooth operation. Lubricate hand tools, such as pliers and wrenches, regularly to prevent seizing.

Sharpening Blades: Regularly sharpen saws, knives, and axes to maintain their cutting efficiency. A dull tool is not only less effective but also more dangerous to use.

Battery Care: For battery-powered tools, ensure batteries are stored in a cool, dry place. Use solar-powered or manual chargers when electricity is limited. Rotate rechargeable batteries regularly to prevent them from losing their charge.

Weather Protection

Waterproof Covers: If tools or equipment need to be stored outside or in areas exposed to moisture, cover them with waterproof tarps or plastic sheeting to protect them from rust and damage.

Ventilation and Humidity Control: Store your tools in a well-ventilated area with low humidity to prevent rusting and mold growth. Using moisture absorbers or dehumidifiers can help maintain a dry environment.

Having the right tools and equipment for shelter construction and maintenance is essential for long-term survival in a post-nuclear environment. A well-rounded collection of basic hand tools, power tools, and specialized equipment will enable you to build, fortify, and maintain a shelter that protects your family from environmental threats, ensures access to clean water and energy, and allows you to manage repairs as they arise.

Proper storage and care of tools extend their usability, ensuring they remain functional throughout the entire two-year survival period. By maintaining a well-organized, well-maintained set of tools, you'll be able to face any challenges that come your way with confidence, keeping your shelter safe, secure, and operational.

Power Generation: How to Keep Your Shelter Powered

In a long-term survival scenario, maintaining a reliable source of power is crucial for ensuring your shelter remains liveable, safe, and functional. Power is necessary for cooking, heating, lighting, communication, and running essential tools. Since access to traditional electricity grids may be unavailable or unreliable, you'll need to explore alternative ways to generate and store energy over the course of two years. This chapter will cover various methods of power generation, the necessary equipment, and strategies to manage and conserve power efficiently.

Key Elements for Power Generation

Assessing Your Power Needs

Solar Power Systems

Wind Power for Long-Term Energy

Generator Options and Backup Power

Battery Storage Solutions

Hand-Crank and Pedal Power

Energy Conservation Strategies

Maintaining Your Power Systems

Assessing Your Power Needs

Before choosing a power generation method, it's important to calculate your shelter's energy needs. Understanding how much power your family will require for lighting, cooking, heating, and essential devices allows you to plan effectively.

Basic Power Requirements

Lighting: LED lights are energy-efficient and require minimal power. Estimate how many lights you need and the total power required.

Cooking: If you're using electric stoves, ovens, or microwave ovens, calculate how much energy they will consume. Consider alternative cooking methods like wood stoves to reduce power use.

Communication: Radios, phones, and emergency communication devices should always be powered. Determine the wattage required to keep these running.

Heating and Cooling: If your shelter is located in a climate that requires heating or cooling, you'll need to account for these in your power calculations. Use energy-efficient heaters or fans to minimize power consumption.

Estimating Total Power Needs

Daily Usage: Estimate how many watts each device consumes per hour and how many hours per day you'll need it. Multiply these figures to get a daily wattage requirement.

Energy Buffer: Plan for occasional power surges or higher usage days by adding an energy buffer to your estimates. Having a surplus in your power generation system ensures you don't run out during critical times.

Solar Power Systems

Solar energy is one of the most reliable and renewable power sources for long-term survival. Solar panels can capture sunlight during the day and store energy in batteries for use at night or on cloudy days.

Solar Panels

Choosing Solar Panels: Select high-efficiency solar panels that can capture sunlight effectively even in low-light conditions. Monocrystalline panels are more efficient but often more expensive than polycrystalline panels.

Installation: Install solar panels in an area that gets maximum sunlight exposure, such as on the roof of your shelter or on a nearby structure. Panels should be mounted at an angle that captures sunlight throughout the day.

Solar Charge Controllers

Purpose: A solar charge controller regulates the voltage and current coming from the solar panels to protect your batteries from overcharging or damage. This is a critical component of your solar setup to ensure battery longevity.

MPPT Controllers: Maximum Power Point Tracking (MPPT) controllers are more efficient than standard charge controllers, especially in conditions where sunlight fluctuates.

Inverters

DC to AC Conversion: Solar panels and batteries produce Direct Current (DC) power, while most household appliances use Alternating Current (AC). An inverter converts DC power to AC, allowing you to run standard appliances off solar energy.

Pure Sine Wave Inverters: These are the most reliable inverters for running sensitive electronics like computers or medical devices. They produce cleaner and more stable power than modified sine wave inverters.

Wind Power for Long-Term Energy

Wind energy can complement solar power, particularly in areas where winds are strong and consistent. Wind turbines generate electricity by harnessing the power of wind to turn a generator.

Choosing a Wind Turbine

Size and Capacity: Select a wind turbine based on the average wind speed in your area and your energy needs. Small turbines (400W to 1kW) are suitable for supplementing solar power, while larger systems (5kW or more) can power an entire shelter.

Installation Location: Install the turbine in an open area where the wind is unobstructed by trees or buildings. Mount it on a tall pole or tower to capture stronger wind currents.

Hybrid Systems

Combining Solar and Wind: A hybrid system that integrates both solar panels and a wind turbine ensures that you have continuous power, even when one energy source is limited. For example, wind turbines can generate power at night or on cloudy days when solar panels are less effective.

Generator Options and Backup Power

A backup generator provides immediate power in emergencies or when your primary system (solar or wind) isn't producing enough energy.

Types of Generators

Gas-Powered Generators: These are reliable but require a consistent supply of fuel, which can be hard to store long-term. Gasoline also degrades over time, so ensure you have stabilizers if you're storing it.

Propane Generators: Propane stores more easily than gasoline and has a longer shelf life, making propane generators a better option for long-term survival. However, they still require regular fuel refills.

Dual-Fuel Generators: These can run on either propane or gasoline, giving you more flexibility with fuel sources. This versatility can be advantageous in a post-nuclear environment where fuel access may be limited.

Portable vs. Stationary Generators

Portable Generators: Portable generators are easier to transport and can be used to power specific devices in different areas. However, they generally have lower power output and may need frequent refuelling.

Stationary Generators: Stationary generators can power your entire shelter but are more expensive and require proper installation. These are ideal as a backup to solar or wind systems.

Battery Storage Solutions

Storing energy in batteries is critical for using power when your solar panels or wind turbines aren't generating electricity. Battery banks store excess energy and release it when needed.

Types of Batteries

Deep-Cycle Batteries: Deep-cycle lead-acid batteries, such as AGM (Absorbent Glass Mat) or gel batteries, are ideal for long-term energy storage. These batteries are designed to discharge slowly over a long period, making them perfect for shelter power needs.

Lithium-Ion Batteries: While more expensive, lithium-ion batteries last longer and are more efficient than lead-acid batteries. They can store more energy in a smaller space and require less maintenance.

Battery Bank Setup

Capacity Planning: Calculate the total storage capacity you need by multiplying your daily energy consumption by the number of days you want backup power (e.g., three days). This will determine how many batteries you need.

Series vs. Parallel Connections: In a series connection, batteries increase the system voltage, while parallel connections increase the system's capacity. Consult an expert or do research to optimize your battery bank for your needs.

Hand-Crank and Pedal Power

Hand-crank and pedal power generators are simple, low-tech solutions for generating small amounts of power without fuel or sunlight. These devices are ideal for charging radios, lights, and small electronics.

Hand-Crank Generators

Uses: Hand-crank generators are compact devices that generate power through manual effort. They are useful for emergency lighting, charging phones, and powering radios.

Effort vs. Output: These devices provide limited power output (often less than 20W), so they should be reserved for small, essential tasks in emergencies.

Pedal Power Generators

Bicycle-Powered Generators: Pedal power generators, often built from modified bicycles, can generate more electricity than hand-crank devices. They can be used to charge batteries or run low-power appliances, making them a valuable backup power option when solar or wind is unavailable.

Exercise and Power: As an added benefit, pedal power generators provide exercise while generating electricity, making them useful for maintaining physical fitness in shelter living.

Energy Conservation Strategies

Even with a reliable power generation system, conserving energy is key to ensuring that you have enough power for essential tasks. Efficient energy use prolongs the life of your batteries and reduces the need for fuel or extensive solar setups.

Using Energy-Efficient Appliances

LED Lighting: Switch to LED lights, which use significantly less energy than incandescent or fluorescent bulbs. Use motion sensors or timers to reduce unnecessary lighting.

Energy-Efficient Devices: Choose appliances and devices labeled as energy-efficient (such as Energy Star-rated products). This includes refrigerators, fans, and heaters.

Timers and Smart Plugs

Timers: Use timers for lights and appliances to ensure they are only on when needed. This prevents unnecessary energy consumption and extends battery life.

Smart Plugs: Smart plugs can help you control devices remotely and turn them off when not in use. These are especially useful for high-energy appliances.

Maintaining Your Power Systems

Regular maintenance of your power generation equipment is essential to ensure long-term functionality. Routine checks and preventive care will extend the life of your solar panels, wind turbines, and batteries.

Solar Panel Cleaning and Care

Cleaning Panels: Dust, dirt, and debris can reduce the efficiency of solar panels. Clean your panels regularly with a soft cloth and mild detergent to ensure they capture as much sunlight as possible.

Inspecting Connections: Regularly check the wiring and connections of your solar panels to ensure there are no loose or corroded parts that could interfere with energy transmission. Tighten or replace any worn-out connectors as needed.

Monitoring Efficiency: Keep track of the energy output of your solar panels and compare it to expected performance. If the output is lower than anticipated, it may indicate a problem with the panels or the charge controller that needs attention.

Wind Turbine Maintenance

Blade Inspection: Inspect the blades of your wind turbine regularly for damage, cracks, or debris. Clean the blades to ensure smooth operation and prevent wear over time.

Lubricating Moving Parts: Wind turbines have moving parts that can wear down without proper lubrication. Follow the manufacturer's guidelines for lubricating the turbine's bearings and other mechanical components to ensure efficient operation.

Checking the Tower: Inspect the tower for structural integrity. Ensure that it remains securely anchored and that guy wires (if applicable) are taut and undamaged. In high winds or storms, double-check the turbine's stability.

Battery Bank Maintenance

Battery Condition: Monitor the health of your battery bank regularly. For lead-acid batteries, check the electrolyte levels (if applicable) and top them off with distilled water as needed. Ensure that terminals are clean and free of corrosion.

Voltage Monitoring: Use a multimeter or charge controller display to monitor the voltage of each battery in your bank. Consistently low voltage readings could indicate a problem with a battery, which may need to be replaced.

Ventilation: Ensure that your battery bank is housed in a well-ventilated area, especially for lead-acid batteries, which can emit gases. Proper ventilation prevents the buildup of harmful gases and prolongs battery life.

Generator Care

Fuel and Oil Changes: If you have a gas or propane generator, change the oil regularly and ensure that fuel is stored properly. Use fuel stabilizers if necessary to prevent the fuel from degrading over time.

Running Tests: Periodically run the generator, even if it's not in regular use, to ensure that it remains operational. Perform these tests under load to check its ability to handle the energy needs of your shelter.

Air Filters and Spark Plugs: Replace air filters and spark plugs as recommended by the manufacturer. Clean air filters regularly to maintain airflow and prevent engine strain.

Keeping your shelter powered for the long term in a post-nuclear environment requires a multi-faceted approach that incorporates solar, wind, and backup generator systems. By understanding your shelter's energy needs, setting up reliable power generation systems, and maintaining your equipment, you can ensure a consistent supply of energy for lighting, cooking, communication, and essential tools. Combining different energy sources and using energy-efficient devices will maximize your power output while conserving resources.

Proper maintenance, routine inspections, and adopting energy-saving practices will extend the life of your power systems and ensure that your family remains safe, comfortable, and prepared to face the challenges of long-term shelter living. With the right combination of power generation methods, you can build a resilient energy infrastructure that will keep your shelter running smoothly for two years or more.

Other Essential Items for Long-Term Shelter Survival

In addition to food, water, clothing, and power, there are several other essential items you'll need to ensure long-term survival in a shelter environment. These items cover a wide range of needs, from medical supplies to tools for communication and hygiene. Being thoroughly prepared with a variety of essential items will ensure that your family remains safe, healthy, and capable of handling any challenges that arise over the course of two years in a survival shelter.

Key Essential Items for Long-Term Shelter Survival

Medical and First Aid Supplies

Hygiene and Sanitation Items

Cooking and Food Preparation Tools

Communication and Information Equipment

Lighting and Illumination Devices

Entertainment and Mental Well-Being Resources

Bartering and Trade Goods

Miscellaneous Survival Tools

Medical and First Aid Supplies

Proper medical care is crucial for long-term survival. Stocking up on a well-rounded first aid kit, along with additional medical supplies, ensures that you are prepared for emergencies, minor injuries, and ongoing health maintenance.

First Aid Kit Essentials

Bandages and Dressings: Include a variety of bandages, gauze, adhesive tape, and antiseptic wipes to treat cuts, scrapes, and minor wounds.

Antibiotics and Antiseptics: Stock up on antibiotic ointments, iodine, and antiseptic sprays to prevent infection in wounds.

Burn Treatments: Include burn creams and gel pads in case of burns from cooking or handling equipment.

Medications

Over-the-Counter Medications: Pain relievers (ibuprofen, aspirin), antacids, anti-diarrheal medications, and antihistamines should be part of your supply.

Prescription Medications: Ensure that you have at least a two-year supply of any necessary prescription medications for chronic conditions like asthma, diabetes, or heart disease.

Antibiotics: If possible, store broad-spectrum antibiotics to treat infections, but only use them when necessary and consult with a medical professional when possible.

Emergency Medical Tools

Splints and Slings: Include materials for immobilizing injured limbs, such as splints, elastic bandages, and slings.

CPR Masks and Gloves: Stock disposable CPR masks and nitrile gloves to protect yourself during emergency procedures or when treating injuries.

Sutures and Wound Closure Kits: A suture kit or wound closure strips (like Steri-Strips) are essential for closing larger wounds if professional medical help isn't available.

Hygiene and Sanitation Items

Maintaining hygiene is critical to prevent the spread of disease, especially in a confined environment. Stock up on sanitation supplies to ensure your shelter remains clean and habitable.

Personal Hygiene Supplies

Soap and Hand Sanitizer: Stockpile plenty of soap, hand sanitizer, and alcohol-based wipes for maintaining cleanliness when water is limited.

Toothbrushes and Toothpaste: Include a supply of toothbrushes, toothpaste, and dental floss for oral hygiene.

Feminine Hygiene Products: Ensure a sufficient supply of feminine hygiene items, including tampons or menstrual cups, for long-term use.

Waste Management

Portable Toilet: If your shelter lacks plumbing, a portable toilet with disposable waste bags is necessary for managing human waste.

Bleach and Disinfectants: Keep a supply of bleach, disinfectants, and cleaning supplies to maintain sanitation in your shelter.

Waste Disposal Bags: Heavy-duty garbage bags are essential for disposing of waste, keeping the environment clean, and preventing contamination.

Cooking and Food Preparation Tools

Even with stored food, you'll need the right tools to prepare meals safely and efficiently. Focus on cooking equipment that doesn't rely on a steady power supply.

Stovetop and Fuel

Portable Gas Stove: A propane or butane-powered camping stove is ideal for cooking when there's no access to electricity. Be sure to stock plenty of fuel canisters.

Rocket Stove: A rocket stove is a small, efficient wood-burning stove that can be used to cook food using minimal fuel, making it ideal for long-term survival.

Cookware and Utensils

Pots, Pans, and Kettles: Include sturdy, multi-purpose pots, pans, and a kettle for boiling water and preparing meals. Choose stainless steel or cast iron for durability.

Manual Can Opener: A reliable manual can opener is essential for accessing canned goods in your food stockpile.

Utensils and Cutlery: Stock up on basic utensils, plates, bowls, and knives. Consider stainless steel items for longevity and ease of cleaning.

Communication and Information Equipment

In a post-nuclear environment, staying informed and connected with the outside world is critical for survival. Ensure you have the necessary communication equipment for updates and emergency messages.

Radios

Battery-Powered or Hand-Crank Radio: A weather radio with AM/FM and NOAA bands will keep you informed about weather conditions, news updates, and emergency alerts. Hand-crank radios are ideal when batteries are scarce.

Two-Way Radios

Walkie-Talkies: Two-way radios are useful for communicating with family members within your shelter or with nearby survivors. Keep extra batteries or solar chargers available.

Emergency Whistles and Signal Mirrors

Whistle: A loud whistle can help attract attention in emergencies, especially if someone gets lost or is in distress.

Signal Mirror: A signal mirror is a low-tech tool that can be used to send visual signals to rescuers or nearby groups.

Lighting and Illumination Devices

Having a reliable source of light is essential for safety, comfort, and performing tasks inside and outside the shelter.

Flashlights and Headlamps

LED Flashlights: LED flashlights are energy-efficient and last longer than traditional bulbs. Keep a supply of high-quality flashlights and extra batteries.

Headlamps: Hands-free lighting is especially useful for working in low-light conditions or at night. Choose headlamps with adjustable brightness settings.

Lanterns and Candles

Battery-Powered Lanterns: Battery-operated lanterns provide ambient lighting for larger areas. Solar-powered lanterns are ideal for long-term use.

Candles: Stockpile long-burning candles and matches as a backup for when battery-powered lighting isn't available.

Entertainment and Mental Well-Being Resources

Surviving long-term in a confined shelter can be mentally and emotionally challenging. Including entertainment and mental well-being resources helps keep morale high.

Books and Puzzles

Books: Stock up on a variety of books, including survival guides, fiction, and educational materials, to provide entertainment and learning opportunities.

Puzzles and Games: Board games, puzzles, and playing cards offer mental stimulation and a distraction from the stress of survival.

Journals and Writing Materials

Notebooks and Pens: Keeping a journal or writing down your experiences can provide emotional relief and help track daily tasks and observations.

Bartering and Trade Goods

In a post-nuclear world, bartering may become a key way to acquire needed goods or services. Stockpile items that are likely to be in demand for trade.

High-Demand Items

Alcohol and Tobacco: Small quantities of alcohol (such as whiskey) and tobacco can be valuable bartering items.

Coffee and Tea: These comfort items can be scarce and highly sought after by other survivors.

Small Tools and Supplies

Lighters and Matches: Stock up on lighters, matches, and fire-starting materials. These will be in high demand for those without reliable fire-starting equipment.

Batteries: Extra batteries, especially AA and AAA, are valuable trade goods due to their wide usage in radios, flashlights, and other small devices.

Miscellaneous Survival Tools

Other tools and supplies can make daily survival tasks easier and more manageable. Stock up on versatile, multi-purpose tools.

Multi-Tools

Swiss Army Knife or Leatherman: A multi-tool with blades, screwdrivers, scissors, and other attachments is a versatile item that can handle a range of survival tasks.

Rope and Paracord

Paracord: Paracord is incredibly strong and can be used for everything from building shelter to securing equipment and creating makeshift repairs.

Duct Tape

Duct Tape: Stock up on duct tape, which is incredibly useful for quick repairs, sealing containers, and constructing makeshift tools or equipment.

Long-term shelter survival requires careful planning and stocking up on essential items that address a wide range of needs. From medical care and hygiene to communication and entertainment, ensuring you have the right tools and supplies helps your family stay safe, healthy, and mentally resilient during extended shelter living. Each of these items plays a crucial role in making survival more manageable and improving the quality of life in a challenging environment. Proper preparation with a diverse collection of essential supplies will keep your shelter functioning smoothly for the duration of your survival period.

Understanding Radiation Levels: What Is Safe for the Long-Term?

Radiation is one of the most critical dangers following a nuclear event, and understanding its effects on the human body and environment is essential for long-term survival. Different levels of radiation exposure can have varying effects, from mild to fatal, and learning how to monitor and manage radiation levels is key to keeping your family safe. This chapter focuses on what radiation levels are considered safe for long-term exposure, how to measure radiation, and strategies for minimizing radiation risks inside and outside the shelter.

Key Concepts of Radiation and Safety

Types of Radiation and Their Effects

Measuring Radiation Levels

Understanding Safe Radiation Exposure

Short-Term vs. Long-Term Radiation Effects

Reducing Exposure in the Shelter

Radiation in the Environment: Contaminated Soil, Water, and Air

When Is It Safe to Leave the Shelter?

Monitoring Radiation Levels over Time

Types of Radiation and Their Effects

There are three main types of radiation to be concerned with after a nuclear event: alpha particles, beta particles, and gamma rays. Each type of radiation behaves differently and poses distinct risks to the human body.

Alpha Particles

What They Are: Alpha particles are large, heavy, and slow-moving. While they cannot penetrate the skin, they are extremely dangerous if ingested or inhaled.

Protection: A basic barrier, such as clothing or a mask, can protect against alpha particles, but precautions must be taken to avoid internal exposure through contaminated food, water, or air.

Beta Particles

What They Are: Beta particles are smaller and faster than alpha particles. They can penetrate the outer layer of skin, causing burns and other skin damage. Prolonged exposure can lead to more severe health issues, including cancer.

Protection: Clothing and gloves offer protection from beta particles, but exposure to contaminated surfaces or materials must be minimized.

Gamma Rays

What They Are: Gamma rays are high-energy waves that can penetrate deep into the body and through most materials, causing severe internal damage. Prolonged exposure to gamma rays increases the risk of acute radiation sickness and cancer.

Protection: Gamma rays are the hardest to protect against. Thick layers of dense materials like lead, concrete, or earth are needed to reduce gamma ray exposure.

Measuring Radiation Levels

To determine radiation safety, you need to be able to measure the radiation levels both inside your shelter and in the surrounding environment. There are several tools for this purpose.

Geiger Counters

How They Work: A Geiger counter detects and measures radiation by counting the number of radioactive particles passing through the sensor. It displays this count in counts per minute (CPM) or microsieverts per hour (μSv/h), which indicates the level of radiation present.

Usage: Keep a Geiger counter on hand to regularly check radiation levels inside and outside the shelter. These readings will help you assess whether it's safe to go outside or if you need to strengthen your shelter's protective measures.

Dosimeters

How They Work: Dosimeters are personal devices that measure an individual's cumulative exposure to radiation over time. They can help track how much radiation a person has absorbed, offering a more personalized risk assessment.

Usage: Each family member should wear a dosimeter to monitor their exposure, especially when going outside or working in potentially contaminated areas. Set limits on acceptable exposure based on the radiation risk and monitor these levels closely.

Understanding Safe Radiation Exposure

There is no "safe" level of radiation, but exposure limits are used to gauge when radiation becomes dangerous for humans. These limits are measured in millisieverts (mSv) per year or per exposure event.

Normal Background Radiation Levels

Typical Levels: In a normal, non-contaminated environment, background radiation levels range between 0.1 to 0.2 μSv/h, which amounts to about 2-3 mSv per year.

What's Safe: At these levels, radiation is not harmful to humans, and people are regularly exposed to these small amounts in everyday life.

Safe Exposure Limits Post-Nuclear Event

Short-Term Safe Levels: Radiation levels between 0.5 µSv/h and 1.0 µSv/h (roughly 5-10 times normal background radiation) are generally considered safe for short-term exposure, but prolonged exposure should be avoided.

Long-Term Safe Levels: For long-term survival, exposure should ideally remain below 0.2 µSv/h, which allows for prolonged habitation with minimal health risks. Living in areas with radiation levels consistently higher than 1.0 µSv/h can increase the risk of long-term health effects like cancer.

Lethal Exposure Levels

Dangerous Levels: Radiation levels above 500 mSv in a short period can cause acute radiation sickness, and exposure to 5,000 mSv or more can be fatal within days or weeks.

Short-Term vs. Long-Term Radiation Effects

Understanding how radiation affects the human body over different periods is essential for managing your exposure during and after a nuclear event.

Short-Term Exposure

Symptoms of Acute Radiation Sickness: Nausea, vomiting, skin burns, and fatigue are early signs of radiation sickness. Exposure to high levels of radiation over a short time can result in immediate damage to cells, tissues, and organs.

Critical Time Period: The first few days and weeks after a nuclear event are the most dangerous in terms of radiation exposure due to high fallout levels. Shelter in place until radiation levels significantly decrease.

Long-Term Exposure

Cumulative Effects: Even low levels of radiation, when experienced over a long period, can lead to an increased risk of cancer, genetic mutations, and cardiovascular diseases.

Children and Pregnant Women: These groups are more vulnerable to long-term radiation exposure due to their rapidly dividing cells. Extra precautions should be taken to minimize their exposure.

Reducing Exposure in the Shelter

Protecting your family from radiation while inside the shelter is a priority. Taking steps to block radiation and prevent contamination from entering the shelter will reduce your long-term risk.

Shelter Materials

Thick Walls: Use materials like concrete, earth, or lead to create thick walls that can block or absorb radiation. A layer of earth or sand at least 3 feet thick can reduce gamma radiation exposure significantly.

Sealing Entry Points: Seal windows, doors, and ventilation systems to prevent radioactive dust from entering the shelter. Use plastic sheeting, duct tape, and weatherstripping to cover cracks and gaps.

Air Filtration

HEPA Filters: Install HEPA (High-Efficiency Particulate Air) filters in your ventilation system to remove radioactive particles from the air. Ensure that your shelter remains airtight to prevent fallout from contaminating indoor air.

Decontamination Area: Create a decontamination zone at the entrance to your shelter where family members can remove clothing and wash exposed skin before entering the living space.

Radiation in the Environment: Contaminated Soil, Water, and Air

Even when radiation levels inside the shelter are safe, the outside environment can remain contaminated for years. Knowing how to deal with contaminated soil, water, and air is critical for long-term survival.

Contaminated Soil

Fallout Deposition: Fallout particles can settle in the soil and remain radioactive for years. Avoid gardening or farming in contaminated soil until radiation levels drop, and consider using raised beds or imported soil.

Decontaminating Soil: Removing the top layer of soil or using phytoremediation (planting certain plants that absorb radioactive particles) can help reduce contamination over time.

Contaminated Water

Water Testing: Regularly test water sources for radiation using a Geiger counter or water testing kit. Use water filters designed to remove radioactive particles or rely on stored water for drinking and cooking.

Boiling Water: Boiling water will kill biological contaminants but does not remove radiation. Use a combination of distillation and filtration methods to purify contaminated water.

When Is It Safe to Leave the Shelter?

Knowing when it's safe to venture outside or permanently leave your shelter is crucial. You'll need to closely monitor radiation levels both inside and outside to determine when exposure risks have dropped to acceptable levels.

The Rule of Sevens and Tens

Radiation Decay: After a nuclear event, radiation levels decrease over time based on the rule of sevens and tens. For every sevenfold increase in time after the explosion, radiation levels decrease by a factor of ten. For example, after 7 hours, radiation is 10% of its initial level, and after 49 hours, it's 1%.

Safe Timeline: Wait at least two weeks before considering venturing outside the shelter. After 14 days, radiation levels typically drop to 1% of their original intensity, but continue to use a Geiger counter to verify that levels are within the safe range.

Monitoring Radiation Levels Over Time

Regular monitoring of radiation levels is essential, especially if you plan to leave your shelter for extended periods or engage in activities like gardening, water collection, or outdoor repairs.

Regular Geiger Counter Readings

Inside the Shelter: Take daily readings inside the shelter to ensure no radiation is seeping in. If levels rise, check for any leaks or compromised areas in the shelter's structure and immediately address them. Focus on entry points like doors, windows, and ventilation systems, as these are common places where radiation can seep in.

Outside the Shelter: Before venturing outside, use your Geiger counter to take readings in various locations around your shelter. Pay close attention to the soil, water sources, and any areas that may have collected fallout (such as roofs or areas with standing water).

Long-Term Tracking

Charting Radiation Decay: Keep a log of radiation readings over time to track how levels are decreasing. This helps you predict when it will be safe to engage in more regular outdoor activities, such as gardening, gathering firewood, or hunting.

Reassessing Activity: Based on your radiation tracking, reassess outdoor activities. For instance, if radiation levels remain consistently low in certain areas, you might be able to spend more time outside, while avoiding areas with higher contamination.

Dosimeter Monitoring

Personal Exposure: Each family member should wear a dosimeter when they leave the shelter. Regularly check these devices to ensure no one exceeds safe exposure levels. If someone's dosimeter shows elevated exposure, have them stay inside the shelter to recover and reduce their cumulative dose.

Understanding radiation levels and how to measure and manage them is essential for long-term survival in a post-nuclear environment. The key to staying safe is consistently monitoring radiation both inside and outside the shelter, using the right protective measures, and limiting your exposure to contaminated environments. By knowing what levels are safe for short-term and long-term exposure, you can make informed decisions about when it's safe to leave the shelter, how to decontaminate areas, and how to protect your family. With the right tools and knowledge, you can navigate the dangers of radiation, ensuring your family remains healthy and secure throughout the two-year shelter period.

Surviving the Psychological Toll of Isolation and Fear

Surviving the psychological toll of isolation and fear is just as important as physical survival in a long-term shelter scenario. Living in a confined space for months or years, cut off from the outside world, can lead to intense feelings of loneliness, anxiety, depression, and even despair. Managing these emotional challenges is key to ensuring your mental well-being and maintaining a positive mindset, which can be as crucial to survival as food and water. In this chapter, we'll explore strategies for coping with isolation, dealing with fear, maintaining mental health, and building emotional resilience during long-term shelter living.

Key Aspects of Psychological Survival

Understanding the Emotional Challenges of Isolation

Managing Fear and Anxiety in a High-Stress Environment

Creating a Daily Routine for Stability

Maintaining Social Connections with Family

Finding Purpose and Goals to Focus On

Managing Conflict and Tension in a Confined Space

Engaging in Mental and Physical Exercise

Coping with the Uncertainty of the Outside World

Understanding the Emotional Challenges of Isolation

Isolation can be mentally and emotionally draining, especially over a prolonged period. Being separated from the outside world, combined with the constant fear of the unknown, can lead to psychological distress. Recognizing these challenges is the first step in managing them.

Loneliness and Boredom

Lack of Social Interaction: The absence of regular social interaction outside your immediate family can lead to feelings of loneliness. Even if you are not physically alone, the sense of separation from the broader community and normal life can weigh heavily.

Boredom and Monotony: Shelter living may involve long stretches of inactivity, especially if it's too dangerous to leave the shelter. The lack of variety in activities can create mental fatigue, making time feel as though it's moving slowly.

Fear of the Unknown

Uncertainty about the Future: Not knowing when it will be safe to leave the shelter, whether the outside world has improved, or if help will ever arrive can create an overwhelming sense of fear and anxiety.

Fear of Exposure: The constant threat of radiation and the possibility of dangerous encounters outside the shelter add to the stress. This persistent fear can lead to feelings of helplessness or dread.

Managing Fear and Anxiety in a High-Stress Environment

Fear and anxiety are natural responses to a dangerous and unpredictable situation, but if left unchecked, they can become debilitating. Learning to manage these emotions is crucial for maintaining mental health in the shelter.

Recognizing the Signs of Anxiety

Physical Symptoms: Anxiety can manifest physically through symptoms like rapid heartbeat, shortness of breath, dizziness, or muscle tension. Recognizing these signs early can help you take steps to calm yourself before the anxiety escalates.

Mental Symptoms: Anxiety often leads to racing thoughts, catastrophizing, or an inability to focus. Mindfulness and mental grounding techniques can help manage these feelings.

Calming Techniques

Deep Breathing: Practicing deep breathing exercises can help slow down your heart rate and bring a sense of calm. Inhale deeply through your nose for four counts, hold for four counts, and exhale slowly for four counts.

Progressive Muscle Relaxation: Tensing and then slowly releasing muscle groups can relieve physical tension caused by stress. Start with your toes and work your way up to your head, focusing on each muscle group.

Visualization: Mentally visualizing peaceful, calming places or memories can reduce stress. Spend a few minutes each day imagining yourself in a peaceful setting to help manage fear.

Creating a Daily Routine for Stability

A daily routine provides structure and stability in an otherwise unpredictable situation. It helps give purpose to your day and keeps your mind occupied, reducing feelings of chaos and uncertainty.

Setting Regular Wake-Up and Sleep Times

Maintaining a Sleep Schedule: Going to bed and waking up at the same time each day helps regulate your body's natural rhythms. A regular sleep schedule can improve both mental and physical health.

Organizing Daily Tasks

Chores and Responsibilities: Assign everyone in the shelter specific daily tasks, such as cooking, cleaning, or maintaining the shelter. These tasks help create a sense of responsibility and accomplishment, breaking up the monotony.

Breaks and Leisure Time: Incorporate scheduled breaks and leisure activities into your routine. Use this time to relax, read, or engage in a hobby.

Maintaining Social Connections with Family

The relationships you have with those in the shelter are vital for emotional support. Maintaining strong connections and fostering communication can alleviate feelings of loneliness and create a more positive atmosphere.

Communication and Emotional Sharing

Open Communication: Encourage family members to express their feelings openly. Listening to one another and validating each other's emotions can strengthen bonds and reduce isolation within the group.

Conflict Resolution: Address conflicts as soon as they arise to prevent them from escalating. Use calm, respectful communication to resolve disagreements and avoid letting tensions build.

Group Activities

Family Meals: Eating meals together as a family creates opportunities for bonding and conversation. Make this time a shared experience rather than a chore.

Group Games and Hobbies: Engage in games or group activities that everyone enjoys. Whether it's playing a board game, solving puzzles, or sharing stories, these activities can help break up the routine and lighten the mood.

Finding Purpose and Goals to Focus On

Having goals to work toward provides a sense of purpose and direction, even in the most difficult situations. Focus on long-term survival goals as well as small daily achievements.

Setting Personal Goals

Skills Development: Use the time in the shelter to learn new survival skills or improve existing ones. Practicing first aid, learning how to grow food, or building something can give you a sense of progress.

Physical Fitness Goals: Set small, achievable fitness goals to maintain your physical health. Regular exercise helps release endorphins, improving mood and reducing anxiety.

Shelter Improvement Projects

Making Improvements: Focus on tasks that improve your shelter's comfort or functionality. This could include organizing supplies, reinforcing walls, or creating a better ventilation system. Completing these tasks can give you a sense of control over your environment.

Managing Conflict and Tension in a Confined Space

Tension and conflict are inevitable when living in close quarters for an extended period. Learning to manage these conflicts constructively is important for maintaining peace and cooperation.

Recognizing the Causes of Conflict

Personal Space Issues: Lack of privacy or personal space can lead to frustration and irritability. Create designated spaces for each person in the shelter to have some alone time.

Stress and Fatigue: Ongoing stress, lack of sleep, and anxiety can exacerbate tensions. Recognizing when someone is under stress and offering support can prevent arguments from escalating.

Conflict Resolution Techniques

Active Listening: When a conflict arises, listen carefully to the other person's point of view without interrupting. Show that you understand their perspective, even if you disagree.

Compromise: Be willing to compromise to resolve conflicts. Find solutions that work for both parties, rather than insisting on being right.

Engaging in Mental and Physical Exercise

Keeping your mind and body active is essential for preventing mental fatigue and physical decline during long-term shelter living.

Mental Stimulation

Reading and Learning: Stimulate your mind by reading books, learning new skills, or solving puzzles. Keeping your brain active helps ward off boredom and keeps your mind sharp.

Creative Outlets: Engage in creative activities like drawing, writing, or crafting. Expressing your thoughts and emotions creatively can be a therapeutic way to manage stress.

Physical Activity

Exercise Routines: Even in a confined space, regular exercise is important. Stretching, bodyweight exercises (like push-ups, squats, and planks), or yoga can help maintain physical health.

Walking and Movement: If you have enough room, take regular walks around the shelter or perform light cardio exercises to keep your body active.

Coping with the Uncertainty of the Outside World

The uncertainty of what's happening outside the shelter can cause anxiety and fear. Learning to accept and manage this uncertainty is critical for long-term psychological well-being.

Focusing on What You Can Control

Shelter Environment: Focus on maintaining and improving your shelter. Taking action to make your environment as comfortable and safe as possible can give you a sense of control.

Daily Tasks: Concentrate on tasks you can complete within the shelter, such as cooking, organizing, or cleaning. These small, achievable goals help you stay grounded.

Limiting Exposure to Negative News

Filtering Information: While staying informed is important, too much negative news can increase stress and anxiety. Limit how often you listen to updates and focus on actionable information rather than speculation or fear-based reports.

Surviving the psychological toll of isolation and fear in a long-term shelter situation requires a combination of mental strength, emotional resilience, and practical coping strategies. By maintaining a structured routine, staying connected with family, managing anxiety, and focusing on achievable goals, you can preserve your mental well-being and create a supportive, stable environment. Managing the psychological challenges of shelter living is just as important as physical survival, and with the right mindset and tools, you can endure the hardships and emerge stronger.

Managing Family Dynamics in a Confined Space

Managing family dynamics in a confined space is one of the biggest challenges in long-term shelter survival. When multiple people share a small living area for an extended period, tensions can arise, leading to conflicts that might hinder cooperation and harmony. It's important to recognize the potential stressors and adopt strategies to foster a positive environment where everyone feels respected and understood. This chapter will explore techniques for maintaining healthy family dynamics, resolving conflicts, and ensuring that everyone contributes to the shelter's survival in a meaningful way.

Key Aspects of Managing Family Dynamics

Understanding the Impact of Close Quarters

Establishing Roles and Responsibilities

Creating Personal Space and Time for Privacy

Effective Communication in a Confined Environment

Resolving Conflicts Constructively

Addressing Power Struggles and Leadership

Fostering Teamwork and Collaboration

Providing Emotional Support for Family Members

Understanding the Impact of Close Quarters

Living in a confined space for months or even years can amplify stress, irritability, and emotional strain. Understanding the psychological and emotional impact of such close quarters is the first step to managing family dynamics effectively.

Physical Proximity and Emotional Strain

Lack of Physical Space: In a shelter environment, personal space is limited. This constant proximity can lead to feelings of claustrophobia or frustration. Over time, even small annoyances can build up into larger issues.

Emotional Amplification: In stressful environments, emotions tend to become heightened. Minor disagreements can escalate quickly, and individuals may struggle to find ways to manage their emotions when personal space is limited.

Children and Adolescents

Special Needs of Children: Younger family members may find it especially difficult to cope with the lack of space and the sudden disruption of their normal routines. They may become restless, bored, or emotionally volatile, requiring extra patience and support.

Teenage Independence: Adolescents may struggle with the loss of independence and personal freedom. Finding ways to give them a sense of autonomy while maintaining safety and order in the shelter is important.

Establishing Roles and Responsibilities

A clear division of labor within the shelter helps create structure and purpose. Assigning specific roles and responsibilities to each family member reduces feelings of chaos and ensures that everyone contributes to the survival effort.

Assigning Roles

Daily Tasks: Assign each family member specific daily tasks, such as cooking, cleaning, maintaining the shelter, or monitoring supplies. This gives everyone a sense of purpose and responsibility.

Rotation of Responsibilities: To prevent boredom or resentment over repetitive tasks, consider rotating responsibilities regularly. This also ensures that everyone gains a broad set of survival skills.

Recognizing Strengths

Leveraging Talents: Consider the unique skills and strengths of each family member when assigning tasks. For example, someone with mechanical skills could maintain equipment, while someone who enjoys cooking could manage meal preparation.

Encouraging Leadership: Encourage family members to take the lead in certain tasks, fostering a sense of ownership and leadership within the group. This can be particularly beneficial for children and adolescents, as it builds confidence.

Creating Personal Space and Time for Privacy

Even in a confined space, finding ways to create personal space and time for solitude is essential for maintaining emotional balance and reducing tension.

Physical Barriers for Privacy

Curtains or Dividers: Use makeshift curtains, dividers, or blankets to create small areas of privacy for sleeping or resting. Even the illusion of separation can provide a psychological break from constant proximity.

Personal Corners: Allow each family member to have a designated personal corner or area where they can store their belongings or retreat for some quiet time. This helps individuals feel more in control of their personal space.

Time for Solitude

Scheduled Alone Time: Schedule specific times during the day when family members can take turns being alone in a designated space for reading, meditating, or simply resting. This allows everyone to recharge emotionally.

Respecting Boundaries: Encourage family members to respect each other's need for solitude. Let them know it's okay to ask for personal space when needed and make sure that others honor that request.

Effective Communication in a Confined Environment

Clear and respectful communication is vital to prevent misunderstandings, resolve conflicts, and ensure that the family works as a team. Developing strong communication habits is essential for long-term survival.

Daily Check-Ins

Family Meetings: Hold regular family meetings to discuss daily plans, address any concerns, and check in on everyone's emotional well-being. These meetings provide a structured opportunity for open communication.

Feedback and Suggestions: Encourage each family member to voice their opinions and suggestions. Create an environment where everyone feels heard and valued, and consider their input when making decisions.

Using Non-Verbal Communication

Body Language and Tone: In close quarters, it's easy for small gestures or tones of voice to be misinterpreted. Encourage family members to be mindful of their body language and tone, as these can often convey frustration or anger without meaning to.

Listening Skills: Teach active listening, where family members focus on understanding each other's viewpoints rather than immediately responding. This fosters empathy and reduces misunderstandings.

Resolving Conflicts Constructively

Conflicts are inevitable in a confined space, but how they are handled makes all the difference. Developing constructive conflict resolution strategies is key to maintaining harmony.

Identifying the Root Cause

Addressing Issues Early: Encourage family members to speak up early when something is bothering them. This prevents minor issues from festering and turning into larger conflicts.

Understanding Triggers: Identify what triggers conflict and take steps to avoid or manage these triggers. For example, if noise is a common irritant, create quiet times during the day.

Mediation and Compromise

Calm Discussions: When a conflict arises, take a break if emotions are running high. Once everyone has calmed down, have a respectful discussion about the issue, focusing on finding solutions rather than assigning blame.

Compromising: Teach family members the art of compromise. No one should feel like they are always giving in, but finding middle ground ensures that everyone's needs are considered.

Addressing Power Struggles and Leadership

Leadership roles may naturally arise within the family dynamic, but power struggles can create tension and disrupt cooperation. Managing leadership and authority in the shelter is essential for maintaining order.

Sharing Leadership Roles

Rotating Leadership: If one person is the designated leader, consider rotating leadership responsibilities for certain tasks. This helps prevent feelings of resentment or overburdening the leader.

Team-Based Decision Making: Make important decisions as a family unit, with input from all members. While someone may take the lead on specific tasks, the overall direction should be a group effort.

Avoiding Dominance

Respecting Contributions: Ensure that no family member dominates the group or consistently overrules others. Foster an environment where everyone's contributions are respected and valued, regardless of their age or skill level.

Fostering Teamwork and Collaboration

Surviving as a group requires strong teamwork. By fostering a collaborative atmosphere, you can reduce tension and ensure that everyone works toward the same goal—survival.

Encouraging Cooperation

Shared Goals: Emphasize that everyone is working toward the same goal: survival and well-being. When disagreements arise, remind the family of the bigger picture to refocus their efforts on collaboration.

Group Projects: Engage in group projects where family members must work together, such as building something for the shelter or preparing a large meal. This reinforces the importance of teamwork and gives a sense of shared accomplishment.

Celebrating Successes Together

Recognizing Achievements: Celebrate small victories together, whether it's completing a difficult task, getting through a tough week, or simply making it through another day. Acknowledging successes boosts morale and strengthens bonds.

Encouraging Positivity: Keep the atmosphere positive by focusing on what's going well and encouraging family members to compliment and thank each other for their contributions.

Providing Emotional Support for Family Members

Emotional well-being is vital for survival, and family members must support one another through tough times. Encouraging open communication about feelings and providing emotional support can help mitigate the emotional strain of long-term shelter living.

Checking in on Each Other

Emotional Check-Ins: Regularly ask family members how they're feeling and encourage them to talk about their emotions. Everyone will have bad days, and knowing they have a support system can make a significant difference.

Validating Feelings: Acknowledge each other's feelings, whether it's frustration, sadness, or fear. Validating these emotions shows empathy and helps family members feel understood.

Mental Health Resources

Journaling: Encourage family members to keep journals where they can express their feelings privately. Writing can be a powerful tool for processing emotions in a confined environment.

Therapeutic Activities: Engage in activities that promote mental health, such as meditation, breathing exercises, or creative outlets like drawing or writing. These activities provide a way to de-stress and relax.

Managing family dynamics in a confined space is one of the most important aspects of surviving long-term in a shelter. By establishing clear roles, maintaining open communication, and fostering a supportive environment, you can reduce tension and conflict while building a cohesive team focused on survival. A well-functioning family dynamic will not only improve day-to-day life in a confined space but also enhance your chances of long-term success by ensuring that everyone feels valued, respected, and emotionally supported.

Understanding the psychological and emotional toll of being in close quarters is crucial for creating a positive environment. Encourage empathy, patience, and respect among family members, as these qualities will go a long way in preventing conflict and promoting cooperation. At the same time, it's essential to create space—both physical and emotional—for individual needs, as well as to balance leadership and shared responsibility so that no one feels overburdened or left out.

By addressing conflicts constructively, ensuring proper communication, and fostering teamwork, your family can maintain a strong sense of unity, even in challenging circumstances. Emotional resilience, mutual support, and a shared focus on survival will help you manage the ups and downs of long-term shelter living. Ultimately, your family's collective strength, adaptability, and cooperation will be key to enduring the trials of isolation and confinement.

The Importance of Physical Fitness in a Fallout Shelter

Physical fitness is a crucial, yet often overlooked, aspect of survival in a fallout shelter. Being in good physical condition helps your body handle the stresses of survival and confinement, maintain overall health, and boost mental well-being. The limited space and resources in a fallout shelter may make it challenging to engage in traditional fitness routines, but it's essential to adapt and find ways to stay active. This chapter will explore why physical fitness is important in a survival scenario, the benefits of maintaining fitness in a fallout shelter, and practical exercises you can do in confined spaces.

Key Reasons for Maintaining Physical Fitness in a Shelter

Strengthening the Immune System and Health

Enhancing Mental Well-Being and Reducing Stress

Preventing Muscle Loss and Joint Stiffness

Building Physical Strength for Survival Tasks

Maintaining Flexibility and Mobility

Improving Cardiovascular Health in a Confined Space

Practical Exercises for a Shelter Environment

Creating a Fitness Routine for Long-Term Shelter Living

Strengthening the Immune System and Health

Maintaining physical fitness strengthens your immune system, which is vital in a confined environment where access to medical care and fresh air may be limited. Regular physical activity helps your body fight off illnesses and infections by boosting circulation, which delivers more oxygen and nutrients to your cells and immune system.

Reducing the Risk of Illness

Improved Circulation: Exercise increases blood flow, which helps remove toxins and deliver vital nutrients to tissues and organs. This enhances your body's ability to fight off infections, which is crucial in a shelter environment where illness could spread quickly.

Boosting Immunity: Physical fitness enhances immune function by promoting the production of antibodies and white blood cells. A strong immune system reduces the likelihood of infections, particularly in a closed environment where ventilation may be poor.

Enhancing Mental Well-Being and Reducing Stress

Exercise has a direct impact on mental health. In a shelter, where stress, anxiety, and depression are common due to isolation, fear, and uncertainty, staying physically active can help alleviate these emotional burdens.

Reducing Anxiety and Depression

Endorphin Release: Physical activity stimulates the release of endorphins, which are natural mood elevators. Regular exercise can reduce feelings of anxiety and depression, helping to maintain mental health in a high-stress environment.

Stress Relief: Exercise helps regulate cortisol levels, the hormone associated with stress. By incorporating fitness into your routine, you can manage stress more effectively and stay calm under pressure.

Preventing Muscle Loss and Joint Stiffness

Inactivity can quickly lead to muscle atrophy (the weakening of muscles) and joint stiffness, especially when confined to a small space for long periods. Without regular movement, your body can lose muscle mass and flexibility, making survival tasks more difficult.

Preventing Muscle Atrophy

Maintaining Strength: Simple bodyweight exercises can help you retain muscle mass and strength, even if you don't have access to traditional exercise equipment. Staying active helps prevent muscle wasting, which can occur after just a few weeks of inactivity.

Joint Mobility: Regular movement helps keep your joints lubricated and mobile. This is important for maintaining your ability to move freely, lift objects, and perform shelter-related tasks like building, cleaning, or repairing.

Building Physical Strength for Survival Tasks

Certain survival tasks, such as carrying supplies, fortifying the shelter, or lifting heavy objects, require physical strength. By staying physically fit, you ensure that your body is ready to handle these tasks when they arise.

Functional Strength

Core and Upper Body Strength: Strengthening your core and upper body is essential for tasks like lifting heavy items, moving debris, or even defending yourself in case of an emergency.

Lower Body Strength: Strong legs are crucial for mobility, especially if you need to climb, crouch, or navigate rough terrain when outside the shelter. Regular squats, lunges, and other lower body exercises can maintain this strength.

Maintaining Flexibility and Mobility

Flexibility and mobility are key to preventing injuries, especially in a confined space where you may have to manoeuvre around obstacles or cramped quarters. Stretching exercises are vital for keeping muscles and joints flexible.

Preventing Injuries

Stretching: Regular stretching helps prevent strains, sprains, and other injuries by keeping your muscles and joints flexible. Flexibility is essential for performing shelter tasks safely and efficiently.

Range of Motion: Maintaining a full range of motion in your joints ensures that you can move comfortably and easily, even in tight spaces. Flexibility exercises such as yoga or dynamic stretches can help.

Improving Cardiovascular Health in a Confined Space

Even in a small shelter, it's important to maintain cardiovascular health. Lack of activity can lead to poor circulation, heart health issues, and increased risk of cardiovascular diseases. Low-impact cardio exercises can help keep your heart strong.

Cardiovascular Endurance

Simple Cardio: Engage in activities like jogging in place, jumping jacks, or high knees to get your heart rate up and improve cardiovascular endurance. These exercises don't require much space but are effective for heart health.

Consistency Over Intensity: You don't need intense cardio workouts to maintain cardiovascular health. Consistent, moderate-intensity exercise is enough to improve circulation and keep your heart healthy.

Practical Exercises for a Shelter Environment

Despite space limitations, there are many exercises you can do inside a shelter to stay fit. Bodyweight exercises are especially useful since they don't require equipment and can be done in a small area.

Bodyweight Exercises

Squats: Squats are excellent for building lower body strength. They work the legs, hips, and core and can be modified for different fitness levels.

Push-Ups: Push-ups target the chest, shoulders, triceps, and core. They can be modified to be easier (knee push-ups) or harder (diamond push-ups) depending on your strength level.

Lunges: Lunges strengthen your legs and improve balance. They don't require much space and can be done as forward or backward lunges.

Planks: Planks engage the entire core and help build endurance and stability. Side planks can also be added for variety and increased difficulty.

Cardio Exercises

Jumping Jacks: Jumping jacks are a simple yet effective cardio exercise that can be done in a small space. They also help improve coordination.

High Knees: High knees are a great way to get your heart rate up and engage your lower body. Simply march or jog in place while lifting your knees as high as possible.

Mountain Climbers: Mountain climbers are a dynamic cardio exercise that works your core, shoulders, and legs. They're ideal for combining strength and cardio in one movement.

Creating a Fitness Routine for Long-Term Shelter Living

A regular fitness routine helps create structure and keeps your body strong. By dedicating time each day to physical activity, you can maintain your health and manage the stress of survival living.

Daily Exercise Schedule

Short Sessions: Aim for short, regular exercise sessions, such as 20-30 minutes a day. This helps prevent fatigue while keeping you active.

Balancing Cardio and Strength: Include a balance of cardio exercises and strength training in your routine. This ensures that you maintain both cardiovascular health and muscle strength.

Incorporating Stretching and Mobility

Morning and Evening Stretches: Start and end each day with a stretching routine to keep your muscles flexible and prevent stiffness. Stretching in the morning prepares your body for the day ahead, while evening stretches help you relax before sleep.

Low-Impact Mobility Work: Incorporate gentle mobility exercises to keep your joints healthy. This can include shoulder rolls, hip circles, and ankle stretches.

Physical fitness is a vital part of survival in a fallout shelter, not only for maintaining physical health but also for boosting mental resilience and overall well-being. By engaging in simple, space-efficient exercises, you can stay strong, flexible, and prepared for the physical demands of survival tasks. A regular fitness routine helps prevent muscle atrophy, joint stiffness, and cardiovascular decline, ensuring that your body remains fit and ready for the challenges ahead.

Incorporating physical activity into your daily routine, even in a confined space, will greatly improve your overall survival experience. You'll feel more energized, less stressed, and better prepared to handle the tasks and demands of long-term shelter living.

Maintaining Morale and Hope During a Nuclear Winter

Maintaining morale and hope during a nuclear winter is perhaps one of the most challenging aspects of long-term survival. While physical survival depends on securing food, water, shelter, and protection from radiation, mental and emotional survival hinges on the ability to foster hope, stay positive, and keep spirits high even in the face of bleak and uncertain conditions. This chapter will explore strategies for maintaining morale, nurturing hope, and fostering resilience in yourself and your family during the long months or even years of nuclear winter.

Key Factors for Maintaining Morale and Hope

Understanding the Emotional Challenges of a Nuclear Winter

Creating a Positive Environment in the Shelter

Establishing and Celebrating Milestones

Keeping a Long-Term Vision: Setting Goals for the Future

Encouraging Group Cohesion and Support

Fostering a Sense of Purpose and Contribution

Engaging in Meaningful Activities and Creativity

Staying Informed Without Overloading on Bad News

Developing Mental and Emotional Resilience

Understanding the Emotional Challenges of a Nuclear Winter

The psychological impact of a nuclear winter can be overwhelming. With limited daylight, harsh weather, and the looming threat of radiation, the emotional burden on survivors is immense. Understanding the emotional challenges that come with isolation and uncertainty is the first step to addressing them.

Fear and Uncertainty

Uncertain Future: The inability to predict when or if life will return to normal can create feelings of fear and anxiety. People may worry about their long-term survival, the state of the world outside, and whether help will ever arrive.

Monotony and Despair: The repetitive nature of shelter living—where every day may feel the same—can lead to boredom and a loss of hope. The lack of variety in daily life, combined with limited activities and interaction, can drain morale.

Isolation

Physical and Emotional Isolation: Being cut off from the outside world can lead to feelings of loneliness, even when surrounded by family. The absence of outside social interactions and normal life can amplify these feelings.

Loss of Control: In a survival scenario, much of life is beyond your control, which can be unsettling. Not being able to leave the shelter, control the weather, or determine how long you'll need to stay confined can lead to frustration.

Creating a Positive Environment in the Shelter

Maintaining a positive atmosphere inside the shelter is crucial for sustaining morale. This requires intentional effort to reduce tension, create comfort, and foster a sense of well-being.

Comfort and Warmth

Physical Comfort: Make your shelter as comfortable as possible by organizing sleeping areas, creating cozy spaces, and ensuring access to warmth, especially during long, cold nights.

Lighting: In a nuclear winter, daylight will be scarce. Use soft, warm lighting like lanterns or battery-powered lamps to create a comforting ambiance. The right lighting can drastically improve mood, especially during long stretches of darkness.

Positive Interaction

Promoting Positivity: Encourage positive conversations, humor, and laughter within the group. While it's important to be realistic, focusing on uplifting topics can help balance the heaviness of the situation.

Avoiding Negativity: Limit negative conversations and complaints. Address concerns constructively but avoid dwelling on the negatives, as this can drain morale.

Establishing and Celebrating Milestones

In an environment where time seems to stretch endlessly, creating and celebrating small milestones gives you something to look forward to and helps mark progress.

Weekly and Monthly Milestones

Setting Small Goals: Break time down into manageable chunks by setting small, achievable goals, such as organizing supplies, completing shelter repairs, or learning new skills. Celebrate the completion of each goal as a team.

Tracking Progress: Keep a visual calendar or log where you can mark off days and see progress. This helps maintain a sense of time and accomplishment, providing a psychological boost.

Celebrating Achievements

Acknowledging Efforts: Recognize and celebrate individual and group accomplishments, whether it's mastering a new skill, solving a problem, or simply surviving another week. Celebrations don't need to be elaborate—acknowledging success is enough to lift spirits.

Keeping a Long-Term Vision: Setting Goals for the Future

Focusing on the future, even if it's uncertain, can help foster hope and give survivors something to work toward. This includes both short-term goals for shelter living and long-term visions for life after the nuclear winter.

Short-Term Shelter Goals

Skill Development: Set goals for learning survival skills, such as first aid, gardening, or repair work. Not only do these skills improve survival chances, but they also provide a sense of accomplishment and progress.

Physical Fitness: Maintain a fitness routine with the goal of staying strong and healthy for when the nuclear winter ends. Exercise improves mental clarity and gives a sense of control over one's body.

Long-Term Vision

Imagining the Future: Regularly talk about life after the shelter. What would you do when it's safe to go outside? Where would you go? What would you want to rebuild? These conversations help keep hope alive and give people something to look forward to.

Preparing for Rebuilding: Start thinking about what will be needed once it's safe to leave the shelter. Planning for the future, even if far off, creates purpose and drive.

Encouraging Group Cohesion and Support

In a confined space, unity is crucial for maintaining morale. Ensuring that everyone feels supported and part of the group prevents feelings of isolation and fosters teamwork.

Group Discussions

Family Meetings: Hold regular group discussions to check in on everyone's emotional well-being, address concerns, and ensure that everyone feels included in decision-making.

Conflict Resolution: Address any tensions or conflicts promptly, using respectful communication to resolve issues. This prevents lingering resentments that could damage group morale.

Supporting Each Other

Emotional Support: Encourage family members to lean on one another for emotional support. Listening and validating each other's feelings during tough moments creates a stronger emotional bond.

Collaborative Activities: Engage in group projects, like organizing the shelter, preparing meals together, or participating in group exercises. Working as a team fosters unity and shared purpose.

Fostering a Sense of Purpose and Contribution

Giving each family member a meaningful role helps create a sense of purpose, reducing feelings of helplessness. Contributing to the survival of the group boosts self-worth and morale.

Assigning Responsibilities

Daily Tasks: Assign everyone important daily tasks, whether it's preparing food, maintaining the shelter, or monitoring supplies. Knowing that their contribution is essential helps foster a sense of importance.

Sharing Leadership: Rotate leadership roles for certain tasks or projects to prevent burnout and ensure that everyone feels involved in the decision-making process.

Encouraging Independence

Personal Projects: Allow each family member to take on a personal project that interests them, whether it's learning a new skill, writing, or creating something. Personal goals give individuals something to focus on beyond survival.

Engaging in Meaningful Activities and Creativity

Keeping busy with meaningful activities can break the monotony of shelter life and provide a much-needed mental escape. Creative outlets are also essential for emotional expression.

Mental Stimulation

Learning and Education: Encourage learning by reading books, practicing survival skills, or teaching each other new things. Keeping your mind engaged prevents boredom and helps time pass more quickly.

Puzzles and Games: Puzzles, board games, or card games provide entertainment and a fun way to stimulate the brain. They also offer a distraction from the harsh realities of shelter life.

Creative Outlets

Writing or Drawing: Creative activities like journaling, writing stories, or drawing help process emotions and offer a form of mental escape. They also serve as a record of your experiences during this time.

Crafting and Building: Engage in crafting projects or building useful items for the shelter. Creative problem-solving boosts morale and provides a sense of accomplishment.

Staying Informed Without Overloading on Bad News

While it's important to stay informed about conditions outside, too much negative information can harm morale. Balancing the intake of news is crucial for mental well-being.

Limiting News Consumption

Moderating Information: Avoid obsessively following the news, especially if updates are consistently negative. Set specific times to check for updates, and focus on actionable information rather than speculation.

Positive News: Whenever possible, look for positive or hopeful news. Even small signs of recovery in the outside world can offer hope and lift spirits.

Developing Mental and Emotional Resilience

Resilience is the ability to adapt and bounce back from difficult situations. Cultivating mental and emotional resilience is crucial for maintaining hope during a nuclear winter.

Practicing Gratitude

Gratitude Journals: Encourage family members to keep a gratitude journal, where they can write down things they are thankful for each day. Focusing on the positive aspects of shelter life, no matter how small, can help shift the mindset away from negativity.

Daily Reflection: At the end of each day, gather as a group and reflect on what went well. This simple practice helps maintain perspective and reinforces the idea that there is always something to be grateful for.

Mindfulness and Meditation Practices: Introduce mindfulness or meditation exercises to help everyone stay grounded and calm. These practices can reduce stress, anxiety, and feelings of helplessness, while improving focus and emotional resilience.

Deep Breathing: Encourage the practice of deep breathing exercises, especially during moments of stress or panic. Taking slow, deliberate breaths can help calm the nervous system and bring a sense of control to the moment.

Guided Meditation: Use guided meditation sessions to help family members focus on positive thoughts, visualize peaceful settings, or simply relax. Even short, daily sessions can improve mental well-being over time.

Fostering a Growth Mindset

Focus on Learning: Encourage the belief that challenges can be opportunities for growth. Reframe difficulties as learning experiences that make everyone stronger and more adaptable.

Recognizing Progress: Celebrate small victories and recognize improvements, whether it's mastering a new skill, improving fitness, or simply managing emotions more effectively. Acknowledging growth helps build confidence and keeps hope alive.

Maintaining morale and hope during a nuclear winter is vital for the long-term mental and emotional survival of your family. While the physical environment may be harsh and isolating, fostering a positive atmosphere within the shelter can provide strength and resilience. By creating a supportive, structured, and emotionally enriching environment, you can help your family cope with the psychological toll of isolation, fear, and uncertainty.

The key to survival in such a scenario lies not only in securing the basic necessities of life but also in nurturing mental and emotional health. Through mindfulness, setting goals, celebrating milestones, fostering creativity, and maintaining group cohesion, you can keep spirits high and ensure that everyone emerges from this experience with hope and determination intact. With a focus on both the present and the future, you can help your family endure the long months of nuclear winter while keeping the light of hope alive.

Communicating with the Outside World: Tools for Survival

In a long-term survival scenario, particularly during a nuclear winter, staying informed and being able to communicate with the outside world is critical for your safety, planning, and mental well-being. Even in a fallout shelter, you'll need reliable tools for communication to monitor what's happening beyond your immediate environment, receive emergency updates, and possibly contact other survivors or rescue teams. This chapter will explore essential communication tools, how to use them effectively, and strategies for maintaining contact with the outside world during a nuclear winter.

Key Aspects of Communication for Survival

The Importance of Communication in a Crisis

Essential Communication Tools for Shelter Living

Radio Systems: Staying Informed and Receiving Alerts

Two-Way Radios for Local Communication

Satellite Phones: Connecting Beyond the Fallout Zone

Emergency Signal Devices

Preserving Power for Communication Devices

Communicating with Rescue Teams or Other Survivors

Monitoring the Outside World: Staying Informed Without Panic

The Importance of Communication in a Crisis

Communication is a lifeline in survival situations, especially when living in isolation during a nuclear winter. Having the ability to reach out for help, receive updates on radiation levels, or simply know what is happening beyond your shelter can make the difference between safety and danger. Additionally, staying informed helps you plan and make decisions, reducing anxiety and uncertainty.

Why Communication is Essential

Staying Informed: The ability to receive news about radiation levels, weather conditions, and government updates will help you know when it's safe to leave the shelter or take other precautions.

Reaching Out for Help: In the event of an emergency, such as medical issues or equipment failure, having communication tools allows you to reach out for help if it's available.

Maintaining Mental Health: The isolation of shelter living can lead to emotional strain, and knowing you have the means to communicate with others provides comfort and a sense of connection to the outside world.

Essential Communication Tools for Shelter Living

In a nuclear survival scenario, traditional communication networks (such as cell phone towers and internet services) may be disrupted or unavailable. Therefore, you'll need alternative communication tools that do not rely on these networks.

Basic Communication Kit

Hand-Crank or Battery-Powered Radios: Radios are essential for receiving news and emergency broadcasts. Make sure they have multiple power options (such as hand-crank, solar, or batteries) to ensure they work even if the power grid is down.

Two-Way Radios (Walkie-Talkies): These are useful for communicating within your group and with nearby survivors.

Satellite Phones: Satellite phones can bypass local communication outages, allowing you to reach emergency services or distant contacts.

Signal Flares and Mirrors: In extreme emergencies, visual signals like flares and mirrors can help attract attention from rescuers or other survivors.

Radio Systems: Staying Informed and Receiving Alerts

Radios are the most critical tool for receiving information during a nuclear winter. They can keep you updated on conditions such as radiation levels, weather patterns, and government-issued instructions. Having multiple radios with different capabilities ensures redundancy in case one system fails.

AM/FM and NOAA Weather Radios

Receiving Broadcasts: Standard AM/FM radios can pick up local and regional broadcasts, while NOAA weather radios provide up-to-date weather alerts and emergency warnings. These radios may broadcast essential survival information during a nuclear winter.

Hand-Crank and Solar-Powered Radios: These radios are designed for use during power outages. Hand-crank models generate their own power, and solar-powered radios can recharge during daylight hours. Always have extra batteries as a backup power source.

Shortwave Radios

Global Communication: Shortwave radios allow you to pick up broadcasts from distant stations, which can be useful if local broadcasting infrastructure is down. You can tune into global news and emergency frequencies, giving you a broader view of the situation.

Monitoring International Updates: During a nuclear winter, international broadcasts may provide useful information about global conditions, rescue efforts, or recovery plans.

Two-Way Radios for Local Communication

Two-way radios, such as walkie-talkies, are an essential tool for communicating with nearby survivors, shelter members in different parts of your shelter complex, or others in your immediate area. These radios don't rely on infrastructure like cell towers, making them reliable in a post-nuclear environment.

Choosing Two-Way Radios

Range and Frequency: Choose radios with a sufficient range for your needs. Standard walkie-talkies typically have a range of 2-5 miles, though the range may vary depending on obstacles such as buildings or terrain.

Multiple Channels: Radios with multiple channels allow you to switch frequencies if the airwaves are crowded or if interference occurs. This can be especially useful in areas with many survivors or in larger shelter communities.

Using Two-Way Radios Effectively

Regular Check-Ins: If you're part of a larger group or community, schedule regular radio check-ins to stay in contact. Agree on a specific channel and time for communication to ensure everyone is on the same page.

Keeping Radios Charged: Like all communication tools, two-way radios must be charged regularly. Keep extra batteries on hand or use hand-crank radios to ensure they remain functional during power outages.

Satellite Phones: Connecting Beyond the Fallout Zone

Satellite phones are a crucial backup for communication, as they don't rely on local infrastructure. They connect directly to satellites in orbit, allowing you to communicate with emergency services or distant family members, even if cell towers and landlines are destroyed.

Why Satellite Phones Are Critical

Global Coverage: Unlike traditional cell phones, satellite phones can work anywhere on Earth, as long as there's a clear line of sight to the sky. This makes them reliable for emergencies, even in remote or isolated locations.

Emergency Calls: In the event of a serious emergency (such as medical issues, equipment failure, or evacuation needs), a satellite phone can be a lifeline to contact authorities, relief agencies, or nearby survivors.

Powering Satellite Phones

Solar Chargers: Keep solar chargers on hand to recharge satellite phones, especially during long periods without grid power. Solar power allows you to maintain functionality without relying on batteries.

Conserving Battery Life: Use satellite phones sparingly to conserve battery life. Keep them turned off when not in use and turn them on only when necessary to make a call or check for updates.

Emergency Signal Devices

Sometimes, more traditional, low-tech methods of communication may be necessary to signal for help or alert others to your location. These tools are particularly useful in situations where radios and phones are unavailable or impractical.

Signal Flares

Signaling for Help: Flares are useful for attracting the attention of nearby aircraft, helicopters, or rescue teams. They are visible from a distance and can be life-saving if you're trying to signal your location in an emergency.

Using Flares Safely: Make sure to store flares in a safe, dry place, and only use them when you're sure help is nearby to avoid wasting them.

Signal Mirrors

Daylight Communication: Signal mirrors can be used during the day to reflect sunlight and send signals to aircraft or distant locations. A well-placed mirror reflection can be seen from miles away.

Effective Use: Practice using a signal mirror before you need it. It requires some skill to accurately direct sunlight to a target.

Preserving Power for Communication Devices

In a long-term survival situation, power is a limited resource. Preserving energy for your communication devices ensures that they remain functional when needed.

Using Power Wisely

Powering Down: Turn off devices when not in use to save battery life. Only turn on radios, phones, or other communication tools when checking for updates or making contact.

Hand-Crank and Solar Chargers: Use hand-crank generators or solar-powered chargers to recharge devices without draining your stored battery supply. These tools are particularly useful for maintaining communication tools without relying on fuel or electricity.

Battery Backup

Spare Batteries: Keep a stockpile of spare batteries for your radios, flashlights, and other devices. Rotate them regularly to ensure they remain functional over time.

Portable Battery Packs: Portable power banks or battery packs can provide additional energy for radios, phones, or other small devices. Recharge them using solar panels or crank generators when possible.

Communicating with Rescue Teams or Other Survivors

Knowing how to communicate effectively with rescue teams or other survivors increases your chances of survival. You'll need to convey your location, condition, and needs clearly and efficiently.

Emergency Contact Protocols

Sharing Critical Information: If you're able to contact emergency services or a rescue team, be prepared to provide important information quickly, such as your location, the number of people in your shelter, and any urgent needs (such as medical supplies or evacuation).

Using Standard Emergency Signals: Familiarize yourself with standard emergency communication protocols, such as distress signals (SOS), that may be used to attract attention or communicate with rescue teams.

Locating Other Survivors

Listening for Other Signals: Keep your radios tuned to emergency channels or general use frequencies to listen for other survivors. In post-disaster scenarios, survivors may use common channels to communicate and form networks.

Establishing a Network: If possible, establish a **communication network** with other survivors. Agree on specific channels and times to check in with each other, share information about local conditions, and coordinate efforts to gather resources or provide mutual assistance. This type of network can also help maintain morale and reduce feelings of isolation.

Monitoring the Outside World: Staying Informed Without Panic

Staying informed is critical for making survival decisions, but it's also important not to overwhelm yourself or your family with too much bad news. Managing information intake can help keep morale high while still ensuring you know what's happening in the outside world.

Setting Limits on News Consumption

Scheduled News Check-Ins: Rather than constantly monitoring news broadcasts, schedule regular times to check for updates on radiation levels, weather conditions, or government alerts. This helps prevent panic from information overload.

Focusing on Relevant Information: Prioritize actionable information, such as updates on safe zones, evacuation routes, or local contamination levels. Avoid dwelling on speculation or worst-case scenarios that can lead to unnecessary fear.

Filtering Out Negative Information

Avoiding Unverified Sources: Stick to official or trusted sources of information, such as government broadcasts, emergency radio stations, or NOAA weather alerts. Be cautious of rumors or unverified reports, which can create confusion and anxiety.

Sharing Information Wisely: When communicating updates with your family or group, focus on facts and practical steps you can take. Reassure others by emphasizing solutions and actions rather than simply relaying negative news.

Effective communication is one of the most important tools for surviving a nuclear winter. With the right equipment and strategies, you can stay informed, connected, and prepared to respond to changing conditions outside your shelter. Radios, two-way communication devices, satellite phones, and emergency signaling tools provide the lifeline needed to monitor the environment, communicate with others, and reach out for help when necessary.

In addition to gathering the right tools, it's crucial to develop communication protocols, preserve power for your devices, and balance staying informed with maintaining mental well-being. By managing communication effectively, you can reduce isolation, enhance cooperation within your group, and improve your chances of long-term survival.

Long-Term Storage: Ensuring Your Food and Water Last

Ensuring that your food and water supplies last throughout a long-term survival scenario, such as a nuclear winter, is a critical part of planning and preparation. Proper storage techniques, rationing, and understanding shelf life are all vital to maintaining your food and water supply over months or even years in a confined space. This chapter will explore strategies for effective long-term food and water storage, tips for avoiding spoilage and contamination, and how to extend the shelf life of your essential resources.

Key Aspects of Long-Term Storage

The Importance of Proper Storage for Survival

Choosing the Right Foods for Long-Term Storage

Water Storage: Keeping Your Supply Clean and Safe

Best Storage Conditions for Food and Water

Managing Rations to Extend Supplies

Protecting Supplies from Contamination and Pests

Preserving Food for Extended Durations

Monitoring Supplies and Rotating Stock

Dealing with Spoilage and Waste

The Importance of Proper Storage for Survival

In a nuclear winter, access to fresh food and clean water will be extremely limited. Therefore, the supplies you have in your shelter must last for the entire duration of your stay, which could be several months or years. Proper storage prevents spoilage, contamination, and waste, ensuring that you and your family have enough resources to survive.

Avoiding Spoilage

Temperature Control: Foods spoil faster in warm, humid environments. Keeping your food at a stable, cool temperature is essential for preventing spoilage.

Moisture and Humidity: Excess moisture can cause food to spoil, mold, or become unsafe to eat. Similarly, moisture in stored water can promote the growth of bacteria or mold. Managing the humidity levels in your storage area is key to preserving your supplies.

Preventing Contamination

Sealed Containers: Using airtight, durable containers prevents contamination from air, moisture, pests, and chemicals, keeping food and water safe for longer periods.

Clean Storage Areas: Regularly cleaning your storage space reduces the risk of contamination and pests, both of which can ruin your supplies and make them unsafe to consume.

Choosing the Right Foods for Long-Term Storage

When planning for long-term survival, the types of food you store will play a major role in how well you can sustain yourself. Focus on foods that are shelf-stable, nutrient-dense, and have long expiration dates.

Non-Perishable Staples

Canned Goods: Canned vegetables, fruits, beans, meats, and soups are excellent for long-term storage because they have a shelf life of several years. Look for low-sodium options and rotate stock regularly to ensure freshness.

Dried Foods: Foods such as rice, pasta, beans, lentils, oats, and powdered milk are lightweight and have long shelf lives if stored in airtight containers. They can also be easily rehydrated when needed.

Freeze-Dried and Dehydrated Foods: Freeze-dried meals, fruits, and vegetables retain their nutrients and can last for years if kept dry. Dehydrated foods also have long shelf lives and are easy to store in compact containers.

High-Energy Foods

Nuts, Seeds, and Nut Butters: These are calorie-dense and provide essential fats and proteins. Store them in sealed containers to prevent oxidation and extend shelf life.

Grains and Cereals: Whole grains, such as quinoa, bulgur, and barley, provide long-lasting energy. Keep them in dry, cool conditions to preserve their freshness.

Energy Bars and Meal Replacement Powders: These items are formulated to provide balanced nutrition in a compact form, making them ideal for long-term storage.

Water Storage: Keeping Your Supply Clean and Safe

Water is the most important resource for survival, and ensuring that your stored water remains safe to drink over the long term is essential. Water can become contaminated with bacteria, viruses, and chemicals, so careful storage is critical.

Storing Water Safely

Water Containers: Use food-grade, BPA-free containers specifically designed for long-term water storage. These containers are durable and prevent chemicals from leaching into the water.

Water Purification Methods: In case your stored water becomes contaminated, have a water purification system ready. Water purification tablets, filters, or boiling can make contaminated water safe to drink.

Rotation of Water Supply: Rotate your water supply every six months to one year to ensure it stays fresh. Label containers with dates so you know when to replace or treat the water.

Emergency Water Sources

Rainwater Collection: If safe, set up a rainwater collection system to supplement your stored water. Make sure to purify the water before drinking, as fallout or contamination may be present.

Water Filtration Systems: Keep portable water filters or gravity-fed filtration systems on hand to purify water from rivers, lakes, or other sources if necessary.

Best Storage Conditions for Food and Water

The conditions in which you store your food and water can have a significant impact on how long they last. Controlling temperature, light exposure, and humidity can extend the shelf life of your supplies.

Temperature Control

Cool and Stable Temperatures: Aim to store food and water in a cool (ideally between 50°F and 70°F), dry location. Higher temperatures can accelerate spoilage, especially for foods like oils and nuts, while freezing temperatures can damage canned goods.

Light Protection

Keep Supplies in the Dark: Direct sunlight can degrade food and water storage containers, causing spoilage. Store supplies in dark areas or use opaque containers to protect them from light exposure.

Humidity and Moisture Control

Low Humidity Levels: Moisture in the air can cause mold growth and degrade dry foods. Use moisture absorbers or desiccants in your storage area to keep humidity levels low.

Airtight Containers: Store food in airtight containers to keep moisture out and preserve freshness. This is particularly important for dried goods like grains and beans.

Managing Rations to Extend Supplies

Properly managing your food and water rations ensures that you don't run out of supplies before it's safe to leave the shelter. Create a rationing system that provides sufficient nutrition while conserving resources.

Setting Daily Rations

Calculating Caloric Needs: Estimate the caloric needs of each person based on their age, gender, and activity level. Adjust rations according to how active you are and the available supply of food.

Pre-Planned Meals: Create meal plans that balance nutrition with rationing. Focus on meals that are filling but don't use excessive amounts of your most valuable resources.

Water Rationing

Daily Water Needs: Plan for at least one gallon of water per person per day for drinking, cooking, and hygiene. In extreme rationing situations, drinking water should take priority over water for hygiene.

Water Conservation Tips: Use water-saving techniques such as sponge baths, limited dishwashing, and reusing gray water (water used for washing) for non-drinking purposes like cleaning.

Protecting Supplies from Contamination and Pests

In a survival scenario, your food and water must be protected from contamination and pests that can ruin your supplies.

Keeping Pests Out

Rodent and Insect Barriers: Use sealed containers to keep pests out of your food and water. Store containers off the ground and inspect them regularly for signs of damage.

Natural Pest Deterrents: Consider using natural pest deterrents such as bay leaves, cloves, or diatomaceous earth to keep insects and rodents at bay.

Avoiding Cross-Contamination

Separate Storage for Chemicals: Store chemicals like cleaning supplies far away from food and water to prevent accidental contamination.

Handling Food Safely: Wash your hands and use clean utensils when handling stored food to avoid introducing bacteria or other contaminants.

Preserving Food for Extended Durations

In addition to buying non-perishable goods, preserving food yourself can help extend the life of fresh foods. Techniques such as canning, dehydrating, and freeze-drying are excellent for long-term storage.

Canning

Home Canning: Canning is an effective way to preserve fruits, vegetables, and meats. Follow proper canning techniques to ensure food is sealed and bacteria-free, which can extend shelf life by years.

Pressure Canning: Use a pressure canner for low-acid foods (like meats and vegetables) to kill harmful bacteria and extend their shelf life.

Dehydrating and Freeze-Drying

Dehydration: Dehydrated foods, such as jerky or dried fruits and vegetables, last longer when stored properly. Keep them in airtight containers to prevent moisture from shortening their shelf life.

Freeze-Dried Foods: Freeze-dried foods retain their nutritional value and have an even longer shelf life than dehydrated foods. Store them in moisture-proof packaging for maximum longevity.

Monitoring Supplies and Rotating Stock

Keeping track of your supplies ensures that nothing goes to waste and allows you to manage your resources effectively.

Inventory Management

Tracking Your Inventory: Keep a detailed inventory of your food and water, including expiration dates. Regularly check this inventory and use items that are nearing the end of their shelf life first.

Rotating Supplies: Practice the "first in, first out" method, using older supplies first before opening newer ones. This ensures that your food and water remain fresh for as long as possible, reducing waste and spoilage. Label your supplies clearly with purchase and expiration dates so you can easily keep track.

Regular Inspections

Checking for Signs of Spoilage: Inspect food and water supplies regularly for signs of spoilage, such as discoloration, off smells, bulging cans, or mold. If any signs of contamination or spoilage are detected, discard the item immediately to prevent it from affecting other supplies.

Ensuring Seals and Containers Are Intact: Periodically check all containers to make sure they are still sealed and intact. Any breaches can lead to contamination from air, moisture, or pests, compromising the safety of your food and water.

Dealing with Spoilage and Waste

No matter how careful you are with storage, there's always a possibility that some food or water will spoil. Knowing how to deal with spoilage and minimize waste is essential to conserving your resources.

Identifying Spoiled Food

Recognizing Spoilage: Familiarize yourself with the signs of spoiled food, such as foul odors, strange colors, unusual textures, or leaking containers. Canned goods with bulging lids or rust are indicators that the contents may no longer be safe to eat.

Tossing Spoiled Items: If any food or water shows signs of contamination or spoilage, discard it immediately. Consuming spoiled food can lead to illness, which could be life-threatening in a survival scenario where medical resources are limited.

Minimizing Waste

Using Perishables Wisely: In a survival situation, it's important to be creative in using every part of your food to avoid waste. For example, save vegetable peels for making broth or use leftover grains in different recipes.

Composting: If space allows, you can set up a small composting system for spoiled food. While you won't be able to compost everything, organic matter like vegetable scraps can be used later to enrich soil if gardening becomes possible.

Ensuring your food and water last through a nuclear winter or any long-term survival scenario requires careful planning, proper storage techniques, and regular monitoring. By choosing non-perishable, nutrient-dense foods and storing them under optimal conditions, you can preserve your supplies for months or even years. Water storage is equally critical, and proper containers, purification methods, and regular rotation will keep your supply safe.

Rationing is key to stretching your supplies for the duration of your shelter stay, and creating a system of regular inspections and stock rotation ensures that nothing goes to waste. By protecting your supplies from contamination, pests, and spoilage, and being prepared to deal with any waste that does occur, you increase your chances of long-term survival. With careful management and proper storage, you and your family can safely endure the challenges of shelter life during a nuclear winter.

What to Do If the Shelter Fails: Emergency Backup Plans

In any survival scenario, especially during a nuclear winter, having an emergency backup plan is critical in case your primary shelter fails. Whether due to structural damage, contamination, equipment failure, or an unexpected emergency, your survival could depend on how well-prepared you are to quickly adapt and relocate. This chapter will cover what to do if your shelter becomes compromised, strategies for emergency evacuation, and how to develop a solid backup plan that ensures your family's safety in a worst-case scenario.

Key Considerations for Emergency Backup Plans

Understanding the Potential Risks to Your Shelter

Signs That Your Shelter is Failing

Immediate Actions to Take When the Shelter Fails

Preparing a Secondary Shelter

Essential Go-Bags for Emergency Evacuation

Finding or Building Temporary Shelter

Navigating Fallout and Radiation During Relocation

Communicating and Coordinating During Evacuation

Post-Evacuation Strategies: Long-Term Survival Without a Shelter

Understanding the Potential Risks to Your Shelter

Even the most carefully constructed shelter can face risks that compromise its safety. Understanding these risks helps you prepare for unexpected situations and identify the early signs of shelter failure.

Structural Damage

Natural Disasters: Earthquakes, extreme weather, or landslides could damage your shelter's structural integrity, making it unsafe to remain inside.

Radiation and Fallout: Fallout particles can seep into your shelter if it's not properly sealed, leading to contamination. The shelter's protective barriers against radiation could also degrade over time.

Equipment Failure

Air Filtration and Ventilation: If your air filtration system fails, it may no longer provide clean air, putting your family at risk of inhaling toxic particles. Lack of ventilation could also lead to dangerous carbon dioxide buildup.

Power Outages: Loss of power can disrupt essential systems such as heating, water filtration, or lighting, making it difficult to maintain a livable environment.

Water or Food Contamination

Spoiled Supplies: If water or food supplies become contaminated by radiation, mold, or pests, you may no longer have access to safe resources for survival.

Water System Failure: If your water purification system or stored water supply is compromised, you may run out of clean water quickly, forcing you to evacuate.

Signs That Your Shelter is Failing

Recognizing early signs of shelter failure allows you to act quickly before the situation becomes life-threatening. Monitoring your shelter regularly for potential issues is key to being prepared.

Structural Warning Signs

Cracks or Leaks: Any signs of cracks in the walls or foundation, water leaks, or damage to seals and doors could indicate that your shelter is no longer providing full protection.

Decreasing Air Quality: If you notice a drop in air quality—such as a musty smell, increased humidity, or difficulty breathing—it could mean that the air filtration system is failing or the shelter is contaminated with outside particles.

System Failures

Malfunctioning Equipment: Unexplained power outages, failure of critical systems like water purification or heating, or malfunctioning electronics could indicate that the shelter is no longer sustainable.

Sudden Illness or Symptoms: If anyone in the shelter begins to show symptoms of radiation sickness (nausea, fatigue, skin burns) or food or water contamination (stomach pain, diarrhea), the shelter may no longer be safe.

Immediate Actions to Take When the Shelter Fails

If your shelter begins to fail, swift and decisive action is necessary to ensure your family's safety. Having a plan in place for an emergency exit can save valuable time when every second counts.

Assess the Situation

Determine the Severity: Assess whether the issue can be fixed or whether you need to evacuate. For example, a small crack in a wall may be temporarily patched, while a collapsed roof or system-wide failure requires immediate evacuation.

Safety First: Before evacuating, make sure the area outside the shelter is safe, particularly in terms of radiation exposure. Use radiation monitoring devices like Geiger counters to assess radiation levels.

Prepare for Evacuation

Gather Essentials: In case of an immediate evacuation, gather your essential supplies (such as water, food, first aid kits, and communication devices) and go-bags quickly. Prioritize survival over non-essential items.

Stay Calm and Coordinated: Remain calm and communicate clearly with your family. Assign tasks to each person to avoid confusion and ensure everyone knows what to do during evacuation.

Preparing a Secondary Shelter

A critical component of your backup plan should be identifying and preparing a secondary shelter. This could be a pre-constructed structure or a temporary shelter you can build in an emergency.

Designating a Backup Shelter Location

Pre-Identify a Safe Location: If possible, choose a secondary shelter site that's far enough from your main shelter to avoid the same risks. It should be within a reasonable distance to reach on foot but far enough away from the immediate threat.

Pre-Built or Makeshift Shelter: If you have access to a second pre-built shelter (such as a reinforced basement in a different location), make sure it's stocked with basic survival supplies. If not, have materials ready for building a makeshift shelter using tarps, heavy-duty plastic sheeting, and sturdy construction tools.

Stocking Your Secondary Shelter

Basic Supplies: Store food, water, tools, clothing, and first aid supplies in the secondary shelter, similar to your primary shelter setup. Keep enough supplies to last for at least a few days.

Essential Equipment: Ensure your secondary shelter has radiation protection, a ventilation system, and basic living necessities. These may not be as extensive as your primary shelter, but they should cover short-term survival.

Essential Go-Bags for Emergency Evacuation

Each family member should have a go-bag prepared with essential survival items in case an evacuation becomes necessary. These bags should be stored near the exit of your shelter and easily accessible.

Go-Bag Essentials

Food and Water: Pack enough food and water for 72 hours, including energy-dense snacks like protein bars, nuts, and jerky. Include water purification tablets or a portable water filter.

First Aid Kit: Include bandages, antiseptic wipes, pain relievers, any necessary medications, and emergency medical supplies.

Clothing and Shelter: Pack thermal blankets, extra clothing layers, and items for building a temporary shelter (such as tarps, ropes, and stakes).

Communication Devices: Include a hand-crank radio, walkie-talkies, a whistle, and a signal mirror to stay informed and communicate with others.

Personal Safety Gear

Radiation Protection: Include N95 masks or other respiratory protection, radiation detection equipment (such as a dosimeter), and protective gloves.

Fire Starting Kit: Pack waterproof matches, a lighter, and fire-starting material like tinder to help keep warm in case of exposure to cold temperatures.

Finding or Building Temporary Shelter

If your secondary shelter is not readily accessible or you need a temporary place to take refuge, knowing how to find or build shelter quickly is essential.

Using Natural Shelter

Caves or Underground Structures: Look for natural caves or underground spaces that provide insulation from radiation and fallout. These spaces should be well-ventilated but also offer protection from the elements.

Forested Areas: Dense forests can provide natural protection from the elements and offer materials for constructing temporary shelters.

Building a Makeshift Shelter

Using Available Materials: If no natural shelters are available, build a temporary shelter using tarps, plastic sheeting, and whatever sturdy materials you can find. Focus on creating a barrier from fallout particles and securing protection from the wind and cold.

Digging In: If necessary, dig a trench or use a natural depression in the landscape to create a shelter. Cover the area with plastic or tarps, using soil or debris to insulate from radiation.

Navigating Fallout and Radiation During Relocation

If you need to relocate during or after a nuclear event, radiation exposure becomes a significant risk. Knowing how to navigate fallout zones and reduce exposure is critical to survival.

Timing Your Movement

Waiting for Fallout to Settle: Fallout levels decrease significantly after the first 48 hours. If possible, wait until radiation levels have dropped before moving outside. Monitor radiation with a Geiger counter or similar device.

Minimizing Time Outside: Plan your route and minimize time spent in open areas. Move quickly but cautiously, using protective gear to reduce radiation exposure.

Wearing Protective Gear

Clothing: Wear long sleeves, pants, and protective gloves. Use a poncho, tarp, or other waterproof layer to protect against radioactive dust.

Respiratory Protection: Use masks (preferably N95 or better) to prevent inhaling radioactive particles. If a mask isn't available, use cloth or clothing to cover your nose and mouth.

Communicating and Coordinating During Evacuation

Clear communication is crucial during an evacuation to ensure everyone stays safe and knows their role. Establishing communication protocols in advance will make the evacuation process smoother.

Setting Communication Plans

Pre-Established Signals: Use pre-determined hand signals, walkie-talkie channels, or whistle codes to communicate silently or across distances during an evacuation. This can be especially helpful if the environment is noisy, or you need to maintain silence to avoid detection by others.

Check-In Points: Set specific check-in points during your evacuation route where everyone in your group must meet. This helps ensure no one gets separated or lost during the chaos of an emergency move.

Staying in Contact

Walkie-Talkies: Use two-way radios to communicate within your group if you are spread out. Make sure each person knows the designated channel and proper radio etiquette for clear and concise communication.

Maintaining Silence if Needed: If you are in an area with potential threats (such as other desperate survivors), maintain radio silence and use non-verbal signals to avoid attracting attention.

Post-Evacuation Strategies: Long-Term Survival Without a Shelter

If you've had to abandon your shelter, you may face a prolonged period without access to your primary resources. Planning for long-term survival after evacuation is essential, as it could take days or weeks before finding a new safe location.

Establishing a New Shelter

Temporary Camps: Once you've evacuated, the first priority is finding or building a temporary shelter. Use tarps, tents, or natural materials to create a structure that shields you from the elements and fallout.

Insulation and Warmth: Focus on staying warm and dry. Use thermal blankets, fire, and natural insulation like leaves or dirt to keep your shelter comfortable.

Securing Water and Food

Water Sources: Immediately identify potential water sources near your new location. Use portable water filters or purification tablets to ensure the water is safe to drink. If no clean sources are available, prioritize collecting rainwater or boiling water.

Foraging and Hunting: If your food supply is running low, look for edible plants, insects, or small animals in the area. Research local foraging techniques before the disaster, so you know which plants are safe to eat.

Maintaining Health and Hygiene

Hygiene Practices: Continue basic hygiene practices even in temporary conditions to prevent illness. Use soap, sanitizers, and wipes if available, and avoid contaminating water supplies with waste.

First Aid: Keep your first aid supplies accessible and replenish them if you find new resources. Be prepared to treat minor injuries, infections, or radiation sickness.

Having a backup plan in place for shelter failure is critical for survival in any long-term crisis, especially during a nuclear winter. Whether it's due to structural damage, contamination, or equipment failure, knowing how to respond swiftly and efficiently can be the difference between life and death. Your emergency response plan should include a secondary shelter, go-bags with essential supplies, clear communication protocols, and strategies for temporary survival without a permanent shelter.

By preparing in advance, you'll be able to react calmly and decisively, protect your family, and continue surviving despite setbacks. Always remember that flexibility and adaptability are key to surviving the unexpected challenges that may arise when your shelter fails.

Advanced Shelter Construction for Extreme Survivalists

For extreme survivalists, advanced shelter construction goes beyond basic survival needs, offering enhanced protection, longevity, and self-sufficiency in extreme environments, such as during a nuclear winter or post-apocalyptic scenario. Building an advanced shelter means designing for long-term survival, incorporating cutting-edge techniques, and ensuring the structure can withstand not only environmental hazards but also human threats and supply shortages. This chapter will explore advanced shelter construction techniques, materials, and systems to ensure long-term sustainability and security.

Key Components of Advanced Shelter Construction

Key Design Principles for an Advanced Survival Shelter

Site Selection and Underground Shelter Options

Reinforced Building Materials for Extreme Durability

Air Filtration and Ventilation Systems for Contaminated Environments

Self-Sustaining Power Systems: Off-Grid Energy Solutions

Water Systems: Collection, Filtration, and Recycling

Food Production: Indoor Growing Systems

Advanced Security Measures Against External Threats

Waste Management and Sanitation for Long-Term Sustainability

Maintaining Psychological Well-Being in a High-Tech Shelter

Key Design Principles for an Advanced Survival Shelter

The foundation of an advanced shelter begins with careful planning, incorporating essential survival features and allowing for adaptability. It must be able to withstand extreme environmental conditions, such as radiation, nuclear fallout, extreme temperatures, and natural disasters.

Redundancy in Critical Systems

Backups for Essential Systems: Every critical system, such as power, water, and air filtration, should have multiple backup solutions to prevent total failure. Redundancy ensures that if one system fails, a backup can immediately take over, minimizing downtime.

Fail-Safe Designs: Consider designing fail-safe mechanisms into the shelter's layout. For example, automated systems that switch to backup power or air filtration without manual intervention.

Modularity and Expansion

Expandable Designs: Advanced shelters should allow for modular expansions, enabling the addition of new rooms or systems as needed. This can be useful for increasing storage capacity or creating additional living spaces if more people join your shelter.

Adaptable Spaces: Multi-purpose rooms that can serve as storage, living quarters, or workspaces enhance shelter flexibility and reduce the need for large, single-purpose rooms.

Site Selection and Underground Shelter Options

Choosing the right location for your advanced shelter is critical. Underground shelters offer superior protection from fallout, radiation, and extreme weather, but they also present unique challenges in terms of construction and maintenance.

Underground vs. Above-Ground Shelters

Underground Shelters: Building your shelter underground provides natural insulation and radiation shielding, protecting you from fallout, temperature extremes, and aerial surveillance. Consider building deep enough to avoid ground-level fallout, but not so deep that ventilation becomes an issue.

Above-Ground Shelters: In areas where underground construction is not feasible, reinforced above-ground shelters can be designed with thick, multi-layered walls and radiation shielding. These shelters must be fortified to resist external threats, including human intrusion.

Choosing the Right Terrain

Natural Barriers: Select a site that offers natural protection, such as hills, forests, or cliffs. These features provide additional defense against environmental threats and make the shelter harder to detect.

Water Table Considerations: If building underground, ensure the water table is deep enough to prevent flooding or moisture problems in the shelter. Consider waterproofing materials and drainage systems to combat seepage.

Reinforced Building Materials for Extreme Durability

The materials used in an advanced shelter should prioritize durability, radiation shielding, and resistance to natural disasters. Advanced materials ensure the shelter remains intact and livable even in extreme conditions.

Radiation-Resistant Materials

Thick Concrete or Lead-Lined Walls: Use concrete reinforced with lead or other heavy materials to block gamma radiation. Concrete walls should be at least 3 feet thick, and lead lining can be added to doors, windows, or key walls for additional protection.

Earth Shielding: For underground shelters, using the surrounding earth as a shield is highly effective. A layer of at least 10 feet of packed earth can reduce radiation exposure significantly.

Blast and Fire Resistance

Reinforced Steel: Using reinforced steel in critical structural areas like support beams or doorframes ensures the shelter can withstand blast shockwaves, as well as resist bending or breakage under stress.

Fireproof Materials: Incorporate fireproof or heat-resistant materials such as ceramic fiber insulation or fire-retardant coatings to protect against fires that may arise from external explosions or internal accidents.

Air Filtration and Ventilation Systems for Contaminated Environments

Air quality in a sealed or underground shelter is crucial for long-term survival, especially if the outside environment is contaminated by radiation or toxic chemicals.

High-Efficiency Filtration Systems

HEPA and Carbon Filters: Install a HEPA filter system to capture 99.97% of airborne particles, including dust, mold, and fallout particles. Carbon filters can remove chemical and gas contaminants from the air.

NBC Filtration: For nuclear, biological, and chemical threats, specialized NBC air filtration systems can filter out harmful agents. These systems should include a manual backup in case power is lost.

Ventilation and Air Circulation

Active Ventilation Systems: Use mechanical ventilation systems to circulate fresh air while filtering out contaminants. Ensure that the system is energy-efficient and that you have a hand-crank or solar-powered backup system to keep air moving in case of power loss.

Air Quality Monitoring: Install air quality sensors to monitor CO2 levels, humidity, and contamination within the shelter. These sensors provide early warnings if filtration fails or if air becomes unsafe to breathe.

Self-Sustaining Power Systems: Off-Grid Energy Solutions

A reliable, off-grid power system is crucial for long-term survival. Advanced shelters require power for air filtration, water systems, heating, lighting, and communication.

Renewable Energy Solutions

Solar Panels: Install solar panels to capture energy from the sun, even during a nuclear winter, when sunlight may be limited. Include a battery bank to store excess power for use during periods without sunlight.

Wind Turbines: Wind power is another renewable energy source. A small wind turbine can generate enough electricity to supplement solar power, especially during windy conditions.

Backup Generators

Diesel or Propane Generators: In the event of extended periods without sunlight or wind, have a backup generator powered by diesel or propane. Keep a supply of fuel that's sufficient for running the generator intermittently over months or years.

Manual Power Generation: Incorporate hand-crank generators or pedal-powered systems to produce small amounts of electricity for emergency use if all other power systems fail.

Water Systems: Collection, Filtration, and Recycling

Water is one of the most critical elements for survival. An advanced shelter should have a reliable system for collecting, filtering, and recycling water to ensure a continuous supply.

Water Collection Systems

Rainwater Harvesting: Even in a nuclear winter, rainwater can be collected, but it must be thoroughly filtered and purified. Set up gutters and a filtration system to capture and treat rainwater safely.

Groundwater Wells: If possible, drill a well to access groundwater, providing a more reliable long-term source. Ensure the well is deep enough to avoid contamination from surface fallout.

Water Filtration and Purification

Multi-Stage Filtration: Use a multi-stage water filtration system that includes pre-filters, activated carbon filters, and UV purification to remove particulates, chemicals, and biological contaminants from your water supply.

Desalination (If Near the Coast): In coastal areas, desalination systems can convert seawater into drinkable water. This requires energy but can be a valuable resource in a pinch.

Food Production: Indoor Growing Systems

Advanced shelters must account for long-term food production, as stored food will eventually run out. Incorporating indoor farming systems ensures a sustainable food source.

Hydroponics or Aquaponics Systems

Hydroponics: Grow plants without soil using nutrient-rich water systems. Hydroponic systems are space-efficient and allow you to grow vegetables, herbs, and fruits indoors year-round.

Aquaponics: Combine fish farming with hydroponics to create a self-sustaining system where fish waste provides nutrients for plants, and plants help filter and clean the water. This system can produce both plant-based and protein-based food sources.

Indoor Gardening

Grow Lights: Install energy-efficient LED grow lights to simulate sunlight for indoor crops. These lights can be powered by your renewable energy systems and provide consistent light for photosynthesis.

Advanced Security Measures Against External Threats

In a post-apocalyptic world, your shelter may be vulnerable to external threats, including other desperate survivors or hostile forces. Advanced security measures are essential for protecting your resources and family.

Fortified Entrances

Blast-Proof Doors: Install thick, blast-resistant doors with reinforced locks. Consider creating multiple layers of doors, with an outer security layer and an inner airtight seal to prevent intruders from breaching the shelter.

Camouflaged Entrances: Hide or disguise the entrance to your shelter to prevent detection by others. Use natural elements like rocks, vegetation, or fake structures to mask your entry points.

Surveillance and Alarms

Surveillance Cameras: Install **surveillance cameras** around the perimeter of your shelter to monitor for potential threats. These can be motion-activated and linked to screens inside the shelter, allowing you to maintain constant awareness of your surroundings without exposing yourself.

Alarm Systems: Set up alarm systems, including motion sensors and tripwires, around the shelter to alert you of any intruders. For additional stealth, consider using silent alarms that notify you inside the shelter without drawing attention from outsiders.

Remote Monitoring: If possible, connect your surveillance system to a remote monitoring device that allows you to check the perimeter from a safe location. This could be through a handheld tablet or secondary monitor.

Waste Management and Sanitation for Long-Term Sustainability

Sanitation is critical for maintaining health in a sealed or underground shelter. Managing human waste, trash, and gray water effectively prevents contamination and keeps the living environment clean.

Human Waste Management

Composting Toilets: Install a composting toilet system that safely breaks down human waste into compost material, which can be used for growing plants (though not edible crops) or safely disposed of. Composting toilets reduce water usage and require minimal maintenance.

Septic or Holding Tanks: For more advanced setups, a septic system can be installed to store and process waste. Ensure your system has proper drainage and ventilation to prevent odors and contamination.

Waste Incineration: In extreme situations, a waste incinerator can burn human waste and other refuse, reducing it to ash. This system requires careful planning for venting smoke and maintaining safety in confined spaces.

Trash Disposal

Recycling Materials: To reduce waste, focus on reusing and recycling materials whenever possible. This includes repurposing containers, using scrap materials for repairs, or composting organic waste.

Waste Storage: If incineration or composting isn't feasible, establish a waste storage area. Ensure this area is well-sealed and ventilated to prevent the spread of bacteria, pests, and odors.

Maintaining Psychological Well-Being in a High-Tech Shelter

Surviving in an advanced, isolated shelter for an extended period can take a toll on mental health. Ensuring that the shelter is comfortable and offers outlets for psychological relief is crucial for long-term survival.

Comfort and Aesthetics

Lighting and Ambiance: Use adjustable lighting to simulate natural daylight cycles and reduce the effects of seasonal affective disorder (SAD). Soft lighting, warm colors, and comfortable furnishings create a more pleasant living space.

Soundproofing: Install soundproofing materials to reduce external noise and create a more peaceful environment. This can help reduce stress and anxiety, especially in a high-stress survival situation.

Entertainment and Leisure

Books, Games, and Media: Stock your shelter with books, board games, card games, and other forms of entertainment that don't require power. These provide a mental escape from the pressures of survival.

Digital Media: If your power system is reliable, consider keeping digital media, such as movies or music, stored on hard drives or other devices. A little entertainment can significantly boost morale during long-term confinement.

Exercise and Physical Activity

Exercise Equipment: In a confined space, it's important to stay physically active. Include compact exercise equipment like resistance bands, weights, or even a treadmill if space allows. Physical fitness not only keeps the body healthy but also improves mental resilience.

Outdoor Time (If Safe): If radiation levels and environmental conditions permit, designate outdoor time in a protected area to get fresh air and sunlight. If not, simulate natural outdoor environments inside with virtual landscapes or lighting.

Advanced shelter construction for extreme survivalists is a comprehensive approach to long-term survival in the harshest conditions. By incorporating reinforced materials, high-tech air and water systems, renewable energy, food production, and robust security measures, you can ensure that your shelter provides the safety and self-sufficiency needed for extended periods of isolation.

Additionally, maintaining mental and physical well-being is essential for surviving not just physically but psychologically. By creating a well-rounded shelter that addresses all aspects of survival—ranging from power and security to comfort and entertainment—you maximize your chances of thriving in even the most extreme survival scenarios.

Preparing for Post-Shelter Life: Rebuilding Society

Preparing for life after emerging from a long-term shelter involves more than just survival; it requires planning for how to rebuild and contribute to the restoration of a functioning society. Once the immediate dangers of a nuclear winter or post-apocalyptic event have subsided, those who survive will face new challenges in terms of rebuilding communities, restoring infrastructure, and ensuring long-term sustainability. This chapter will explore the essential steps for preparing for post-shelter life, from planning how to safely re-enter the world to helping reestablish a safe, functioning society.

Key Aspects of Post-Shelter Life and Rebuilding

Assessing Environmental Safety Before Leaving the Shelter

Physical and Mental Health After Extended Confinement

Essential Tools and Skills for Rebuilding

Forming and Rebuilding Communities

Restoring Infrastructure: Water, Power, and Communication

Agriculture and Food Security in a Post-Apocalyptic World

Law, Order, and Security in Rebuilt Societies

Trade and Economic Systems After a Collapse

Maintaining Mental and Emotional Well-Being in a New World

Creating a Vision for the Future: Sustainability and Growth

Assessing Environmental Safety Before Leaving the Shelter

After a nuclear winter or similar catastrophic event, the outside world may still pose significant dangers, including radiation, unstable infrastructure, or contaminated resources. Before you can begin rebuilding, it's critical to assess the environment and determine when it's safe to leave the shelter.

Monitoring Radiation Levels

Geiger Counters and Dosimeters: Use radiation detection devices to assess the safety of the environment outside the shelter. Radiation levels may vary depending on the location and how much time has passed since the nuclear event. Ensure levels are safe for long-term exposure before venturing out.

Testing Soil and Water: Collect samples of soil and water from your immediate surroundings and test them for contamination. Clean water and soil are essential for survival and agricultural growth, so understanding contamination levels is critical.

Observing Wildlife and Plant Life

Signs of Recovery: Look for signs that the natural environment is recovering, such as the return of plant life or animal activity. However, be cautious—some wildlife may be more resilient to radiation than humans, so their presence doesn't necessarily mean the area is safe.

Air Quality Monitoring: In addition to radiation, air quality should be tested for pollutants, toxic chemicals, or fallout particles. You may need to continue using air filtration masks until the atmosphere stabilizes.

Physical and Mental Health After Extended Confinement

After a prolonged period in a confined space, your physical and mental health may be affected. Before embarking on the task of rebuilding, you must ensure that both you and your family are in good shape to handle the challenges ahead.

Physical Readiness

Exercise and Movement: Begin a regimen of light physical activity to reacclimate your body to outdoor conditions, particularly if your shelter life was sedentary. Strengthen muscles, improve endurance, and gradually expose yourself to outdoor conditions.

Nutrition: Focus on rebuilding your strength by eating nutrient-dense foods. Ensure you have access to enough food to meet the increased caloric demands of working and rebuilding in the new environment.

Mental Health Considerations

Psychological Adjustment: Re-emerging into a drastically changed world can be mentally overwhelming. Plan to address the emotional toll of leaving the shelter, facing a new reality, and potentially encountering loss and devastation.

Coping with Trauma: After witnessing or experiencing major destruction, trauma may be prevalent. Establish support systems within your group or community to discuss emotions and reduce the psychological burden.

Essential Tools and Skills for Rebuilding

Rebuilding society requires a broad range of practical skills and tools. Advanced preparation for this process can give you a significant advantage when the time comes to rebuild.

Essential Tools for Post-Shelter Life

Hand Tools: Ensure you have durable hand tools such as hammers, saws, shovels, axes, and wrenches for building structures and managing resources.

Repair Kits: Have access to toolkits for repairing mechanical devices, generators, and other equipment that might be essential for power generation or water systems.

Gardening Tools: For food production, gardening tools like spades, hoes, and watering cans will be essential to plant and maintain crops.

Skills for Rebuilding

Basic Construction Skills: Understanding how to construct or repair buildings will be key to rebuilding homes and communal spaces. If possible, learn basic carpentry, masonry, and plumbing skills before leaving the shelter.

Medical Knowledge: Having first aid and basic medical knowledge will be vital in the early stages of rebuilding society, where healthcare systems may not exist. Knowing how to treat injuries, illness, and radiation poisoning can save lives.

Agricultural Skills: Basic farming and gardening skills will be crucial for growing food. Knowledge of crop rotation, seed saving, and soil regeneration will be essential to food security.

Forming and Rebuilding Communities

No one can rebuild society alone. Once it's safe to leave the shelter, forming communities and re-establishing cooperative systems will be necessary for survival and progress.

Finding Other Survivors

Communication Tools: Use radios or communication devices to seek out other survivors in your area. Establishing contact with other groups can lead to the formation of cooperative communities.

Reaching Out Safely: Approach new groups with caution, as resources may still be scarce, and desperation can drive conflict. Establish trust through mutual goals, like food production and safety.

Creating a Cooperative Structure

Leadership and Organization: When forming a community, create a leadership structure to manage responsibilities and make decisions. Consider rotating leadership roles or forming councils to ensure everyone has a voice.

Division of Labor: Assign tasks based on individual skills—some may focus on construction, others on farming or defense. A well-organized division of labor ensures that all essential tasks are covered.

Restoring Infrastructure: Water, Power, and Communication

Rebuilding infrastructure is one of the most challenging and essential tasks after leaving the shelter. Water, power, and communication systems must be restored to create a functioning society.

Water Infrastructure

Building Wells and Purification Systems: Identify clean water sources and build wells if necessary. Install water filtration systems to ensure that water is safe for drinking and agriculture.

Establishing Water Distribution: Set up a system to store and distribute water within your community. Rainwater harvesting and groundwater collection systems can be useful in the early stages.

Power Generation

Renewable Energy Systems: Consider setting up renewable energy systems such as solar panels or wind turbines for small-scale power generation. These systems can power essential devices like water pumps or communication systems.

Energy Conservation: Until energy production increases, encourage conservation practices by limiting the use of non-essential electronics and lighting.

Agriculture and Food Security in a Post-Apocalyptic World

Food security will be a primary concern in post-shelter life. Establishing reliable food sources is critical for survival, and long-term planning will be necessary to sustain your community.

Starting an Agricultural System

Seed Banks: If possible, leave the shelter with a collection of seeds to start growing crops. Focus on fast-growing vegetables, high-calorie crops, and perennials that can produce food over multiple seasons.

Livestock: If feasible, raising livestock such as chickens, goats, or rabbits provides a steady source of protein, dairy, and manure for fertilizer.

Food Preservation Techniques

Canning and Drying: Learn how to preserve food through canning, drying, and fermentation to ensure that excess harvests can be stored for winter or times of scarcity.

Greenhouses: If the environment is still too harsh for outdoor farming, build greenhouses to grow food in a controlled environment, protecting crops from fallout and extreme weather.

Law, Order, and Security in Rebuilt Societies

Maintaining law and order is essential in a post-apocalyptic world, where resources may be limited, and human desperation can lead to conflict.

Creating Fair Governance

Community Rules: Establish clear rules within your community to avoid conflicts and ensure the fair distribution of resources. Create a system for resolving disputes and maintaining peace.

Security Teams: Organize volunteers or security teams to patrol and protect the community from external threats or potential violence from within.

Defending Against External Threats

Fortifications: Reinforce key areas of your community to protect against intruders or attacks. This could include building fences, watchtowers, or simple alarm systems.

Nonviolent Conflict Resolution: When possible, establish communication with other groups to prevent misunderstandings or violent encounters. Negotiation over resources may be necessary for peaceful coexistence.

Trade and Economic Systems After a Collapse

In a newly rebuilt society, bartering and trade will likely replace conventional currency systems, at least initially.

Bartering Goods and Skills

Trade Networks: Form networks with other communities to exchange goods and services. Valuable items for trade include food, tools, medical supplies, and skilled labor.

Resource Management: Keep track of your community's resources, ensuring that you produce more than you consume. This surplus will be useful for trade or during times of scarcity.

Building an Economy

Skills as Currency: In addition to physical goods, skills such as medical knowledge, construction, and farming will become valuable commodities in the early stages of rebuilding society. People with specialized skills can trade their labor for food, tools, or other necessities, creating a barter-based economy that helps distribute essential resources and services.

Establishing Value: Over time, communities may develop more formal systems of value, such as using scarce goods (e.g., ammunition, fuel, precious metals) as a medium of exchange. Developing an economy based on trust and mutual benefit will be essential for cooperation between groups.

Maintaining Mental and Emotional Well-Being in a New World

Rebuilding after a catastrophic event is as much a psychological challenge as it is a physical one. Maintaining mental and emotional health for yourself and your community is key to long-term survival and cohesion.

Dealing with Trauma and Loss

Grief Counseling: Survivors may face significant grief and loss, especially if loved ones, homes, or familiar surroundings were destroyed. Establish regular community gatherings or group discussions where people can share their experiences and emotions.

Finding Purpose: Rebuilding society offers a sense of purpose. Encourage individuals to take on meaningful roles in the community, whether through farming, building, teaching, or organizing. A sense of contribution helps combat despair and depression.

Creating a Sense of Normalcy

Rituals and Celebrations: Even in difficult times, maintaining traditions, rituals, or creating new celebrations can bring comfort and a sense of normalcy. Birthdays, holidays, or the completion of communal projects can be celebrated to foster unity and lift spirits.

Education and Leisure: If possible, prioritize education and recreational activities for both children and adults. Teaching new skills, reading, playing games, and creating art can provide a much-needed mental escape and ensure long-term cultural and intellectual growth.

Creating a Vision for the Future: Sustainability and Growth

Once the immediate needs of survival are met, communities can start planning for a more sustainable and prosperous future. Establishing a long-term vision for growth and sustainability ensures that the next generations can thrive.

Sustainable Practices

Permaculture and Sustainable Farming: Adopt farming methods that focus on sustainability and ecosystem balance, such as permaculture, crop rotation, and natural pest control. These techniques reduce the reliance on finite resources and ensure long-term food security.

Renewable Energy Expansion: As resources allow, expand your renewable energy systems to increase the community's energy independence. Solar, wind, and water-based energy sources should be prioritized to avoid dependence on fossil fuels, which may be in short supply.

Education and Skill Building for the Next Generation

Passing down Knowledge: Ensure that critical survival skills, farming practices, and medical knowledge are passed down to younger generations. Formal or informal education systems can help maintain and grow your community's knowledge base.

Technology Recovery: Once the community is stable, focus on recovering lost technologies or developing new ones that improve living conditions. This could include reestablishing communication systems, rebuilding medical facilities, or improving agricultural techniques.

Building for the Future

Expanding the Community: Plan for gradual expansion of your community, ensuring that as it grows, it remains self-sustaining. Balance resource management with population growth to avoid overextending your capabilities.

Forming Alliances with Other Communities: Strengthen relationships with neighboring communities through trade, communication, and mutual defense agreements. A network of allied communities can share resources, knowledge, and labor, making everyone more resilient to future threats.

Rebuilding society after a nuclear winter or post-apocalyptic event is a daunting but achievable task for those who survive. Preparing for post-shelter life involves more than just emerging from your shelter—it requires planning, cooperation, and vision for a sustainable future. By assessing the environment, securing essential resources, rebuilding infrastructure, and forming new communities, you can help lay the foundation for a new, functioning society.

Equally important is maintaining mental and emotional well-being, passing down knowledge, and creating a system of governance and economy that promotes fairness and cooperation. With resilience, foresight, and collaboration, rebuilding after a catastrophe offers a chance to not just survive, but thrive in a new world.

Preparing Children for Nuclear Survival: A Family Guide

Preparing children for nuclear survival is one of the most challenging and delicate aspects of survival planning. Children are particularly vulnerable during a nuclear event due to their physical and emotional needs. However, with the right approach, they can be equipped with knowledge, skills, and the psychological resilience needed to navigate the challenges of survival in a fallout shelter and beyond. This chapter provides a family guide to preparing children for nuclear survival, covering key strategies for education, psychological support, and practical preparedness.

Key Areas to Address When Preparing Children for Nuclear Survival

Explaining the Situation: Educating Without Causing Panic

Teaching Essential Survival Skills to Children

Helping Children Cope with Fear and Anxiety

Maintaining Routines for Stability and Comfort

Creating Age-Appropriate Responsibilities in the Shelter

Ensuring Physical Safety and Health of Children

Mental and Emotional Support for Children During Isolation

Post-Shelter Adaptation for Children: Re-Entering the Outside World

Family Unity and Teamwork: Strengthening Bonds for Survival

Long-Term Strategies for Raising Resilient Children

Explaining the Situation: Educating Without Causing Panic

When preparing children for a nuclear survival situation, it's important to be honest but not overwhelming. Children need to understand what's happening in a way that makes them feel safe rather than afraid.

Age-Appropriate Explanations

Young Children (Ages 4-7): Keep explanations simple, focusing on safety. Explain that you need to go to a special place (the shelter) to stay safe, just like going inside during a storm.

Older Children (Ages 8-12): Provide more details about what nuclear fallout is and why it's dangerous, but emphasize the safety measures you've taken. Encourage questions and be ready to explain how the shelter protects them.

Teens (Ages 13-18): Be more direct with teenagers. Discuss the scientific and practical aspects of nuclear survival and explain how they can contribute to the family's safety.

Reassuring Safety

Focus on Protection: Reinforce the idea that the shelter is a safe place where the family can stay protected from harm. Emphasize the steps you've taken to make the shelter comfortable and secure.

Answering Questions: Allow children to ask questions and answer them as honestly as possible, but without unnecessary detail that might provoke fear. Your calm attitude will help set the tone for how they process the situation.

Teaching Essential Survival Skills to Children

Children can play an important role in the family's survival, but they need to be taught the skills required in a way that's appropriate for their age and abilities. Teaching children these skills not only makes them more capable but also gives them a sense of control and responsibility.

Basic Survival Skills

Safety Drills: Practice shelter-in-place drills with your children, teaching them how to quickly and safely get to the shelter. Make sure they understand the importance of following instructions in an emergency.

Water and Food Storage: Teach children how to help with basic tasks like gathering and storing food and water. Older children can learn how to use water filtration systems or prepare simple meals from stored supplies.

First Aid: Depending on their age, teach children basic first aid skills, such as treating cuts, scrapes, or burns, and how to recognize signs of illness or radiation sickness. Even simple lessons on handwashing and hygiene are important in a shelter.

Self-Sufficiency Skills

Helping with Chores: Give children responsibilities, such as helping with cleaning the shelter, organizing supplies, or checking that everything is in order. These tasks help children feel useful and contribute to the family's well-being.

Communication Skills: Teach older children how to use communication devices like radios or walkie-talkies. In an emergency, this knowledge can help them stay in contact with family members if they are separated.

Helping Children Cope with Fear and Anxiety

Children will inevitably feel fear or anxiety during a nuclear event or while living in a shelter. It's crucial to provide emotional support to help them cope with these feelings in a healthy way.

Validating Their Emotions

Acknowledge Their Fears: Let children know it's okay to feel scared or upset. Validate their emotions and reassure them that you understand their feelings.

Offering Comfort: Provide comfort through physical closeness, such as hugs, and by maintaining a calm demeanor. When children see their parents remaining composed, they are more likely to feel secure.

Creating a Safe Space to Talk

Encouraging Open Communication: Make time for family discussions where everyone can talk about their feelings. Encourage children to share their thoughts and fears, and listen without judgment.

Providing Reassurance: Frequently reassure your children that you are all safe in the shelter and that you have prepared for this situation. Reinforce that you will face challenges together as a family.

Maintaining Routines for Stability and Comfort

Routines provide a sense of normalcy, even in survival situations. For children, familiar routines offer structure and comfort, helping them adapt to life in a shelter.

Daily Schedules

Morning and Bedtime Routines: Maintain morning and bedtime routines as much as possible. These rituals, like brushing teeth, changing clothes, and reading bedtime stories, help children feel secure and create a sense of normalcy.

Structured Activity Times: Include periods of learning, play, and rest in the daily schedule. Structure keeps children engaged and prevents boredom, which can lead to anxiety.

Educational and Play Activities

Continuing Education: Keep children's minds active by incorporating educational activities into their day. Even if formal schooling isn't possible, teach practical skills, math, reading, or science through hands-on lessons in the shelter.

Time for Play: Encourage creative play to allow children to express their emotions and use their imaginations. Toys, games, and puzzles can also provide distraction and reduce stress.

Creating Age-Appropriate Responsibilities in the Shelter

Giving children responsibilities helps them feel empowered and useful, especially during long stays in the shelter. These tasks should be age-appropriate and designed to build confidence.

Chores and Shelter Maintenance

For Young Children: Assign simple tasks like organizing small items, cleaning up their space, or helping distribute food. These tasks help younger children feel involved without overwhelming them.

For Older Children and Teens: Older children can take on more significant roles, such as assisting with food preparation, helping manage supplies, or learning basic shelter maintenance skills like checking filters or monitoring water systems.

Building Confidence through Contribution

Praise and Encouragement: Acknowledge your children's efforts and praise them for their contributions. This positive reinforcement builds confidence and encourages them to continue taking an active role in the family's survival.

Ensuring Physical Safety and Health of Children

Children's health and safety are paramount in a nuclear survival scenario. Ensuring their well-being requires both physical care and education on health risks.

Monitoring Physical Health

Nutrition and Hydration: Ensure children are eating balanced meals and drinking enough water. Ration food and water wisely but be mindful of their nutritional needs, especially during growth stages.

Sleep and Rest: Adequate sleep is crucial for children's health. Ensure they have a comfortable sleeping space and maintain a consistent sleep schedule.

Radiation and Contamination Protection

Clothing and Masks: If children need to go outside the shelter, ensure they are properly dressed in protective clothing and masks to avoid exposure to radioactive fallout. Teach them how to put on and take off protective gear.

Hygiene and Cleanliness: Teach children the importance of cleanliness in the shelter, such as washing hands frequently, keeping living spaces tidy, and properly disposing of waste.

Mental and Emotional Support for Children During Isolation

Long-term shelter living can lead to feelings of isolation and boredom, which are particularly hard on children. Providing mental and emotional support helps them cope.

Creative Outlets

Art and Writing: Encourage children to express themselves through drawing, painting, or writing. These creative outlets help them process their emotions and provide a distraction from the monotony of shelter life.

Journaling: For older children, keeping a journal can be a therapeutic way to document their experiences and emotions.

Family Bonding and Activities

Family Games: Spend time playing games together as a family. Board games, card games, or group storytelling can foster connection and alleviate stress.

Regular Check-Ins: Have daily family meetings where everyone can discuss how they are feeling, share updates, and voice any concerns. This reinforces the idea that everyone is in this together.

Post-Shelter Adaptation for Children: Re-Entering the Outside World

Once it's safe to leave the shelter, children will face new challenges in adapting to a changed world. Helping them navigate this transition is crucial for their long-term emotional and psychological health.

Gradual Re-Introduction to the Outside World

Safety First: Ensure children are protected from lingering dangers, such as radiation or contaminated water and soil, before venturing outside. Monitor their health closely for signs of distress.

New Normals: Help children understand that the world may look different than before and that adapting to these changes is part of the recovery process. Encourage them to focus on the positives of rebuilding.

Helping Children Process Changes

Talking About the New Reality: Once outside, have age-appropriate conversations with your children about the changes they will see. Be honest about the destruction they may witness, but emphasize the opportunity to rebuild and create a new, better future.

Encouraging Hope: Focus on hope and resilience. Encourage children to think about the future and how they can contribute to rebuilding. Help them understand that they are part of something important and that their efforts matter.

Family Unity and Teamwork: Strengthening Bonds for Survival

Family unity is crucial in survival situations, and preparing children for nuclear survival should emphasize teamwork and mutual support. A strong family bond will help you all navigate the challenges of shelter living and the transition back to the outside world.

Encouraging Cooperation

Shared Responsibilities: Make sure that everyone, including children, understands their role in the family's survival. Assign tasks that suit each person's abilities, and emphasize the importance of working together to ensure everyone's well-being.

Collaborative Decision-Making: Involve children in family decisions whenever possible. This helps them feel empowered and teaches them important lessons about cooperation and leadership in times of crisis.

Building Stronger Relationships

Quality Time: Spend quality time together in the shelter, engaging in activities that strengthen family bonds. These moments of connection help relieve stress and reinforce the idea that you are a team.

Open Communication: Keep lines of communication open between all family members. Encourage children to express their feelings, ask questions, and share their concerns. This openness fosters trust and emotional resilience.

Long-Term Strategies for Raising Resilient Children

Raising resilient children is a long-term process that goes beyond the immediate survival scenario. Teaching them how to cope with adversity, adapt to changing circumstances, and thrive in difficult situations will benefit them for life.

Fostering Resilience

Problem-Solving Skills: Teach children how to solve problems on their own by encouraging critical thinking and creativity. When faced with challenges, guide them to come up with solutions and praise their efforts.

Emotional Regulation: Help children learn how to manage their emotions, especially in stressful situations. Teach them techniques like deep breathing, mindfulness, or journaling to cope with anxiety, fear, or frustration.

Instilling a Growth Mindset

Learning from Challenges: Frame challenges and difficulties as learning opportunities. Encourage children to see setbacks as part of the journey toward success, helping them develop a growth mindset that will serve them in all areas of life.

Building Confidence: Praise children not just for what they achieve, but for their effort and persistence. This helps them build confidence and understand that they can overcome difficulties through hard work and determination.

Preparing children for nuclear survival is a delicate balance of providing them with the skills and knowledge they need to stay safe while ensuring they remain emotionally resilient and mentally healthy. By explaining the situation in age-appropriate ways, teaching them essential survival skills, maintaining routines, and offering emotional support, you can help children not only survive but thrive in a challenging environment.

As a family, your unity, teamwork, and communication will be key to ensuring that everyone emerges from the crisis stronger and more connected. By focusing on resilience, adaptability, and hope, you'll give your children the tools they need to face an uncertain future with confidence and courage, both in the shelter and in the world beyond.

How to Identify Safe Zones after Leaving the Shelter

Identifying safe zones after leaving the shelter is one of the most critical tasks for ensuring your survival in the aftermath of a nuclear event. The world outside your shelter will likely be drastically altered, with radiation, contaminated water, and unstable infrastructure posing significant dangers. Knowing how to evaluate the environment and locate areas that are free from harmful radiation, contamination, and other hazards will improve your chances of long-term survival. This chapter will cover the essential strategies and tools for finding and identifying safe zones after you exit the shelter.

Key Factors to Consider When Identifying Safe Zones

Monitoring Radiation Levels: Tools and Techniques

Testing Water and Soil for Contamination

Assessing Shelter and Infrastructure Stability

Identifying Natural Barriers and Geographical Features for Protection

Safe Distance from Impact Zones and Fallout Paths

Assessing Air Quality and Wind Patterns

Locating Clean Water Sources

Determining Safe Zones for Agriculture and Food Production

Identifying and Avoiding Hazardous Zones

Establishing Safe Zones for Community Rebuilding

Monitoring Radiation Levels: Tools and Techniques

The first step in identifying safe zones is to monitor radiation levels, which can remain dangerously high for weeks, months, or even years after a nuclear event. Radiation hotspots can form in unexpected areas due to wind patterns, precipitation, and topography.

Using a Geiger Counter

Geiger Counter Basics: A Geiger counter is essential for detecting radiation levels in the environment. Familiarize yourself with how to use it before leaving the shelter. Ensure the battery is fully charged or have extra batteries on hand.

Measuring Safe Levels: Safe radiation levels are generally below 0.2 microsieverts per hour (μSv/h). Anything higher may pose long-term health risks, particularly for children, the elderly, and those with weakened immune systems. If radiation levels exceed this, avoid spending long periods in that area.

Radiation Monitoring Techniques

Regular Readings: Take regular radiation readings at multiple locations as you travel. Radiation levels can vary dramatically over short distances. Record these readings to identify areas that remain consistently low in radiation.

Avoiding Hotspots: Stay away from areas with significantly elevated radiation levels, especially near urban centers, factories, or industrial sites that may have been hit by a nuclear strike. These areas are likely to have high fallout concentrations.

Testing Water and Soil for Contamination

Safe zones must have access to clean water and soil for agriculture. Testing these resources is vital to ensuring that you are not exposed to harmful contaminants or radiation.

Water Testing

Portable Water Filters: Use portable water filters designed to remove radiation particles, heavy metals, and bacteria. Even if water appears clear, it may still be contaminated. Filters like those with activated carbon can help eliminate dangerous substances, but always check the filter's effectiveness against radioactive contaminants.

Water Testing Kits: Carry a water testing kit to detect common contaminants, including heavy metals (like lead or mercury), radiation, and biological hazards. These kits provide an initial assessment of whether a water source is safe for drinking.

Soil Testing

Testing for Radiation: Use a Geiger counter to test soil for radiation. Areas with low radiation levels in the soil may be suitable for planting crops. However, high readings suggest the soil has absorbed fallout and may not be safe for agricultural use.

Chemical Contamination: In addition to radiation, the soil may be contaminated with chemicals from industrial fallout. Avoid planting crops in areas near factories, refineries, or urban centers, as these regions are more likely to be polluted.

Assessing Shelter and Infrastructure Stability

Once you begin exploring outside your shelter, it's important to assess the stability of nearby buildings and infrastructure to determine if they are safe to use or occupy.

Structural Integrity

Checking Buildings: Before entering abandoned buildings, examine them for signs of structural damage such as cracks in the foundation, collapsed roofs, or unstable walls. If a building shows significant damage, avoid it, as it may collapse further.

Utilizing Safe Shelters: Look for sturdy, well-built structures that could provide temporary shelter. Reinforced concrete buildings, underground bunkers, or basements may still offer protection from fallout and provide a secondary safe zone.

Avoiding Unstable Infrastructure

Bridges and Roads: Be cautious when crossing bridges or traveling on roads, as these may have been damaged by explosions or natural disasters. Unstable bridges or roads may collapse under weight, especially if they haven't been maintained.

Power Lines and Gas Pipelines: Stay clear of downed power lines or damaged pipelines. These pose additional dangers, including electrocution or gas leaks that could lead to fires or explosions.

Identifying Natural Barriers and Geographical Features for Protection

Natural geography can offer protection from radiation and fallout, making certain areas more suitable for safe zones than others.

Mountains and Hills

Shielding from Fallout: Mountains, hills, and cliffs can serve as natural barriers, shielding you from radioactive fallout carried by the wind. Areas on the opposite side of a hill or mountain from a fallout zone may have lower radiation levels.

Higher Elevation: Higher elevations tend to have less fallout accumulation because fallout particles settle more heavily in low-lying areas. However, ensure the area has access to clean water sources, as higher elevations may be drier.

Forested Areas

Dense Vegetation: Dense forests can offer protection from fallout, as trees and vegetation can absorb radiation to some extent. The canopy may prevent fallout particles from reaching the ground, but always test the soil before settling.

Natural Resources: Forested areas provide natural resources, such as wood for shelter, fire, and foraging opportunities. However, be cautious of potentially contaminated wildlife.

Safe Distance from Impact Zones and Fallout Paths

Identifying how far you are from the original impact zone is crucial for safety. Fallout zones are typically more dangerous closer to the detonation site.

Staying Far from Ground Zero

Distance from Blast Sites: If you are aware of the location of nuclear detonation sites, maintain a distance of at least 10 to 20 miles, depending on the size of the blast. Radiation levels drop significantly the farther you are from the epicenter.

Wind and Fallout Patterns: Understand how wind patterns have spread fallout. Areas directly downwind of a detonation site are more likely to be contaminated, even at a significant distance. Moving crosswind or upwind may help reduce radiation exposure.

Assessing Air Quality and Wind Patterns

Clean air is essential for survival, and monitoring air quality after leaving the shelter will help you determine if the area is safe for long-term habitation.

Wind and Fallout Dispersion

Monitoring Wind Direction: Wind plays a major role in dispersing radioactive particles. Use wind direction to your advantage by moving in directions that avoid fallout clouds or regions where the wind has carried radioactive material.

Air Quality Testing: Use portable air quality testers to monitor for radioactive particles, chemicals, or pollutants in the air. If air quality is poor, continue using masks or air filtration systems until it improves.

Locating Clean Water Sources

Finding clean water is critical for survival, and identifying uncontaminated water sources should be a top priority after leaving the shelter.

Safe Water Sources

Springs and Underground Wells: Look for natural springs, as underground water sources tend to be less contaminated than surface water. Wells that tap into deep aquifers are also more likely to provide clean water.

Rainwater Harvesting: In areas where surface water may be unsafe, consider setting up rainwater collection systems. Use filtration and purification methods to ensure the rainwater is drinkable.

Avoiding Contaminated Water

Stagnant Water: Avoid drinking from stagnant water sources like ponds or ditches, as these are more likely to contain fallout particles and contaminants. Running water (rivers, streams) is generally safer but should still be tested and purified.

Determining Safe Zones for Agriculture and Food Production

Growing your own food will be essential for long-term survival, but it's important to identify areas where soil and water are safe for agriculture.

Low-Contamination Areas

Testing Soil: Before planting crops, thoroughly test the soil for radiation and chemical contamination. Areas with low radiation levels and no industrial contamination are best for growing food.

Sunlight and Water Access: Ensure the area receives sufficient sunlight for crop growth and has access to clean water for irrigation.

Greenhouse Agriculture

Controlled Environment: If soil contamination is an issue, consider building greenhouses where you can grow crops in a controlled environment. Use clean soil and filtered water to avoid introducing contaminants.

Identifying and Avoiding Hazardous Zones

While searching for safe zones, you will inevitably encounter hazardous areas that should be avoided due to high radiation levels, contamination, or other dangers.

Industrial Zones and Urban Areas

Avoid Industrial Sites: Areas near factories, chemical plants, or refineries are likely to be heavily contaminated with both radiation and toxic chemicals. These areas are not safe for long-term habitation or food production.

Ruined Cities: Urban areas, especially those that have been hit by nuclear strikes, pose multiple dangers, including radiation, structural collapse, and limited resources. If you must pass through, limit your exposure and avoid settling there.

Floodplains and Low-Lying Areas

Risk of Contaminated Water: Low-lying areas and floodplains are particularly vulnerable to the accumulation of fallout particles and other contaminants that can be carried by water. Avoid settling in these areas, as they are more likely to experience flooding that can spread radioactive or chemical pollutants over a wide area.

Standing Water: Stagnant water in low-lying areas often becomes a collection point for contaminants. Even if the water appears clear, it may be hazardous to drink or use for irrigation.

Establishing Safe Zones for Community Rebuilding

Once you've identified a safe area, it's important to consider how to establish a base for long-term survival and potentially rebuild a community. This requires assessing the area's resources, defensibility, and suitability for sustainable living.

Key Criteria for a Long-Term Safe Zone

Access to Resources: Ensure the safe zone has access to essential resources such as clean water, fertile land for agriculture, and materials for building shelters. Proximity to natural resources like forests for timber and rivers for water will be important for sustaining a community.

Defensibility: Choose a location that is easy to defend against potential threats, whether from wildlife or other humans. Natural barriers like cliffs, hills, or dense forests can provide protection and make it harder for outsiders to approach unnoticed.

Space for Expansion: If you're planning to build or expand a community, make sure the area has enough space for future growth. This includes not only living spaces but also room for agriculture, livestock, and storage facilities.

Building Infrastructure

Shelters and Housing: Start by establishing temporary shelters using materials like tarps, tents, or scavenged building supplies. As time goes on, more permanent structures can be built using wood, stone, or reinforced concrete.

Water and Waste Systems: Design water collection and filtration systems to ensure a continuous supply of clean water. Additionally, build waste management systems to keep the area sanitary and prevent contamination of water sources.

Building Community Cohesion

Cooperation with Others: If you encounter other survivors, working together can increase your chances of long-term survival. A larger group can share skills, resources, and knowledge, making the task of rebuilding society more manageable.

Communication: Establish communication systems with other safe zones or communities to share information about hazards, resources, or mutual defense strategies. Two-way radios, signal flares, and even couriers on foot can be used to maintain contact.

Identifying and establishing a safe zone after leaving the shelter is essential for long-term survival in the aftermath of a nuclear event. By carefully monitoring radiation levels, testing water and soil for contamination, and selecting areas with natural protection and access to resources, you can create a safe environment for yourself and others. Once a suitable safe zone is found, you can begin the process of rebuilding, focusing on the development of shelter, agriculture, and infrastructure that will support long-term survival. Cooperation, resource management, and careful planning will help ensure that your chosen safe zone not only provides immediate safety but also fosters a sustainable future for you and your community.

Foraging for Food in a Post-Nuclear World

In a post-nuclear world, foraging for food becomes a crucial survival skill, especially when supplies run low or you lack access to traditional agriculture. The environment after a nuclear event may be significantly altered, with contamination posing serious risks. However, with proper knowledge and caution, it's still possible to find edible plants, animals, and other sources of sustenance. This chapter will guide you through safe foraging practices, identifying edible plants and animals, and understanding the risks associated with foraging in a contaminated environment.

Key Aspects of Foraging for Food in a Post-Nuclear World

Understanding Contamination Risks and Safety Precautions

Identifying Edible Plants and Safe Locations for Foraging

Foraging for Wild Edible Plants and Roots

Hunting and Trapping Small Game

Fishing and Aquatic Food Sources

Edible Insects as a Survival Food

Mushroom Foraging: Risks and Benefits

Preserving Foraged Foods for Long-Term Use

Avoiding Poisonous Plants, Animals, and Fungi

Creating a Foraging Strategy for Long-Term Survival

Understanding Contamination Risks and Safety Precautions

The biggest challenge in foraging after a nuclear event is avoiding contamination from radioactive fallout, chemicals, or other pollutants that may have settled on or near plants and animals. Taking the right precautions will help reduce the risks of consuming contaminated food.

Radiation Testing

Use a Geiger Counter: Always carry a portable Geiger counter to check radiation levels in the areas where you plan to forage. If radiation levels are high, avoid foraging in that location.

Test Food Sources: Whenever possible, use radiation detection tools to check the food you gather, especially plants, water sources, and animals. Contaminated food can accumulate radioactive particles, making it dangerous to consume.

Avoiding Fallout-Contaminated Areas

Downwind from Blast Sites: Stay away from areas downwind of nuclear impact zones or areas that have been heavily affected by fallout. Radiation levels will be higher, and plants and animals in these areas are more likely to be contaminated.

Surface Contamination: Be mindful of areas where fallout may have settled, especially on plants or in water sources. Fallout particles can accumulate on the leaves and surfaces of plants and in the top layers of soil, so avoid consuming plants that have been exposed.

Identifying Edible Plants and Safe Locations for Foraging

Certain plants are better suited for foraging in a post-nuclear environment because they may be less prone to contamination or are more abundant. Identifying safe locations and edible species is key to successful foraging.

Prioritize Low-Contamination Areas

Higher Elevations: Plants growing at higher elevations may be less affected by fallout, as fallout particles tend to settle in low-lying areas. Search for food in these areas, but still test plants for contamination before consuming.

Sheltered Areas: Look for plants growing under dense tree cover or in areas protected by natural barriers like hills or cliffs. These locations may have shielded plants from the direct impact of fallout.

Common Edible Wild Plants

Dandelions: Dandelions are resilient, easy to identify, and fully edible. The leaves, flowers, and roots can be used in salads, teas, or cooked as greens.

Wild Garlic and Onion: These are easily recognizable by their smell and can be found in many parts of the world. They're useful for flavouring food and providing essential nutrients.

Cattails: Cattails grow near water sources and have multiple edible parts. The young shoots, roots, and pollen can be harvested for food, making them a versatile survival resource.

Chickweed: This common wild green grows in many climates and can be eaten raw or cooked. It's rich in vitamins and can be used in soups or salads.

Foraging for Wild Edible Plants and Roots

In addition to common plants, many wild plants and roots offer important nutrients and can be vital for survival in a post-nuclear environment.

Roots and Tubers

Burdock Root: Burdock is a hardy plant whose root is edible and packed with nutrients. It can be boiled or roasted to add bulk to meals.

Jerusalem Artichoke: This tuber is found in many temperate regions and can be eaten raw or cooked. It's a good source of carbohydrates and is easy to dig up.

Wild Carrots (Queen Anne's Lace): The roots of wild carrots can be eaten, but they should be harvested before the plant flowers. However, be cautious, as wild carrot can be confused with poisonous hemlock.

Leaves and Greens

Plantain: Common plantain is a nutritious green that can be found in disturbed soil. It's edible raw or cooked and is rich in vitamins and minerals.

Nettle: Stinging nettles can be a good source of food once boiled to remove their sting. Nettles are high in protein, vitamins, and minerals, making them an excellent survival food.

Hunting and Trapping Small Game

Small game such as rabbits, squirrels, and birds can provide much-needed protein. However, hunting in a contaminated environment requires extra caution.

Safe Hunting Practices

Trapping Instead of Hunting: Setting traps allows you to conserve energy and reduce exposure to the outside environment. Use snares, deadfalls, or other simple traps to catch small game.

Monitoring Radiation in Animals: Animals can absorb radiation through the food they eat, so it's essential to test any game you hunt for radiation. If you have no way to test, prioritize animals in areas with lower radiation levels, such as those farther from fallout zones.

Common Small Game

Rabbits: Rabbits are abundant and reproduce quickly, making them a reliable source of food. Use traps or snares near their burrows or along trails they frequent.

Squirrels: Squirrels are another common target for small game hunting. They are often found in wooded areas, and their meat can be used in stews or roasted.

Pigeons and Doves: These birds are commonly found in urban and rural areas alike. They can be caught using traps or by hunting with makeshift weapons like slingshots.

Fishing and Aquatic Food Sources

Fishing is an important way to supplement your food supply, but water contamination poses significant risks in a post-nuclear world. You must be careful about where and how you fish.

Assessing Water Safety

Testing Water for Radiation: Before consuming fish or other aquatic life, test the water for radiation. Fish that live in contaminated water are more likely to absorb radioactive particles.

Prioritize Flowing Water: Fish in rivers and streams are less likely to be contaminated than those in stagnant ponds or lakes, as flowing water can help disperse fallout particles.

Fish and Aquatic Life to Target

Small Fish: Smaller fish tend to accumulate less contamination than larger, predatory fish. Stick to small species like trout, bass, or perch when foraging for fish.

Shellfish: In areas where water contamination is low, shellfish like crayfish and clams can be foraged. However, they should be thoroughly cooked to reduce any potential toxins.

Edible Insects as a Survival Food

Insects are abundant, easy to gather, and packed with protein, making them an ideal survival food in extreme conditions.

Safe Insects to Eat

Grasshoppers and Crickets: These insects are easy to find and high in protein. Remove the legs and wings before cooking, and roast or boil them for a quick meal.

Ants: Ants can be eaten raw or cooked and are rich in nutrients. You can collect them by disturbing their nests and gathering the workers or larvae.

Beetles: Many beetle larvae, such as mealworms or June bug grubs, are edible and provide substantial nutrition.

Preparing and Cooking Insects

Boiling and Roasting: Always cook insects to kill any parasites or bacteria they may be carrying. Roasting or boiling are the most effective methods for preparing insects.

Grinding into Flour: Insects can also be dried and ground into a fine flour, which can be mixed with other foraged foods or stored for later use.

Mushroom Foraging: Risks and Benefits

Mushrooms can be a valuable food source, but they also come with significant risks, especially in a contaminated environment.

Identifying Safe Mushrooms

Edible Varieties: Some common edible mushrooms include morels, chanterelles, and puffballs. However, proper identification is crucial, as many toxic varieties closely resemble edible ones.

Avoiding Toxic Mushrooms: Never eat mushrooms unless you are 100% certain of their identification. Mistakes can lead to poisoning, which may be fatal without medical treatment.

Mushroom Contamination

Absorbing Contaminants: Mushrooms are highly absorbent and can take in contaminants from the soil, including radiation. Avoid foraging mushrooms in areas where soil contamination is suspected.

Preserving Foraged Foods for Long-Term Use

Once you've gathered food, preserving it ensures you have a supply during times when fresh resources are scarce.

Drying and Smoking

Drying Food: Drying is one of the simplest and most effective ways to preserve foraged foods. Lay fruits, vegetables, and herbs in a dry, well-ventilated area out of direct sunlight, or use low heat near a fire to speed up the process. Ensure the food is fully dried before storing to prevent mold growth.

Smoking Meat and Fish: Smoking meat and fish not only preserves them but also adds flavor. Hang the meat or fish over a low, smoky fire and let the smoke dry it out. This method can preserve food for weeks or even months, making it an essential skill in survival scenarios.

Pickling and Fermenting

Pickling Vegetables: Pickling vegetables in vinegar, salt, and water is an effective way to preserve them for long periods. Use a clean, sterilized container to store pickled items and ensure they remain sealed to prevent spoilage.

Fermenting Foods: Fermentation can be used to preserve vegetables like cabbage (sauerkraut) or fruits. This method relies on natural bacteria to create lactic acid, which preserves the food and adds nutritional benefits. Be mindful of contamination and use clean containers.

Salting

Curing with Salt: Salt is one of the oldest and most effective preservation methods for meat. Rub the meat thoroughly with salt, then store it in a cool, dry place. This method helps prevent bacterial growth and extends the shelf life of meat significantly.

Brining: You can also preserve meat and vegetables by submerging them in a saltwater brine. The high salt content prevents bacteria from spoiling the food and keeps it edible for longer.

Avoiding Poisonous Plants, Animals, and Fungi

In a survival situation, it's critical to be cautious and avoid potentially toxic foods. Misidentification of plants, animals, or fungi can lead to illness or even death.

Common Poisonous Plants

Hemlock: Hemlock is one of the most toxic plants and can easily be mistaken for wild carrots or parsley. All parts of the plant are deadly, so avoid foraging plants that look similar if you're not certain.

Oleander: This plant, often found in gardens, is extremely poisonous. Even a small amount can cause serious health problems, so avoid eating any part of it.

Deadly Nightshade (Belladonna): Belladonna produces small, dark berries that may look tempting but are highly toxic. Avoid foraging for berries unless you are confident in their identification.

Dangerous Animals and Insects

Venomous Species: Be cautious of venomous animals such as snakes, spiders, or scorpions. These creatures can be dangerous if mishandled or consumed. Learn to identify which species are safe for hunting and which should be avoided.

Contaminated Animals: Animals that appear sick or have visible sores should not be eaten. They may be contaminated by radiation or disease. Always inspect game before preparing it for consumption.

Poisonous Mushrooms

Toxic Fungi: Many poisonous mushrooms resemble edible varieties, making them especially dangerous. Death cap mushrooms, for instance, are fatal if consumed. Only forage for mushrooms if you are absolutely certain of their identification, or avoid them altogether to minimize risk.

Creating a Foraging Strategy for Long-Term Survival

Foraging alone cannot sustain you indefinitely unless you develop a sustainable strategy that incorporates safe locations, seasonal availability, and food preservation.

Mapping Foraging Areas

Identify Low-Risk Zones: Use your Geiger counter and other testing tools to map out areas with the least contamination where foraging is safe. These areas should be visited regularly to gather food.

Seasonal Planning: Plan your foraging efforts around seasonal growth patterns. For example, wild greens and herbs may be abundant in spring, while roots and berries become available in late summer or fall.

Diversifying Food Sources

Avoid Over-Harvesting: Be mindful not to over-harvest from any one area, as this can deplete resources and make it harder to gather food in the future. Rotate your foraging locations to give plants time to recover and regrow.

Incorporating Protein: Combine plant-based foraging with hunting, trapping, and fishing to ensure a balanced diet that includes necessary proteins and fats. This is essential for long-term health in a survival situation.

Building a Foraging Knowledge Base

Education and Research: Before you leave the shelter, spend time studying local plants, animals, and fungi to become familiar with what's edible and what isn't. Carry a field guide if possible, and continue learning as you explore the environment.

Training Family Members: Teach your family or group how to forage safely. This way, the responsibility for gathering food can be shared, and you reduce the risk of someone accidentally consuming something harmful.

Foraging for food in a post-nuclear world is a vital survival skill that requires knowledge, caution, and adaptability. While the environment may be dangerous and resources scarce, understanding how to identify safe plants, animals, and water sources can provide sustenance for the long term. By taking proper safety measures to avoid contaminated areas, poisonous plants, and dangerous animals, and by learning effective food preservation methods, you can extend the availability of foraged food and improve your chances of surviving in a challenging environment. Building a sustainable foraging strategy that incorporates safe locations, seasonal patterns, and a diverse range of food sources will allow you to endure the uncertainties of a post-nuclear landscape and support the survival of your group or community.

First Aid and Medical Supplies: What You'll Need to Survive

In a post-nuclear world, access to medical care will be extremely limited, and it's crucial to be prepared to handle a wide range of medical issues on your own. Having a well-stocked first aid kit and essential medical supplies can make the difference between life and death in a survival scenario. This chapter will guide you through the must-have first aid items, medications, and medical knowledge you'll need to ensure your family's health and safety in a fallout shelter and beyond.

Key Medical Supplies and First Aid Considerations

The Importance of First Aid Knowledge and Training

Assembling a Comprehensive First Aid Kit

Medications and Prescription Drugs for Long-Term Survival

Dealing with Radiation Exposure: Potassium Iodide and Other Treatments

Treating Common Injuries: Wounds, Burns, and Fractures

Managing Illnesses and Infections without a Doctor

Handling Radiation Sickness Symptoms

Caring for Chronic Conditions and Special Needs

Hygiene and Sanitation in a Survival Situation

Mental Health and Emotional Support During Isolation

The Importance of First Aid Knowledge and Training

Before assembling your first aid kit, it's important to recognize that having the right supplies is only part of the equation. Knowing how to use them effectively is just as crucial. In a post-nuclear world, professional medical help may not be available for a long time, so having first aid training and a basic understanding of emergency care is essential.

First Aid Training

Take a First Aid Course: Before any survival situation, ensure that you and your family members have taken a basic first aid and CPR course. Knowing how to stop bleeding, perform CPR, and treat burns or fractures could save a life.

Emergency Medical Guides: Include a first aid manual or survival medical guide in your kit. This will help you manage medical situations you may not have trained for, offering step-by-step instructions for dealing with various injuries and illnesses.

Self-Assessment Skills

Recognizing Symptoms: Learning to recognize symptoms of common injuries or illnesses, such as infection, dehydration, or shock, is important for early intervention. The sooner you identify a problem, the better your chances of managing it effectively.

Assembling a Comprehensive First Aid Kit

Your first aid kit should cover a wide range of potential injuries and health issues. While many pre-made kits are available, building a custom one that addresses long-term survival needs is ideal.

Essential Supplies for Injury Treatment

Bandages and Dressings: Stock a variety of adhesive bandages, sterile gauze pads, and medical tape. Include butterfly closures for deeper cuts that need to be held together but don't require stitches.

Antiseptic Wipes and Ointments: Use antiseptic wipes to clean wounds and prevent infection. Include antibiotic ointments like Neosporin for wound care. Hydrogen peroxide or iodine can also be used for cleaning wounds.

Elastic Bandages and Splints: Elastic bandages (like ACE bandages) can help stabilize sprains or fractures. Include splints for immobilizing injured limbs and a triangular bandage for slings.

Supplies for Emergency Situations

Tourniquets: In the event of severe bleeding, a tourniquet may be necessary to prevent life-threatening blood loss. Learn how to apply a tourniquet properly to avoid additional injury.

Burn Treatment: Include sterile burn dressings or burn gel to treat minor to moderate burns. Burn injuries may be more common after a nuclear event due to fires or explosions.

Trauma Shears: Trauma shears are essential for cutting through clothing or materials during emergencies. They are designed to cut thick fabric, making it easier to access wounds quickly.

Pain Relief and Anti-Inflammatory Medications

Over-the-Counter Pain Relievers: Include ibuprofen (Advil), acetaminophen (Tylenol), and aspirin. These can help manage pain, reduce inflammation, and control fevers.

Cold and Heat Packs: Reusable cold and heat packs can be used to reduce swelling, relieve muscle pain, or treat injuries like sprains and bruises.

Medications and Prescription Drugs for Long-Term Survival

Access to pharmacies and hospitals may be non-existent in a post-nuclear environment, so it's essential to stockpile critical medications in advance.

Prescription Medications

Extended Prescriptions: If you or a family member relies on prescription medications for chronic conditions (e.g., heart disease, diabetes, asthma), work with your doctor to obtain a longer supply of these drugs, typically 90 days or more. Store them properly to extend their shelf life.

Backup Supplies: Inquire with your physician about alternatives or generic versions of essential medications in case you run out. Consider natural remedies as supplements for managing conditions when pharmaceuticals are unavailable.

Essential Over-the-Counter Medications

Anti-Diarrheal and Antacids: Gastrointestinal issues can arise from contaminated food or water. Include anti-diarrheal medications like loperamide (Imodium) and antacids for stomach relief.

Antihistamines: Include antihistamines like Benadryl for treating allergic reactions, insect bites, or hay fever. They also help with skin irritations and swelling.

Cough and Cold Medications: Prepare for respiratory issues by including decongestants, cough syrups, or lozenges in your kit. Nuclear fallout may cause respiratory distress in the form of coughs or colds due to poor air quality.

Dealing with Radiation Exposure: Potassium Iodide and Other Treatments

Radiation exposure is one of the primary health risks after a nuclear event. Specific medical treatments can help mitigate the effects of radiation on the body.

Potassium Iodide (KI)

Blocking Radiation Absorption: Potassium iodide pills are a critical treatment in a nuclear fallout situation. KI helps protect the thyroid gland by blocking the absorption of radioactive iodine. Have an adequate supply for each family member, and ensure you understand the proper dosage for adults and children.

Radiation Detoxification Treatments

Chelation Therapy: In some cases, chelation therapy drugs may help remove certain radioactive particles from the body. These medications are often used to treat heavy metal poisoning and can be part of your long-term radiation response plan.

Activated Charcoal: Activated charcoal can absorb toxins in the digestive system, which may be useful if contaminated food or water is consumed. Always follow the proper dosage and instructions.

Treating Common Injuries: Wounds, Burns, and Fractures

In a survival scenario, injuries like cuts, burns, or broken bones are more likely to occur. Knowing how to treat them effectively is essential for preventing complications.

Wound Care

Cleaning and Dressing Wounds: Always clean wounds thoroughly with antiseptic wipes or a saline solution. Apply antibiotic ointment to prevent infection, and cover the wound with a sterile dressing.

Sutures and Skin Glue: For deep cuts, you may need to apply sutures or use skin glue to close the wound. Practice using these supplies before an emergency, as they require skill to apply correctly.

Burn Care

Treating Minor Burns: For first- and second-degree burns, cool the area with clean water and apply burn gel or aloe vera. Cover the burn with a sterile, non-stick dressing.

Handling Severe Burns: For third-degree burns or burns covering a large area, medical attention is critical. In a survival situation, keep the burned area clean, prevent shock, and provide fluids to the victim.

Fracture Management

Immobilizing a Broken Bone: Use splints, bandages, or any rigid material to immobilize a broken limb. Keep the injured area elevated if possible to reduce swelling and pain.

Pain Management: Provide pain relievers and apply cold packs to the affected area to reduce swelling.

Managing Illnesses and Infections Without a Doctor

Without access to antibiotics or professional medical care, managing infections becomes more challenging. Early detection and treatment are crucial for preventing serious complications.

Antibiotics

Stockpiling Antibiotics: If possible, stock antibiotics like amoxicillin, doxycycline, or ciprofloxacin for treating bacterial infections. Speak with your doctor about obtaining these medications in advance for emergency use.

Natural Alternatives: In situations where antibiotics aren't available, consider natural antiseptics such as garlic, honey, or tea tree oil for treating mild infections. These are not replacements for antibiotics but can help in a pinch.

Signs of Infection

Monitoring for Symptoms: Watch for signs of infection, such as redness, swelling, pus, or fever. Treating infections early with antiseptics, proper wound care, and hydration is critical for preventing the spread of bacteria.

Handling Radiation Sickness Symptoms

Radiation sickness occurs when you are exposed to large amounts of radiation over a short period. Knowing how to recognize and manage symptoms is essential in a post-nuclear environment.

Early Symptoms

Nausea and Vomiting: Radiation sickness often begins with nausea, vomiting, and diarrhea. Manage these symptoms with anti-nausea medications and ensure the person stays hydrated.

Fatigue and Weakness: Extreme fatigue is common. Encourage rest and, if possible, provide electrolyte solutions to help with energy levels.

Long-Term Radiation Sickness Management

Hair Loss and Skin Damage: Radiation exposure can cause hair loss, burns, and skin damage over time. Treat skin damage with burn ointments or aloe vera gel, and keep the affected areas clean to prevent infection. Hair loss is a common symptom, but it is not life-threatening.

Immune System Weakness: Radiation exposure weakens the immune system, making the body more susceptible to infections. Ensure that the affected individual maintains proper hygiene, eats nutritious food if available, and avoids contaminated areas to reduce further exposure.

Consult Medical Guides: Having a radiation sickness treatment guide or handbook is essential, as this condition requires careful management over time. While the immediate symptoms may pass, long-term care will focus on preventing infection and managing the overall health of the person.

Caring for Chronic Conditions and Special Needs

In a long-term survival situation, caring for individuals with chronic conditions such as diabetes, asthma, or heart disease can be difficult without proper supplies. Planning ahead and stockpiling essential medications and equipment will help mitigate these risks.

Managing Chronic Conditions

Diabetes: For individuals with diabetes, managing blood sugar levels is critical. Stockpile insulin, oral diabetes medications, and syringes. Include glucose monitoring tools such as test strips and lancets. Have high-sugar snacks (e.g., glucose tablets or candy) available for emergency hypoglycaemia.

Asthma: Keep a supply of inhalers and any prescribed asthma medications. If medications are unavailable, attempt to avoid triggers such as smoke, dust, or allergens, which may be more prevalent in a post-nuclear world.

Heart Disease: Ensure you have a stockpile of heart medications, such as beta-blockers, nitroglycerin, or blood pressure medications. Monitor for symptoms of heart attacks or other cardiovascular issues and avoid strenuous activity that could exacerbate these conditions.

Special Needs Supplies

Mobility Devices: If a family member requires mobility devices such as wheelchairs, walkers, or canes, ensure these are in good working condition and accessible in the shelter. Keep extra parts or repair kits on hand if possible.

Infant and Elderly Care: For infants, include diapers, formula, and any essential baby supplies. For elderly family members, consider medications for conditions like arthritis or dementia, as well as adult diapers, hearing aids, or other assistive devices.

Hygiene and Sanitation in a Survival Situation

Maintaining good hygiene and sanitation practices is critical to preventing disease, especially in the confined space of a shelter or in a contaminated environment.

Hygiene Supplies

Soap and Hand Sanitizer: Stock plenty of soap and hand sanitizer to ensure that hands are kept clean, especially before preparing food or treating wounds. Alcohol-based hand sanitizers are effective for killing germs when water is scarce.

Toilet Paper and Wet Wipes: Toilet paper is essential for hygiene, but wet wipes or baby wipes can also be used for cleaning and maintaining personal hygiene when water is limited.

Feminine Hygiene Products: Keep a supply of sanitary pads, tampons, or reusable menstrual products. These are essential for maintaining hygiene in long-term survival situations.

Waste Management

Human Waste Disposal: In a shelter or survival environment, setting up a safe system for disposing of human waste is critical. This could include portable toilets, composting toilets, or digging latrines far from the living area to avoid contamination.

Garbage Disposal: Keep waste sealed in plastic bags or containers to prevent attracting pests and to avoid contamination. Dispose of garbage far from the shelter or bury it if necessary.

Mental Health and Emotional Support During Isolation

Living in a post-nuclear world can take a heavy psychological toll, especially in extended isolation or high-stress environments. Ensuring mental health care is an often-overlooked but essential aspect of survival.

Recognizing Mental Health Challenges

Anxiety and Depression: Isolation, fear, and uncertainty can lead to anxiety and depression. Recognize the signs early, such as withdrawal, lack of interest, irritability, or feelings of hopelessness. Create an environment where family members can talk openly about their feelings and provide emotional support.

Post-Traumatic Stress: The trauma of experiencing a nuclear event, losing loved ones, or dealing with the aftermath can lead to PTSD. Be aware of symptoms like flashbacks, nightmares, or extreme emotional reactions. Patience, understanding, and communication are key to managing PTSD symptoms.

Coping Mechanisms for Mental Health

Routine and Structure: Maintaining a daily routine helps create a sense of normalcy, reducing anxiety and stress. Incorporate regular activities such as meals, exercise, and relaxation times into your routine.

Social Interaction: Encourage family members to spend time together, playing games, reading, or talking. Isolation can worsen mental health, so fostering a sense of community within your group or family is crucial.

Entertainment and Distraction: Having books, games, or other forms of entertainment can help pass the time and keep the mind occupied. Laughter and relaxation are important for maintaining a healthy mental state in survival situations.

First aid and medical preparedness are critical to surviving a post-nuclear event, where access to professional healthcare may be impossible. By assembling a comprehensive first aid kit, stocking essential medications, and gaining the necessary medical knowledge, you can manage a wide range of injuries, illnesses, and health conditions.

In addition to physical health, attention must be given to hygiene, sanitation, and mental well-being, as these factors are just as important for long-term survival. Managing chronic conditions, handling radiation exposure, and providing emotional support in the face of trauma will improve your chances of enduring the harsh realities of a post-nuclear world. A well-prepared first aid kit and the right knowledge can keep your family safe, healthy, and resilient in even the most challenging circumstances.

Managing Waste and Sanitation in a Confined Shelter

In a post-nuclear world, especially within the confines of a shelter, waste management and sanitation are critical to maintaining health and avoiding disease. Without proper disposal methods, human waste, garbage, and other forms of refuse can quickly become breeding grounds for bacteria, pests, and illness, leading to serious health risks. This chapter will cover the essential strategies for managing waste and maintaining sanitation within a confined space, ensuring a clean and healthy environment for long-term survival.

Key Considerations for Waste Management and Sanitation

The Importance of Waste Management in a Confined Shelter

Setting Up a Human Waste Disposal System

Managing Garbage and Household Waste

Dealing with Hazardous Waste and Contaminated Materials

Personal Hygiene and Its Role in Sanitation

Water Conservation and Sanitation

Preventing Disease and Infection in Confined Spaces

Dealing with Pest Control in the Shelter

Managing Air Quality in a Confined Environment

Long-Term Waste Disposal Solutions in a Survival Situation

The Importance of Waste Management in a Confined Shelter

In a confined shelter, waste can accumulate quickly, leading to health hazards if not properly managed. Without access to outdoor facilities or regular garbage disposal, you must have a system in place for dealing with both human waste and household refuse. Poor sanitation can lead to the spread of diseases such as cholera, dysentery, or respiratory infections, which can be fatal in a survival situation where medical care is limited.

Health Risks of Poor Sanitation

Infection and Disease: Improper waste disposal can lead to contamination of water, food, and living areas, creating a breeding ground for bacteria and viruses. These pathogens can cause severe gastrointestinal illnesses or respiratory issues.

Pest Infestations: Unmanaged waste attracts pests like rodents and insects, which carry diseases. Controlling waste is crucial to preventing infestations and keeping the shelter clean and pest-free.

Setting Up a Human Waste Disposal System

Proper management of human waste is essential to maintaining hygiene and preventing the spread of disease in a shelter. There are several options for waste disposal that can be adapted to a confined environment.

Portable Toilets and Waste Containers

Portable Toilets: In a confined space, a portable toilet or chemical toilet is a convenient option for waste management. These toilets use chemicals to break down waste and neutralize odors. They require regular emptying, and waste must be disposed of properly, either in a designated pit or sealed in bags for later disposal.

Composting Toilets: A composting toilet can be an effective long-term solution. These toilets separate liquid and solid waste, allowing solid waste to decompose over time with the addition of materials like sawdust or peat moss. This method reduces odor and produces compost that can be used for non-food purposes, such as fertilizing non-edible plants.

Bucket and Bag Method

Emergency Waste Disposal: If more advanced options are unavailable, the bucket and bag method can be used. Line a sturdy bucket with heavy-duty plastic bags, add absorbent materials like sawdust or kitty litter, and seal the bag after use. This method requires careful disposal of the sealed bags to prevent contamination.

Hygiene and Odor Control: Add a layer of sawdust, kitty litter, or baking soda after each use to control odor and absorb liquid. Keep the waste container covered when not in use, and empty it regularly to avoid overflow.

Establishing a Safe Disposal Area

Designating a Disposal Site: If possible, designate an area outside the shelter (such as a latrine or burial pit) for waste disposal. Digging a latrine pit far from your water source can provide a long-term solution for disposing of human waste in a controlled manner.

Burial or Sealed Disposal: Waste should be buried at least 6 to 8 inches deep to prevent contamination of the surrounding environment. Alternatively, sealed waste bags can be stored in a separate area until it is safe to dispose of them properly.

Managing Garbage and Household Waste

In addition to human waste, other forms of garbage—such as food scraps, packaging, and non-biodegradable materials—must also be managed to prevent the spread of bacteria and attract pests.

Segregating Waste

Organic vs. Inorganic Waste: Separate organic waste (food scraps, biodegradable materials) from inorganic waste (plastic, metal, glass). Organic waste can be composted or buried, while inorganic waste must be sealed in bags for proper disposal.

Recycling Materials: Consider reusing or repurposing materials like plastic bottles, cans, and packaging. These items can be used for storage, water collection, or even shelter repairs.

Composting Food Scraps

Compost Pile: If space allows, create a small compost pile for organic waste. This helps manage food scraps and reduces the amount of waste in the shelter. Compost can eventually be used for gardening if you have the means to grow food outside.

Sealed Containers: Use sealed containers to store organic waste until it can be disposed of or composted. This helps control odors and prevents attracting pests to your living space.

Dealing with Hazardous Waste and Contaminated Materials

In a post-nuclear environment, you may encounter hazardous waste, including radioactive materials, contaminated clothing, or chemical waste. Properly handling and disposing of these materials is crucial to maintaining a safe environment.

Radioactive and Chemical Waste

Handling Contaminated Items: Use gloves and masks when handling materials that may have been exposed to radiation or chemicals. Contaminated clothing or tools should be stored in sealed, marked containers until it is safe to dispose of them or decontaminate them.

Decontaminating Surfaces: Clean any surfaces that have come into contact with hazardous materials using decontamination solutions, such as water and bleach. Ensure proper ventilation when dealing with chemicals.

Safe Disposal Practices

Sealing Hazardous Waste: Contaminated waste should be sealed in heavy-duty, airtight containers and stored away from living areas. If disposal outside the shelter is not possible, designate a specific area for temporary storage until it is safe to handle the waste.

Personal Hygiene and Its Role in Sanitation

Maintaining personal hygiene is critical to preventing the spread of disease, especially in a confined space. Regular cleaning of the body, clothing, and living environment helps reduce the risk of infection.

Bathing and Cleaning

Sponge Baths: In a situation where water is scarce, use a damp cloth or sponge to clean your body. Focus on areas where bacteria can accumulate, such as the hands, face, and underarms.

Handwashing: Wash hands regularly, especially after using the toilet, handling waste, or preparing food. If water is limited, use hand sanitizer with at least 60% alcohol.

Laundry and Clothing Maintenance

Cleaning Clothes: Dirty clothes can harbor bacteria, leading to skin infections or other illnesses. Wash clothing periodically using small amounts of water and biodegradable soap. Hang clothes to dry in a well-ventilated area.

Alternate Sets of Clothing: Rotate clothing to allow worn items to air out and prevent the buildup of sweat and bacteria.

Water Conservation and Sanitation

In a survival scenario, conserving water is essential, but you also need enough for basic sanitation practices. Striking a balance between hygiene and water conservation is key.

Gray Water Reuse

Reusing Water: Water used for washing clothes or dishes (gray water) can be reused for other purposes, such as flushing waste or cleaning surfaces. Ensure that gray water does not contain harmful chemicals before reuse.

Water Filtration: If possible, filter water before reuse to remove particulates or contaminants. This helps prolong the usability of your water supply for hygiene purposes.

Minimizing Water Use

Dry Cleaning Methods: Use dry shampoo, wet wipes, or minimal water techniques for personal hygiene when water is in short supply. These methods allow for cleanliness without depleting your water reserves.

Preventing Disease and Infection in Confined Spaces

In a confined shelter, disease can spread quickly due to close quarters. Taking steps to prevent infection is crucial to maintaining the health of everyone in the shelter.

Quarantine for Sick Individuals

Isolating the Sick: If someone in the shelter becomes ill, isolate them from the rest of the group to prevent the spread of infection. Use separate bedding, clothing, and hygiene supplies for the sick individual.

Disinfection: Regularly disinfect common areas, particularly surfaces that are frequently touched, such as door handles, tables, and toilets. Use bleach or alcohol-based solutions to clean and disinfect.

Boosting Immunity

Nutritional Health: Maintain a balanced diet to support the immune system. If fresh food is limited, take multivitamins or supplements to ensure you're getting essential nutrients.

Dealing with Pest Control in the Shelter

Pests such as rodents and insects can carry diseases and contaminate food and water supplies. Managing pest control in a confined space is crucial for maintaining hygiene and preventing illness.

Preventing Pests

Sealed Food Storage: Store all food in sealed containers to prevent pests from accessing it. Avoid leaving food out in the open, as this will attract rodents and insects.

Eliminating Attractants: Clean up food scraps and waste immediately to avoid attracting pests. Regularly clean the shelter, especially areas where food is prepared or consumed, to eliminate any crumbs or spills that could draw rodents or insects.

Setting Traps and Barriers

Rodent Traps: Set up traps around the shelter to catch mice, rats, or other small rodents. Use bait traps, snap traps, or glue traps as necessary, and check them regularly. Dispose of any captured pests safely and sanitize the area.

Insect Barriers: Install insect traps or use natural repellents such as essential oils (e.g., peppermint, eucalyptus, or citronella) to keep bugs at bay. Mosquito nets or screen barriers can help prevent insects from entering the shelter.

Pest-Proofing the Shelter: Seal any cracks, holes, or gaps in walls, windows, and doors to prevent pests from getting inside. Use materials like steel wool, caulk, or weatherstripping to block access points.

Managing Air Quality in a Confined Environment

Good air quality is essential for maintaining health in a confined shelter, especially if the space is sealed off from the outside world. Poor ventilation can lead to a buildup of harmful gases, mold, or airborne pathogens.

Ventilation Systems

Air Filtration: If possible, set up an air filtration system to remove dust, pathogens, and potential radiation particles from the air. High-efficiency particulate air (HEPA) filters are particularly effective in trapping small particles.

Natural Ventilation: If no air filtration system is available, ventilate the shelter periodically to bring in fresh air. However, be cautious of external air contamination, and only open vents or windows if radiation levels are safe.

Preventing Mold and Mildew

Humidity Control: Mold and mildew thrive in damp environments, so controlling humidity is key. Use moisture-absorbing materials, such as silica gel or desiccant packs, to reduce humidity in the shelter.

Cleaning and Disinfecting: Regularly clean and disinfect areas prone to moisture, such as corners, bathrooms, and around windows, to prevent mold growth. Use bleach or mold-killing products if mold appears.

Long-Term Waste Disposal Solutions in a Survival Situation

For long-term survival, managing waste becomes even more important as resources diminish and the space becomes more confined. Establishing a sustainable system for waste disposal is essential for maintaining health and cleanliness over time.

Composting and Recycling

Composting Human Waste: If you have access to a composting system, human waste can be safely composted over time using natural materials like sawdust or leaves. Make sure to follow proper composting procedures to avoid contamination and ensure that the composted material is only used for non-food plants.

Recycling Inorganic Waste: Where possible, recycle materials such as plastic, metal, or glass for other uses. Broken containers can be repurposed for storage, and scrap materials may be useful for repairing the shelter.

Safe Disposal Methods

Burying Waste: If it is safe to go outside, dig a pit away from your shelter and water sources for waste disposal. Burying waste helps keep the shelter clean and minimizes the risk of attracting pests or spreading disease.

Incineration: If conditions allow, burning waste may be an option for reducing the volume of garbage. However, this method should be used with caution, as it can produce smoke and odors that may attract unwanted attention or pose health risks in confined spaces.

Managing waste and sanitation in a confined shelter is essential for long-term survival. Without access to regular waste disposal services, you must develop systems to handle human waste, household garbage, and hazardous materials while maintaining personal hygiene and preventing the spread of disease. By staying vigilant about sanitation practices, controlling pests, ensuring proper air quality, and creating a sustainable waste disposal plan, you can ensure a clean, safe environment for yourself and your family. Proper sanitation is not only about cleanliness but is a critical part of protecting health in an already challenging survival situation.

Dealing with Desperation: How to Stay Calm and Focused

In a survival situation, especially after a nuclear event, maintaining calm and focus can be as crucial as having food, water, and shelter. Desperation can lead to panic, poor decision-making, and a breakdown in group dynamics, which can increase the likelihood of danger. This chapter will cover strategies for managing desperation, keeping your mental state in check, and fostering calmness and focus in high-pressure situations. Learning how to stay calm under pressure not only enhances survival but also improves your ability to lead and protect others.

Key Strategies for Staying Calm and Focused in a Survival Situation

Understanding the Impact of Desperation on Decision-Making

Practical Techniques for Managing Panic and Anxiety

Focusing on What You Can Control

Breaking Down Tasks to Stay Focused and Productive

Creating a Routine to Maintain a Sense of Normalcy

Breathing Exercises and Meditation for Stress Management

Using Visualizations to Stay Mentally Strong

Staying Connected with Others to Reduce Isolation and Fear

Recognizing and Managing Group Dynamics

Fostering a Mindset of Resilience and Adaptability

Understanding the Impact of Desperation on Decision-Making

Desperation often arises when you feel overwhelmed, trapped, or uncertain about the future. In survival situations, this can lead to rash decisions, emotional outbursts, or even dangerous behavior. Understanding how desperation affects your thinking can help you recognize these feelings early and take steps to counteract them.

Fight-or-Flight Response

Adrenaline and Stress: Desperation triggers the fight-or-flight response, flooding the body with adrenaline. This can lead to increased heart rate, rapid breathing, and a heightened sense of anxiety. While this response is useful in emergencies, it can cloud judgment and make it harder to think clearly.

Recognizing Desperation: Acknowledge when you or others are feeling desperate. Common signs include agitation, frustration, difficulty concentrating, and impulsive behavior. Recognizing these feelings early allows you to address them before they spiral out of control.

The Importance of Rational Thinking

Slowing Down Decisions: When feeling desperate, slow down the decision-making process. Take a moment to pause, breathe, and assess the situation before making critical choices. This helps prevent mistakes caused by panic and ensures that your actions are more deliberate and thoughtful.

Practical Techniques for Managing Panic and Anxiety

Panic is one of the most dangerous mental states in a survival situation. It leads to impulsive actions that can make the situation worse, whether by wasting resources, causing injuries, or making poor decisions. Learning to manage panic is crucial to staying calm and focused.

Grounding Techniques

Focusing on the Present: One way to ground yourself is to focus on your immediate surroundings. Engage your senses—what can you see, hear, feel, or smell? This technique helps bring your mind back to the present moment and reduces the overwhelming feeling of panic about the future.

Counting Backward: Slowly counting backward from 10 or 20 can help interrupt a panic attack and give your mind a chance to reset. Pair this with slow, deep breaths to calm your body's fight-or-flight response.

Progressive Muscle Relaxation

Tensing and Releasing: Practice tensing and then relaxing each muscle group in your body, starting from your toes and moving upward. This technique helps release physical tension caused by stress and can help calm your mind in moments of desperation.

Focusing on What You Can Control

In a survival situation, it's easy to feel overwhelmed by everything that's going wrong or by the uncertainty of the future. However, focusing on what you can control helps you regain a sense of agency and reduces feelings of helplessness.

Immediate Actions vs. Future Worries

Control the Present: Focus on tasks that are within your immediate control, such as securing water, organizing supplies, or caring for your family. These are actions you can take right now, and they give you a sense of accomplishment that helps mitigate feelings of desperation.

Letting Go of Uncertainty: Accept that there will be many unknowns in a survival situation, such as how long it will last or what challenges lie ahead. While it's normal to feel anxious about the future, recognize that worrying about the unknown won't change it. Direct your energy toward what you can actively improve.

Breaking Down Tasks to Stay Focused and Productive

When the world around you feels chaotic, staying productive by focusing on small, manageable tasks helps keep your mind grounded and gives you a sense of control.

Prioritizing Tasks

Immediate Needs First: Focus on your most immediate survival needs: water, food, shelter, and safety. Break these down into small tasks, such as collecting and filtering water, organizing your food supplies, or reinforcing your shelter.

Tackle One Task at a Time: Avoid trying to do everything at once. Choose one task, complete it, and then move on to the next. This method helps prevent feeling overwhelmed and maintains focus.

Accomplishing Small Wins

Celebrate Progress: Even small achievements, like securing clean drinking water or completing a minor repair, can boost your morale. Celebrate these victories with your group or family, as they remind you that you are making progress toward survival.

Creating a Routine to Maintain a Sense of Normalcy

Establishing a daily routine helps create a sense of stability in the midst of uncertainty. Routines provide structure, reduce stress, and prevent idleness, which can lead to increased feelings of desperation.

Daily Schedules

Set Regular Tasks: Create a daily schedule that includes basic survival activities (e.g., collecting water, cooking, maintaining the shelter) along with time for relaxation and social interaction. Having a predictable routine helps reduce anxiety about the unknown.

Adapting as Needed: While a routine provides structure, be flexible enough to adapt to changing conditions. Don't be afraid to adjust your schedule if new challenges arise or if you need to focus on urgent tasks.

Breathing Exercises and Meditation for Stress Management

Simple breathing exercises and meditation techniques can help you stay calm and focused, even in high-stress situations. These methods lower your heart rate, reduce anxiety, and give your mind a break from constant worry.

Deep Breathing Techniques

Box Breathing: Inhale deeply for four counts, hold your breath for four counts, exhale for four counts, and hold again for four counts. Repeat this cycle several times to calm your nervous system and regain control over your emotions.

4-7-8 Breathing: Inhale for a count of 4, hold for 7, and exhale for 8. This method helps reset your body's stress response and brings your mind to a more relaxed state.

Mindful Meditation

Focusing on the Breath: Take five minutes to sit quietly and focus on your breathing. If your mind wanders, gently bring it back to your breath. This practice helps train your mind to stay present and reduces the likelihood of panic.

Gratitude Meditation: Spend a few minutes focusing on things you are grateful for, even in difficult times. This shifts your focus away from fear and helps foster a more positive mental state.

Using Visualizations to Stay Mentally Strong

Visualization techniques can help you maintain a sense of calm and mental strength in a survival situation. By mentally rehearsing positive outcomes or peaceful scenarios, you can train your mind to focus on resilience rather than fear.

Positive Visualization

Envisioning Success: Picture yourself successfully navigating the challenges you face, whether it's finding food, reaching a safe zone, or reuniting with loved ones. Visualizing success can build confidence and reinforce a sense of control.

Calming Visualizations: In moments of high stress, imagine yourself in a peaceful, safe environment, such as a beach or forest. This mental escape helps reduce anxiety and provides your mind with a brief reprieve from the harshness of the situation.

Staying Connected with Others to Reduce Isolation and Fear

Isolation can intensify feelings of desperation, while social interaction can provide comfort, support, and a sense of solidarity. Staying connected with others helps reduce fear and reminds you that you are not facing the situation alone.

Supporting Each Other

Open Communication: Encourage regular communication among your group or family. Sharing your concerns, frustrations, or fears with others can help release pent-up emotions and foster a sense of unity.

Assigning Roles: Giving everyone in the group a role or responsibility not only reduces feelings of helplessness but also reinforces the importance of teamwork. Shared tasks help individuals feel valued and reduce isolation.

Recognizing and Managing Group Dynamics

Group dynamics can be challenging in a survival situation, as stress and desperation may lead to conflict. Knowing how to manage group dynamics and diffuse tension helps maintain harmony and ensures that everyone stays focused on survival.

Addressing Conflicts

Calm Conflict Resolution: When conflicts arise, address them calmly and without judgment. Allow each person to express their point of view, and work together to find a solution that benefits the group. Avoid escalating arguments, as they waste energy and can fracture the group's cohesion.

Empathy and Understanding: Recognize that everyone is dealing with stress and fear. Showing empathy toward others' struggles can defuse tense situations and foster a sense of camaraderie.

Fostering a Mindset of Resilience and Adaptability

Survival requires resilience—the ability to adapt to changing circumstances, bounce back from setbacks, and keep going even in the face of adversity. Developing a resilient mindset helps you stay focused and push through difficult times.

Adopting a Growth Mindset

Viewing Challenges as Opportunities: Instead of seeing obstacles as insurmountable, view them as opportunities to grow stronger and more resourceful. Each challenge you overcome reinforces your ability to adapt and survive.

Learning from Mistakes: In survival situations, mistakes happen. Instead of dwelling on failures, learn from them. Every mistake is a lesson in how to improve and avoid similar issues in the future.

Staying Optimistic

Maintaining Hope: Even in the darkest moments, maintaining hope is critical to survival. Focus on small victories and remind yourself that each day survived brings you closer to safety or a better situation.

In any survival scenario, dealing with desperation is as much about mental strength as it is about physical preparedness. Staying calm and focused in the face of adversity allows you to make sound decisions, maintain morale, and avoid the pitfalls of panic and fear. By employing practical techniques like deep breathing, breaking down tasks, and staying connected with others, you can foster resilience and adaptability, ultimately improving your chances of survival. The ability to manage stress, maintain hope, and keep a level head in the midst of chaos will serve as your greatest tool for overcoming the challenges of a post-nuclear world.

Rebuilding the Economy: Currency, Trade, and Resources

In a post-nuclear world, rebuilding the economy will be one of the most important and challenging tasks for survivors. Traditional currency systems may collapse, and resources that were once taken for granted will become precious commodities. In such a scenario, new forms of trade and currency will emerge, and the value of goods and services will be based on their immediate utility and scarcity. This chapter will explore how to rebuild an economy through barter, resource management, and the creation of new systems of trade and currency.

Key Considerations for Rebuilding the Economy

The Collapse of Traditional Currency Systems

The Rise of Barter and Trade: Exchanging Goods and Services

The Value of Resources in a Post-Nuclear World

Creating a New Currency: What Could Replace Money?

The Importance of Skills and Knowledge as Currency

Managing Scarce Resources: Food, Water, and Medical Supplies

Establishing Trade Networks Between Survivor Groups

Protecting Trade Routes and Resources from Theft or Conflict

Building Trust and Fairness in Economic Transactions

Long-Term Economic Sustainability: Growing Beyond Survival

The Collapse of Traditional Currency Systems

In the immediate aftermath of a nuclear event, traditional currency systems—such as paper money, coins, or digital banking—are likely to lose their value. The infrastructure that supports modern financial systems will be disrupted, making it difficult to use or rely on currency for purchasing goods and services. Instead, the economy will shift toward a barter system where tangible goods and resources are traded directly for their value.

Loss of Faith in Currency

Devaluation of Money: In a post-nuclear world, people may no longer trust paper money, and banks may not be accessible or functioning. Without government backing or economic stability, currency may become worthless, leading people to seek out physical goods for survival.

Shift to Physical Commodities: Gold, silver, and other precious metals might retain some value as they are physical commodities with a long history of being used as currency. However, in the short term, basic resources like food, water, and medicine will be far more valuable than gold or silver.

What Happens to Wealth?

Redistribution of Wealth: The collapse of traditional currency systems may lead to a redistribution of wealth. Those who were previously wealthy in money but lack practical resources or skills may find themselves at a disadvantage, while individuals with stockpiles of supplies, farming knowledge, or survival skills will become the new "wealthy."

The Rise of Barter and Trade: Exchanging Goods and Services

In the absence of money, barter will become the primary means of exchange. People will trade goods and services based on immediate needs, and the value of items will fluctuate depending on scarcity and demand.

Bartering Essentials

Trading Food and Water: Food and clean water will be among the most valuable commodities in the early stages of rebuilding. Those with access to surplus supplies can trade for other essential items like tools, clothing, or medicine.

Skills for Survival: Barter doesn't just apply to goods; skills will also be valuable. Those who know how to repair tools, treat injuries, grow crops, or build shelters can trade their expertise for goods they need.

Barter Networks

Creating Local Trade Hubs: Survivor groups or communities can establish local trade hubs where people gather to exchange goods and services. This can be as simple as a weekly market or an organized barter network between multiple groups.

Standardizing Trade: As bartering grows, communities may develop standardized values for certain items. For example, a gallon of clean water could be worth a specific amount of food, and these values will be adjusted based on supply and demand.

The Value of Resources in a Post-Nuclear World

The value of resources in a post-nuclear economy will be vastly different from the pre-crisis world. What was once considered cheap or easily accessible will now be precious, while other items may lose their value altogether.

High-Value Resources

Water: Clean water will be one of the most valuable resources, especially in areas where water sources have been contaminated by radiation. People will trade significant amounts of other goods for access to clean water.

Food: Non-perishable food, seeds for planting, and hunting supplies will be critical. Canned goods, dried food, and hunting or fishing equipment will be highly prized in trade.

Medicine and First Aid Supplies: Medical supplies, from bandages to antibiotics, will be extremely valuable. Access to healthcare may be limited, so any type of medication or treatment supply will become a powerful bargaining tool.

Low-Value or Useless Resources

Luxury Items: Items like jewelry, designer clothing, or electronics will lose much of their value, as they are not essential for survival. These items may only regain some worth once society begins to stabilize and people can focus on more than basic needs.

Paper Money and Bank Cards: As mentioned earlier, paper money and bank cards will likely lose all value, as there will be no functioning system to support them. These items will become meaningless unless a new financial system is established.

Creating a New Currency: What Could Replace Money?

As society begins to stabilize and people start thinking beyond immediate survival, new forms of currency may emerge. These currencies will likely be based on tangible, valuable commodities or even alternative systems like labor or community credits.

Commodities as Currency

Precious Metals: Gold, silver, and other metals may re-emerge as a form of currency once survival needs are met and people start to value items for long-term investment. These metals have a history of being used as currency and are universally accepted.

Food and Resources as Currency: In some cases, food, fuel, or tools may serve as currency. For example, a set amount of grain could be traded as a standard currency in a local economy.

Bartering for Time or Labor

Labor Exchange: People may develop a system where labor itself becomes a form of currency. For instance, hours of work can be exchanged for goods, allowing for a more flexible and sustainable economy. This form of currency relies heavily on trust and community cooperation.

The Importance of Skills and Knowledge as Currency

In a post-nuclear economy, practical skills and knowledge will become as valuable as any physical good. Those who have the ability to contribute to the rebuilding of society through specialized skills will be in high demand.

Survival Skills

Farming and Food Production: Those who know how to grow food, tend livestock, or forage safely will have a significant advantage. Communities will rely on individuals with farming expertise to rebuild food systems.

Medical Knowledge: People with medical training, whether doctors, nurses, or first aid specialists, will be incredibly valuable. Their knowledge can be exchanged for essential resources.

Technical and Repair Skills

Mechanics and Engineers: As machinery and tools wear down over time, mechanics and engineers will become vital for repairing or repurposing equipment. The ability to fix generators, vehicles, or water filtration systems will make these individuals highly sought after.

Construction and Building: Rebuilding infrastructure will require people skilled in construction, carpentry, and masonry. These skills will be indispensable in creating long-term shelters, fortifications, and community structures.

Managing Scarce Resources: Food, Water, and Medical Supplies

Managing resources wisely will be critical to long-term survival. Communities will need to develop systems for rationing, distributing, and conserving essential resources.

Rationing Systems

Fair Distribution: Establishing a fair rationing system ensures that resources like food and water are distributed evenly within a community. This can help prevent hoarding, theft, and conflicts over scarce supplies.

Stockpile Management: Communities should designate individuals responsible for managing stockpiles of food, water, and medical supplies. Regular inventory checks can help prevent shortages and ensure that resources are used efficiently.

Water Conservation

Purification and Recycling: Water sources may be limited, so communities will need to purify and recycle water as much as possible. Setting up rainwater collection systems and water filtration methods will help extend the available supply.

Establishing Trade Networks between Survivor Groups

As survivors begin to organize into groups or communities, trade networks will develop between them. These networks will allow the exchange of surplus goods and resources, helping communities access what they cannot produce locally.

Specialization and Trade

Resource Specialization: Different groups may specialize in producing certain resources, such as one group focusing on farming while another focuses on manufacturing tools. This specialization allows for a more efficient and mutually beneficial trade system.

Traveling Traders: Individuals or small groups may become traveling traders, carrying goods between communities in exchange for food, shelter, or protection. These traders can play an essential role in connecting isolated communities.

Protecting Trade Routes and Resources from Theft or Conflict

With resources scarce, theft and conflict over goods will be a constant threat. Communities will need to protect their resources and trade routes from bandits or rival groups.

Defense and Security

Securing Resources: Communities should establish defenses around key resources, such as water supplies, food stockpiles, or trade routes. Fortifying these areas can help prevent theft and protect against outside threats.

Escorting Trade Caravans: Armed escorts or security details may be needed to protect trade caravans from theft or ambush while traveling between communities. Cooperation between groups to maintain secure trade routes will help build trust.

Building Trust and Fairness in Economic Transactions

In the absence of formal laws or government oversight, trust becomes the foundation of any economic system. Communities will need to develop systems of accountability and fairness in trade to prevent disputes and ensure cooperation.

Community Trust Systems

Honest Trading: Establishing a reputation for honesty and fairness in trade will be essential. Those who are known to be trustworthy will become valued members of the trading network, while those who cheat or deceive will be ostracized.

Conflict Resolution: Set up systems for resolving trade disputes fairly and peacefully. Appointing community leaders or councils to mediate conflicts can help prevent tensions from escalating into violence.

Long-Term Economic Sustainability: Growing Beyond Survival

As society begins to stabilize, the economy can shift from mere survival to sustainability and growth. Communities can work toward rebuilding a more formal economy based on mutual cooperation, the development of industries, and resource management.

Growing Food and Resources

Rebuilding Agriculture: Once immediate survival needs are met, focus on long-term food production through sustainable farming methods. Crop rotation, permaculture, and livestock management will help ensure a stable food supply.

Energy and Manufacturing: Communities may begin to rebuild small-scale industries, such as toolmaking, renewable energy production, or textile manufacturing. These industries can help create new goods for trade and support economic growth.

Creating New Markets

Expanding Trade Networks: Over time, new markets may emerge as communities become more connected. These markets can serve as hubs for exchanging not only essential goods but also luxury items and non-essential goods as society rebuilds.

Rebuilding the economy in a post-nuclear world requires innovation, cooperation, and adaptability. As traditional currency systems collapse, survivors will rely on barter, trade, and resource management to meet their needs. Skills, knowledge, and tangible resources will become the new currency, while trust and fairness will form the backbone of economic interactions. By establishing sustainable practices, securing trade networks, and fostering a resilient mindset, communities can gradually rebuild their economies and create a stable future beyond survival.

The Importance of Education and Knowledge in Post-War Society

Education and knowledge are vital to the rebuilding and future stability of any post-war society, particularly after a nuclear event. While the immediate priorities may be survival—food, shelter, and security—the long-term success of a community depends on its ability to pass on critical skills, rebuild infrastructure, and foster innovation. Education not only ensures the continuation of essential skills but also provides hope and a sense of purpose to future generations. In this chapter, we'll explore how education can drive recovery and why knowledge is one of the most important resources for rebuilding society.

Key Considerations for Education and Knowledge in Post-War Society

Preserving Essential Survival Skills for Future Generations

Rebuilding Society through Practical and Technical Education

Passing Down Knowledge: Oral Tradition, Books, and Technology

Education as a Means of Fostering Hope and Stability

Specialized Skills for Rebuilding Infrastructure

Teaching Leadership and Governance in a New World

Protecting and Expanding Medical Knowledge

Adapting Education to New Realities and Limited Resources

The Role of Innovation and Problem-Solving in Recovery

Creating a Knowledge-Based Economy for Long-Term Growth

Preserving Essential Survival Skills for Future Generations

In the immediate aftermath of a nuclear conflict, survival skills will be the foundation of education. Skills such as growing food, hunting, purifying water, and making shelters will be essential for everyday survival. However, these skills must not be limited to the current generation—they need to be taught to children and future generations so that they, too, can thrive in this new world.

Survival Training for All Ages

Basic Survival Education: Children and young adults must learn essential skills such as food preservation, fire-making, basic first aid, and how to secure water sources. These skills will be the foundation of early education.

Community Teaching: Organize group education sessions where experienced adults pass on their knowledge to younger members of the community. This ensures that critical skills are not lost and that future generations can continue to build on them.

Preserving Knowledge through Apprenticeships

Hands-On Learning: In a post-nuclear society, formal education systems may no longer exist. Instead, focus on apprenticeships where young people learn directly from skilled adults, whether it's farming, engineering, or medical care.

Rebuilding Society through Practical and Technical Education

While survival is the immediate concern, communities will eventually need to rebuild their infrastructure, communication systems, and industries. This requires practical and technical education that focuses on rebuilding and advancing the community.

Technical Skills for Rebuilding

Engineering and Mechanics: Basic engineering knowledge will be crucial for repairing damaged infrastructure and creating new systems. Mechanics will be needed to fix machines, generators, and vehicles that can help with rebuilding efforts.

Agriculture and Sustainability: As communities grow, sustainable farming methods must be taught to ensure long-term food production. Techniques such as crop rotation, permaculture, and irrigation systems should be part of the educational focus.

Crafting and Manufacturing

Tool Making and Construction: Rebuilding society requires tools, and individuals with skills in blacksmithing, carpentry, and metalworking will be essential. Passing on knowledge of how to create and maintain tools is vital for a functional society.

Small-Scale Industry: Basic manufacturing, such as making clothing, shelter materials, or medical supplies, can be taught using available resources. This ensures that communities can produce necessary goods even with limited technology.

Passing Down Knowledge: Oral Tradition, Books, and Technology

In the absence of formal schools or digital systems, preserving and passing down knowledge becomes a communal responsibility. Oral traditions, books, and even remnants of technology can help keep knowledge alive for future generations.

Oral Tradition and Storytelling

Verbal Education: In societies where books or digital devices may be scarce, oral tradition will play a critical role. Elders and knowledgeable individuals should share their experiences and skills through storytelling, teaching, and verbal instruction.

Community Gatherings: Regular gatherings where knowledge is shared, whether in the form of stories, lessons, or instructions, can ensure that valuable information isn't lost.

Preserving Written Knowledge

Books and Manuals: Collect and preserve any books, manuals, or written records that can be used for teaching. Books on farming, medicine, engineering, and even history can become invaluable educational tools.

Repurposing Technology: If possible, salvage and repurpose technology like solar-powered devices or old computers to store digital knowledge. Even small caches of digital information can help preserve important technical data.

Education as a Means of Fostering Hope and Stability

In times of crisis, education provides more than just practical skills—it fosters hope and gives people a sense of purpose. By investing in education, communities can create a foundation of stability and forward-thinking that helps them not just survive but thrive.

Creating Purpose Through Learning

Hope for the Future: Educating the younger generation gives people something to strive for beyond mere survival. It fosters a sense of hope, purpose, and progress, helping to prevent despair and fatalism.

Continuity and Progress: Education allows communities to envision a future where rebuilding is possible. It reminds them that they are not just reacting to the present but building something better for future generations.

Specialized Skills for Rebuilding Infrastructure

Rebuilding infrastructure—whether it's roads, bridges, or communication systems—requires specialized knowledge. Education in civil engineering, construction, and logistics becomes crucial as communities move from survival to growth.

Basic Engineering and Construction

Rebuilding Roads and Bridges: Instruction on basic engineering principles, such as how to construct stable roads, bridges, or temporary shelters, can help communities rebuild transportation networks and infrastructure.

Water and Energy Systems: Teach individuals how to create or repair basic water and energy systems, including wells, wind turbines, and solar panels. These skills are essential for creating sustainable living conditions.

Advanced Technical Knowledge

Electrical Engineering: As communities stabilize, the need for basic electrical systems will arise. Those with knowledge of generators, batteries, and circuits can train others to repair or rebuild electrical systems, ensuring that power is restored to vital areas.

Teaching Leadership and Governance in a New World

Leadership skills and governance structures are essential for maintaining order, fairness, and collaboration in a post-war society. Teaching the next generation about leadership, conflict resolution, and governance helps ensure the long-term success of rebuilding efforts.

Fostering Leadership Skills

Decision-Making and Conflict Resolution: Teach individuals how to make decisions for the group's benefit and manage conflicts peacefully. These skills are crucial in preventing internal disputes from derailing community progress.

Shared Leadership Models: Consider implementing shared leadership or council-based governance models where education focuses on collaboration, transparency, and fairness.

Justice and Fairness

Establishing Laws: Teach young leaders the importance of justice, fairness, and accountability in governance. Establishing fair laws and systems of punishment helps maintain peace and order in growing communities.

Protecting and Expanding Medical Knowledge

In a post-nuclear world, access to modern medical care may be limited or nonexistent. Preserving and expanding medical knowledge is essential for treating injuries, managing diseases, and ensuring long-term health.

Medical Training

First Aid and Emergency Care: Teach basic first aid skills to as many people as possible, including how to treat wounds, fractures, burns, and common illnesses. This knowledge can be life-saving in the absence of professional medical care.

Midwifery and Childcare: Ensuring safe childbirth and care for infants is essential for the survival of the population. Midwifery, infant care, and knowledge of child nutrition should be taught to those responsible for the health of mothers and children.

Herbal Medicine and Natural Remedies

Alternative Treatments: With limited access to modern medicine, communities can turn to traditional herbal remedies and alternative treatments. Teaching the medicinal uses of plants and how to identify, harvest, and use them is vital.

Adapting Education to New Realities and Limited Resources

In a world with limited resources, education systems must be adapted to the new realities. This might mean forgoing traditional classrooms in favor of informal learning environments, where the focus is on practical skills and hands-on experience.

Flexible Learning Methods

Learning by Doing: Instead of structured lessons, emphasize hands-on experience where individuals learn by directly engaging in tasks such as farming, building, or crafting. Practical application is often more valuable than theoretical knowledge in a survival situation.

Mentorship and Apprenticeship: Pair less experienced individuals with mentors who can pass on their skills. This system allows for one-on-one learning that is tailored to the specific needs of the community.

The Role of Innovation and Problem-Solving in Recovery

While the initial focus will be on survival, long-term recovery requires innovation and problem-solving. Communities that encourage creative thinking and experimentation will be better equipped to adapt to changing circumstances and rebuild successfully.

Encouraging Innovation

Problem-Solving Education: Teach critical thinking and problem-solving skills, encouraging people to come up with creative solutions to the challenges they face. Whether it's developing new farming techniques or finding ways to purify water, innovation will be essential for long-term survival.

Experimenting with New Ideas: Allow room for experimentation in education. Encourage individuals to try new approaches, whether in engineering, agriculture, or medicine. This fosters a culture of learning and growth.

Creating a Knowledge-Based Economy for Long-Term Growth

As society begins to stabilize, education will be the foundation of long-term economic growth. A knowledge-based economy, where skills and innovation drive progress, will help communities move beyond mere survival and toward prosperity.

Building Industries on Knowledge

Specialization and Expertise: Encourage individuals to specialize in areas where they can excel, such as medicine, engineering, or agriculture. This will allow for the development of industries that are essential for rebuilding a functioning economy.

Education as an Investment: Viewing education as an investment in the community's future will lead to a more prosperous and resilient society. By nurturing the next generation of leaders, workers, and innovators, communities can ensure their survival and eventual growth.

Education and knowledge are the foundation of any society's recovery, especially in the wake of a nuclear conflict. By preserving essential skills, passing down knowledge, and fostering innovation, communities can rebuild and create a stable future. Whether through practical skills, leadership training, or medical expertise, education will guide the next generation in not only surviving but thriving in a post-war world. A focus on teaching and learning ensures that the lessons of the past are not lost and that future generations are prepared to rebuild society stronger than before.

Reintroducing Animals and Wildlife to a Nuclear-Scarred World

Reintroducing animals and wildlife to a world devastated by nuclear conflict is a delicate but vital part of rebuilding ecosystems and ensuring long-term sustainability for humans and the environment. In a post-nuclear landscape, much of the natural world may be affected by radiation, habitat destruction, and the loss of biodiversity. However, restoring wildlife can help rejuvenate ecosystems, promote biodiversity, and create sustainable food sources for humans. This chapter will explore the challenges and strategies for reintroducing animals and wildlife into a nuclear-scarred world, ensuring a balance between human survival and ecological restoration.

Key Considerations for Reintroducing Wildlife

The Impact of Nuclear Fallout on Wildlife and Ecosystems

Assessing the Safety of the Environment for Animal Reintroduction

Selecting Species for Reintroduction: Prioritizing Resilient and Native Species

The Role of Wildlife in Ecological Balance and Human Survival

Establishing Safe Habitats: Rehabilitating Contaminated Land

Breeding Programs and Conservation Efforts for Endangered Species

Managing Predators and Prey in a Fragile Ecosystem

The Role of Pollinators in Rebuilding Agriculture

Monitoring Wildlife Health and the Effects of Radiation

Balancing Human Needs with Wildlife Conservation

The Impact of Nuclear Fallout on Wildlife and Ecosystems

The aftermath of nuclear warfare will have devastating effects on wildlife and ecosystems, as radioactive fallout spreads over large areas, contaminating land, water, and air. The severity of the impact will depend on the scale of the conflict, the proximity to ground zero, and the types of animals and plants exposed to radiation.

Immediate Effects on Wildlife

Radiation Poisoning: Animals living close to the nuclear detonation sites may suffer from acute radiation poisoning, leading to sickness, birth defects, and death. Birds, small mammals, and fish are particularly vulnerable to high radiation exposure.

Habitat Destruction: Nuclear blasts, fires, and subsequent fallout can destroy entire ecosystems, leaving animals without food or shelter. Forests, rivers, and grasslands may become uninhabitable due to contamination and physical destruction.

Long-Term Ecological Damage

Reduced Biodiversity: Radiation can cause mutations in animals, leading to reproductive issues and a decline in populations. Over time, this can reduce biodiversity and disrupt the balance of ecosystems, which are vital for long-term environmental recovery.

Impact on Food Chains: The loss of key species, such as top predators or primary consumers, can collapse food chains, affecting the entire ecosystem. If pollinators, herbivores, or predators are wiped out or severely reduced, the entire balance of the ecosystem could be disrupted.

Assessing the Safety of the Environment for Animal Reintroduction

Before reintroducing wildlife, it is crucial to assess the environment to determine if it is safe and suitable for animals to thrive. This involves testing soil, water, and air for radiation levels, as well as evaluating the availability of food and shelter for various species.

Radiation Testing

Monitoring Radiation Levels: Use Geiger counters and other radiation monitoring devices to measure contamination levels in the environment. Areas with low to moderate radiation may still support certain animal species, while highly contaminated zones will remain off-limits for decades.

Mapping Safe Zones: Identify and map out regions where radiation levels are low enough to safely support wildlife. These areas will serve as the first places to begin reintroduction efforts.

Assessing Habitat Conditions

Food and Water Sources: Ensure that there are adequate food and water sources in the reintroduction areas. Contaminated water sources and barren landscapes may not support the return of wildlife, so these need to be addressed before any animals are introduced.

Vegetation Recovery: Examine the condition of plant life in the area. Healthy vegetation is crucial for herbivores, while carnivores will need prey species to sustain themselves.

Selecting Species for Reintroduction: Prioritizing Resilient and Native Species

The choice of which species to reintroduce first will depend on their ability to adapt to the post-nuclear environment. Prioritizing resilient species, particularly those native to the region, will increase the likelihood of successful reintroduction and ecological restoration.

Resilient Species

Hardy, Fast-Reproducing Animals: Start with species that are known for their resilience to harsh conditions and ability to reproduce quickly. These animals can help stabilize ecosystems and provide food sources for other species.

Herbivores and Omnivores: Reintroducing herbivores such as deer, rabbits, and small mammals can help kick-start ecosystems by facilitating plant growth and providing prey for carnivores.

Native Species

Preserving Local Biodiversity: Focus on reintroducing species that are native to the region, as they are better adapted to the local environment and will have a better chance of thriving. Non-native species may disrupt the balance of the ecosystem or compete with native species for resources.

The Role of Wildlife in Ecological Balance and Human Survival

Reintroducing wildlife serves more than just ecological purposes—it plays a role in human survival as well. Healthy ecosystems provide food, clean water, and raw materials that humans rely on, and animals are a key component of these ecosystems.

Wildlife as a Food Source

Hunting and Gathering: In a post-nuclear world, reintroduced wildlife can provide a sustainable food source for humans through hunting or controlled harvesting. Animals such as deer, fish, and small game can help supplement food supplies.

Regulating Populations: Introducing animals that help control plant growth and balance ecosystems can prevent overgrowth and maintain fertile lands for agriculture.

Ecosystem Services

Soil Fertility: Animals such as herbivores and insects play a vital role in maintaining soil fertility through grazing and decomposition. Their presence supports plant growth, which is essential for rebuilding agriculture and natural habitats.

Establishing Safe Habitats: Rehabilitating Contaminated Land

To successfully reintroduce animals, contaminated land must be rehabilitated or managed to reduce radiation exposure. This may involve a combination of natural recovery and human intervention.

Soil Remediation

Phytoremediation: Planting specific types of vegetation, such as sunflowers, that can absorb radioactive particles from the soil may help reduce radiation levels over time. These plants can be harvested and safely disposed of to prevent further contamination.

Soil Replacement: In some cases, the top layer of contaminated soil may need to be removed and replaced with clean soil to support vegetation growth and create a safe habitat for animals.

Water Purification

Decontaminating Water Sources: Efforts to purify contaminated water through filtration systems or chemical treatments will be essential for supporting both animal and human populations. Clean water is critical for wildlife to thrive in newly reintroduced habitats.

Breeding Programs and Conservation Efforts for Endangered Species

Some species may be too endangered or vulnerable to reintroduce into the wild immediately. In these cases, captive breeding programs and conservation efforts will be necessary to rebuild their populations before releasing them into rehabilitated habitats.

Captive Breeding Programs

Repopulating Endangered Species: For species that are on the brink of extinction, breeding programs can help bolster their numbers in controlled environments before they are released back into the wild. This will help restore biodiversity over time.

Genetic Diversity: Maintaining genetic diversity in breeding programs is important to ensure healthy populations. Careful management can prevent inbreeding and improve the species' chances of surviving in the wild.

Reintroduction Strategies

Phased Reintroductions: Gradually reintroduce species in phases, starting with those that are more resilient to the post-nuclear environment. As ecosystems stabilize and more resources become available, more delicate or endangered species can be introduced.

Managing Predators and Prey in a Fragile Ecosystem

Maintaining a balance between predators and prey is essential for the long-term stability of ecosystems. Too many predators can decimate prey populations, while too few can lead to overpopulation of herbivores, resulting in habitat destruction.

Careful Predator Reintroduction

Introducing Predators Slowly: Predators such as wolves or large cats should be introduced only after prey populations have stabilized. This helps prevent the collapse of prey populations before they can support themselves.

Maintaining Ecological Balance: Monitor predator-prey dynamics closely to ensure that neither group becomes dominant or endangered. This requires careful tracking of animal populations and food availability.

Managing Human-Wildlife Conflict

Protecting Livestock and Crops: As wildlife is reintroduced, humans may face challenges with animals encroaching on farmland or livestock areas. Teaching communities how to protect their resources while coexisting with wildlife is essential for maintaining balance.

The Role of Pollinators in Rebuilding Agriculture

Pollinators such as bees, butterflies, and birds are crucial for the recovery of plant life and agriculture. Reintroducing pollinators helps regenerate vegetation, supports food production, and restores ecological diversity.

Pollinator Habitat Restoration

Providing Safe Zones for Pollinators: Create habitats that support pollinators by planting native flowers, shrubs, and trees. These plants provide food and shelter for bees and other pollinators, ensuring their survival and promoting plant growth.

Monitoring Pollinator Health: Keep an eye on the health of pollinator populations, as they are sensitive to environmental changes and can be indicators of ecosystem recovery. Ensuring clean water and low radiation levels in pollinator habitats is essential.

Monitoring Wildlife Health and the Effects of Radiation

Reintroducing animals into a post-nuclear environment requires close monitoring of their health to ensure that radiation exposure is not causing harm. Wildlife health can also serve as a bellwether for human safety and environmental recovery.

Tracking Animal Health

Radiation Effects: Monitor animals for signs of radiation sickness, such as tumors, reproductive issues, or birth defects. These signs can provide important information about the safety of the environment for both wildlife and humans.

Population Studies: Conduct regular population surveys to assess the success of reintroduction efforts and detect any negative trends, such as population declines or abnormal behaviors.

Balancing Human Needs with Wildlife Conservation

Reintroducing wildlife and rebuilding ecosystems must be balanced with the needs of human survival. Communities will need to cooperate in managing resources, protecting habitats, and ensuring that wildlife and human populations can coexist peacefully.

Sustainable Resource Management

Shared Resources: Manage resources like water, land, and food in a way that benefits both humans and wildlife. Sustainable farming practices and wildlife corridors can help reduce competition for resources between humans and animals.

Education and Awareness

Teaching Coexistence: Educate communities on the importance of wildlife conservation and the benefits of a balanced ecosystem. Teaching people how to coexist with wildlife, including avoiding dangerous encounters with predators, will help reduce conflicts.

Reintroducing animals and wildlife to a nuclear-scarred world is a critical step in restoring ecosystems and ensuring the long-term survival of both human and animal populations. Through careful planning, habitat rehabilitation, and phased reintroductions, wildlife can help regenerate damaged environments and create a sustainable future. Managing the balance between predators and prey, protecting pollinators, and ensuring the safety of habitats are key to rebuilding the natural world. By prioritizing resilience, biodiversity, and ecological balance, communities can foster a thriving environment where both humans and wildlife can coexist in a post-nuclear world.

The Legal and Ethical Dilemmas of Survival in a Nuclear Winter

In a post-nuclear winter, the laws and ethics that once governed society may be significantly challenged or even collapse altogether. The breakdown of governmental structures, societal norms, and access to resources will create unprecedented legal and ethical dilemmas for survivors. What was once considered normal behavior or the rule of law may no longer apply in a world where survival takes precedence over established legal systems. This chapter will explore the complex legal and ethical issues that may arise during a nuclear winter, and how individuals and communities can navigate these challenges while maintaining a moral compass in a lawless world.

Key Legal and Ethical Dilemmas in a Nuclear Winter

The Collapse of Legal Systems and the Need for New Governance

Ethical Choices in Resource Distribution and Ownership

The Morality of Self-Defense and Protecting Resources

Dealing with Desperate People: When to Show Mercy or Force

The Ethics of Leadership and Decision-Making in Survival Situations

Personal Rights vs. Community Needs: Balancing Freedom and Survival

Medical Ethics in a Resource-Scarce World

Property Rights in a Post-Apocalyptic World

Rebuilding Justice Systems: Fairness and Accountability in a New Era

Survivor Justice: Dealing with Crime, Punishment, and Retribution

The Collapse of Legal Systems and the Need for New Governance

In the aftermath of a nuclear event, the collapse of governments and legal institutions is highly likely. Without formal law enforcement or judicial systems, survivors will face the challenge of creating new forms of governance that are both functional and fair. This brings up the question of who will enforce laws and how justice will be administered in a world where resources are scarce and survival is the top priority.

Loss of Centralized Authority

Absence of Government: With governments likely to collapse, survivors may be left without formal legal systems or law enforcement. This creates a vacuum in which individuals or groups may seek to impose their own rules.

Local Governance: Communities may form their own governance structures, but these will vary based on the values and priorities of the individuals involved. Some groups may adopt democratic principles, while others may lean toward authoritarian control to maintain order.

The Role of Leadership

Choosing Leaders: New leaders will need to emerge in survivor communities. The ethical dilemma lies in determining who has the right to lead and how leaders should be chosen—whether through democratic processes, strength, or other means.

Accountability: In the absence of formal systems, maintaining accountability becomes a challenge. Leaders may be tasked with making life-or-death decisions, and ensuring they act justly and fairly will be crucial for long-term survival.

Ethical Choices in Resource Distribution and Ownership

Access to resources such as food, water, and shelter will be limited, raising ethical questions about how these should be distributed. In a world where everyone is struggling to survive, deciding who gets what can become a matter of life and death.

Resource Scarcity

Rationing: Should resources be rationed equally, or should they be distributed based on need, contribution, or another factor? These decisions can create tensions within a group and pose significant ethical challenges.

Ownership of Resources: In many cases, survivors may take over abandoned property or supplies. The ethical dilemma lies in determining whether these resources should be considered personal property, community property, or up for grabs by anyone in need.

Sharing vs. Hoarding

Community Survival: Ethical questions will arise around whether individuals should hoard resources for themselves or share them for the greater good. Hoarding may ensure personal survival, but at the cost of community solidarity.

Enforcing Fairness: Without formal laws, communities will need to decide how to enforce fairness in resource distribution. This may involve collective decision-making or leaders enforcing rules on resource usage.

The Morality of Self-Defense and Protecting Resources

In a world where law and order have collapsed, self-defense becomes a critical issue. People may be forced to defend their resources, their families, and their own lives from those who would take them by force.

Use of Force

Self-Defense: When is it justified to use force, even lethal force, to protect oneself or one's resources? In the absence of law enforcement, individuals must make ethical decisions about how far they are willing to go to ensure their survival.

Protecting Others: The ethical question extends beyond self-defense to defending others. Should individuals risk their lives to protect their families or communities, or does personal survival take precedence?

Moral Limits

Excessive Force: While self-defense is often seen as justified, the use of excessive force can blur ethical lines. Survivors must weigh the consequences of their actions and whether taking a life is truly necessary in a given situation.

Vigilantism: In the absence of formal justice, individuals may take the law into their own hands, leading to vigilantism. While this may seem necessary for survival, it risks escalating violence and creating a cycle of retribution.

Dealing with Desperate People: When to Show Mercy or Force

In a nuclear winter, desperation will drive many people to take extreme actions to survive, including stealing, looting, or even attacking others. Deciding how to respond to such acts—whether with mercy or force—creates a significant ethical dilemma.

Mercy and Compassion

Showing Leniency: Should survivors show mercy to desperate individuals, understanding that they are acting out of necessity? Offering help or sharing resources may reduce conflict, but it could also deplete limited supplies.

Trust and Betrayal: Showing compassion to desperate people may carry the risk of betrayal. While offering aid can foster cooperation, it could also make survivors vulnerable to theft or violence from those they try to help.

Tough Decisions

Turning People Away: In some cases, survivors may be forced to turn away those in need to protect their own limited resources. The ethical question lies in how much responsibility individuals have to help others versus ensuring their own survival.

Choosing Force: When faced with potential threats, individuals may need to use force to protect themselves. The ethical dilemma here is whether force should be the first option or the last resort, and how to balance the need for safety with compassion.

The Ethics of Leadership and Decision-Making in Survival Situations

Leaders in survival situations face complex ethical decisions, particularly when it comes to prioritizing the needs of the group over the rights of individuals. These leaders must strike a balance between enforcing rules for the good of the group and respecting individual autonomy.

Utilitarian vs. Individual Rights

The Greater Good: Leaders may be forced to make utilitarian decisions, prioritizing the survival of the group over the desires or needs of individuals. This can create ethical conflicts, especially if it involves sacrificing personal freedoms for the collective well-being.

Respecting Personal Autonomy: Balancing the rights of individuals with the needs of the group is an ongoing challenge. Leaders must navigate situations where individuals may disagree with decisions that affect their lives, such as rationing, curfews, or enforced labor.

Informed Consent

Decision-Making Transparency: Leaders must strive to make decisions transparently and involve the group in decision-making when possible. While emergency situations may call for swift action, involving others in the process builds trust and ensures that decisions are ethical.

Personal Rights vs. Community Needs: Balancing Freedom and Survival

In a survival situation, the tension between personal freedom and the needs of the community often comes into sharp focus. Individuals may be asked to give up certain freedoms for the sake of the group's survival, which can lead to ethical dilemmas about autonomy and consent.

Curfews and Movement Restrictions

Freedom of Movement: Restrictions on movement, such as curfews or lockdowns, may be necessary to maintain safety or control resources. However, these restrictions infringe on personal freedoms and raise questions about how much control leaders should have over individual actions.

Consent to Rules: Should individuals be forced to comply with group rules, or should they have the right to opt out, even if doing so endangers the community? This dilemma is particularly challenging when people disagree with the imposed regulations.

Forced Labor and Contributions

Mandatory Work: In some cases, communities may require everyone to contribute labor to ensure survival, such as growing food or maintaining defenses. Ethical dilemmas arise when individuals refuse to contribute or feel they are being exploited for the benefit of others.

Fair Distribution of Labor: Ensuring that everyone shares the burden of work fairly can prevent resentment and build solidarity within the community. However, leaders must also consider the capacities of individuals, such as the elderly, sick, or disabled, when assigning tasks.

Medical Ethics in a Resource-Scarce World

In a nuclear winter, medical supplies will be limited, and difficult decisions about who receives care may become a regular occurrence. Medical professionals and community leaders must navigate these ethical challenges while upholding the values of compassion and fairness.

Triage and Resource Allocation

Who Gets Treatment?: In a resource-scarce world, medical professionals may be forced to triage patients based on their likelihood of survival. Ethical questions arise about who should receive limited medical resources and whether factors like age, health, or social contribution should influence these decisions.

Withholding Treatment: When resources are scarce, withholding treatment may be necessary. However, this raises ethical questions about how to decide who will not receive care and the emotional toll it takes on medical providers and communities.

End-of-Life Decisions

Euthanasia and Pain Management: In extreme cases, the question of euthanasia may arise, particularly for those suffering from terminal conditions or severe injuries with no chance of recovery. The ethics of euthanasia and palliative care will be heavily debated in survival scenarios.

Respecting Dignity: Ensuring that individuals die with dignity, even in harsh survival conditions, is an ethical responsibility. Communities must decide how to handle death respectfully, whether through burial, cremation, or other means, even with limited resources.

Property Rights in a Post-Apocalyptic World

The collapse of legal property rights raises significant ethical questions about ownership in a post-nuclear world. Without formal laws to govern property ownership, communities must decide how to handle issues such as squatting, resource ownership, and abandoned property.

Squatting and Abandoned Property

Taking Over Abandoned Property: With homes and land abandoned after a nuclear event, ethical dilemmas arise about whether it is acceptable to occupy or use these resources. Should abandoned property be free for anyone to claim, or should there be rules about who has rights to it?

Community vs. Individual Ownership: Survivors may need to decide whether property ownership should be communal or individual. Communal ownership can foster cooperation, but it also raises concerns about fairness and distribution.

Protecting Personal Property

Looting and Theft: With law enforcement absent, theft and looting may become common. Survivors must decide how to protect their property and whether it is ethical to take resources from others if they are essential for survival.

Rebuilding Justice Systems: Fairness and Accountability in a New Era

As survivors begin to rebuild communities, they will need to establish systems of justice to maintain order and fairness. Without traditional legal systems, the challenge will be creating a justice system that holds people accountable without resorting to tyranny or excessive punishment.

Establishing New Laws

Creating Fair Laws: Communities will need to create new laws that reflect their values and priorities. These laws must balance the need for order with fairness, ensuring that they do not disproportionately punish or exclude certain individuals.

Community Involvement: Involving the entire community in the creation of laws can help ensure that they are seen as legitimate and just. This helps prevent the concentration of power in the hands of a few and ensures that laws reflect the needs of the group.

Judgment and Punishment

Punishments for Crimes: Deciding on appropriate punishments for crimes, such as theft, violence, or betrayal, will be a difficult ethical issue. Communities must balance the need for deterrence with the value of rehabilitation and mercy.

Restorative vs. Retributive Justice: Some communities may opt for restorative justice, focusing on repairing harm and reintegrating offenders into the group, while others may adopt more punitive measures to maintain order.

Survivor Justice: Dealing with Crime, Punishment, and Retribution

Crime in a post-nuclear world will be a major challenge, as survivors may resort to theft, violence, or even murder out of desperation. Without formal law enforcement, survivors must decide how to deal with crime and what form of justice is appropriate.

Vigilante Justice

Community Enforcement: With no formal police force, communities may rely on collective enforcement of rules. This can lead to vigilantism, where individuals take the law into their own hands, raising ethical concerns about fairness and due process.

Escalating Violence: Without proper justice systems, retribution can lead to cycles of violence. Survivors must decide how to avoid retaliatory violence and maintain peace within their communities.

Exile and Banishment

Exiling Offenders: In some cases, communities may choose to exile or banish offenders rather than punish them physically. While this may seem less violent, it raises ethical questions about the rights of the individual to survival and whether exile is a humane punishment in a hostile world.

Surviving a nuclear winter will require difficult legal and ethical decisions, particularly in the absence of formal laws and governance. Navigating the collapse of legal systems, managing scarce resources, and dealing with crime will present significant challenges to individuals and communities. By striving to maintain fairness, accountability, and compassion in decision-making, survivors can rebuild a society that reflects their values and promotes cooperation, even in the face of extreme hardship. While the dilemmas may be complex, addressing them thoughtfully and ethically will be key to creating a just and functional post-apocalyptic world.

Surviving the First Year: Lessons Learned in Isolation

Surviving the first year after a nuclear event is an immense challenge, marked by isolation, scarcity of resources, and the breakdown of societal structures. This initial period will test both physical endurance and mental resilience. The first year is crucial for laying the foundation for long-term survival, rebuilding, and adjusting to a world that has dramatically changed. In this chapter, we will explore the lessons learned during the first year of isolation, focusing on survival strategies, mental fortitude, and the adaptation to a harsh and uncertain environment.

Key Lessons for Surviving the First Year

Adjusting to a New Reality: Accepting Loss and Change

Prioritizing Immediate Needs: Water, Food, and Shelter

Learning to Manage Limited Resources

The Importance of Mental and Emotional Resilience

Building a Sustainable Routine for Long-Term Survival

Overcoming Isolation: Staying Connected with Others

Adapting Skills and Knowledge to a New World

Staying Healthy in a Resource-Scarce Environment

Handling the Psychological Effects of Loneliness and Fear

Preparing for Year Two: Planning for the Future

Adjusting to a New Reality: Accepting Loss and Change

The first and most important lesson of survival is accepting the reality of the situation. The world, as it was before the nuclear event, is gone, and survivors must come to terms with the loss of normalcy. This acceptance is key to moving forward and focusing on survival.

Accepting Loss

Grieving the Old World: Survivors may experience profound grief over the loss of loved ones, homes, and the familiar world. Accepting and processing this grief is essential to avoid being paralyzed by despair.

Embracing Change: Recognizing that life has changed irrevocably can help survivors focus on adapting rather than clinging to what once was. Acceptance of change enables individuals to think creatively and flexibly about new ways to live and thrive.

Reframing Goals

Survival as the Priority: Instead of focusing on regaining the past, shift the goal to surviving the present and planning for the future. This new mindset will help direct energy toward practical tasks rather than yearning for what is lost.

Prioritizing Immediate Needs: Water, Food, and Shelter

The first year of survival is marked by a constant focus on securing basic necessities. Without stable access to water, food, and shelter, survival is impossible, making these the top priorities.

Water as the Most Urgent Need

Securing Clean Water: Finding and purifying water will be a daily priority. Whether through collecting rainwater, purifying contaminated sources, or tapping into underground supplies, water management is essential to avoid dehydration and illness.

Rationing: Learning to ration water carefully is critical, especially in the early months when access to clean water may be unpredictable. Conserving and reusing water whenever possible becomes a necessary habit.

Food and Nutrition

Preserved Foods: In the first few months, relying on stored or scavenged food will be the norm. Learning to stretch supplies and avoid waste can extend food stores significantly.

Growing Food: As the months progress, survivors must shift toward growing their own food, whether through small-scale gardens, hydroponics, or foraging. Learning which plants are safe to eat and how to cultivate them becomes critical for long-term survival.

Shelter and Protection

Building or Fortifying Shelter: Ensuring that your shelter can protect you from radiation, extreme weather, and potential threats is crucial. Whether it's reinforcing an existing building or constructing a makeshift shelter, it should provide safety, warmth, and privacy.

Learning to Manage Limited Resources

Scarcity will define the first year, and learning to manage resources wisely will be a key survival skill. Wastefulness can lead to disaster, while careful management of supplies can mean the difference between life and death.

Inventorying Supplies

Know What You Have: Early on, take stock of all available resources, from food and water to tools and clothing. Knowing exactly what is available helps in planning rationing and future needs.

Setting Priorities: Prioritize what needs to be used immediately and what can be saved for later. Perishable items should be consumed first, while non-perishables and essentials like medicine should be used sparingly.

Recycling and Repurposing

Reuse Everything: In a resource-scarce environment, learning to repurpose and reuse materials is essential. Old clothing can become bandages or insulation, while broken tools can be repaired or used for parts.

The Importance of Mental and Emotional Resilience

Survival is not just a physical challenge but a mental one. The isolation, fear, and constant uncertainty will take a toll on mental health. Emotional resilience will be just as critical as securing physical resources.

Staying Mentally Strong

Finding Purpose: Having a sense of purpose, even in small tasks, helps maintain mental strength. Focusing on daily goals, such as securing food or improving shelter, provides a structure that can prevent despair.

Controlling Anxiety: Practice mindfulness techniques such as deep breathing, meditation, or journaling to manage anxiety. These practices can provide emotional relief and help maintain a clear head.

Dealing with Trauma

Processing Trauma: The trauma of witnessing or surviving a nuclear event can be overwhelming. Finding ways to process these feelings, whether through talking to others or personal reflection, is crucial for maintaining emotional stability.

Building a Sustainable Routine for Long-Term Survival

Routines provide a sense of normalcy in an otherwise chaotic world. Establishing daily habits and schedules can bring structure and help with both physical and mental survival.

Daily Survival Tasks

Routine Tasks: Create a daily schedule that prioritizes essential survival tasks such as securing water, managing food, and maintaining shelter. This helps break down overwhelming challenges into manageable tasks.

Time for Rest: Building in time for rest and relaxation is crucial for preventing burnout. Maintaining your energy reserves is as important as rationing food and water.

Self-Sufficiency Planning

Long-Term Plans: Begin thinking about how to become more self-sufficient over time. This could mean starting a garden, learning to hunt or fish, or mastering a new skill that will help in long-term survival.

Overcoming Isolation: Staying Connected with Others

While isolation may be a reality for some, it is essential to seek out human connection where possible. Maintaining social bonds is vital for mental health and can provide support in difficult times.

Finding Other Survivors

Reaching Out: If possible, find other survivors to share resources, knowledge, and emotional support. Forming small communities can ease the burden of survival and create a support system for the future.

Building Trust: While caution is necessary when interacting with others in a survival situation, building trust and cooperation can improve chances of long-term survival. Mutual aid and collaboration are powerful tools in isolation.

Communicating Safely

Signals and Messages: Establish methods of communication, such as signal fires, radio transmissions, or written messages, to stay in contact with others. Communication with other survivors can lead to resource sharing and emotional support.

Adapting Skills and Knowledge to a New World

Surviving the first year requires adapting old skills and learning new ones. The ability to learn, adapt, and innovate will be a crucial factor in overcoming the challenges of a nuclear winter.

Learning New Skills

Survival Skills: Whether it's purifying water, growing food, or making basic repairs, learning practical survival skills will be necessary for long-term survival. Identify areas where you are lacking knowledge and seek out ways to learn or practice.

Innovating with Limited Resources: Creativity will be necessary for problem-solving. Whether it's figuring out how to generate power, make tools from scavenged parts, or create clothing from repurposed materials, innovation will be a daily part of life.

Staying Healthy in a Resource-Scarce Environment

Maintaining physical health is a priority during the first year, especially when access to healthcare is limited. Preventing illness and injury is far easier than dealing with their consequences in isolation.

Preventing Illness

Hygiene: Even in resource-scarce environments, maintaining basic hygiene helps prevent the spread of disease. This includes regular handwashing, cleaning wounds, and safely disposing of waste.

Food Safety: Properly preparing and storing food is crucial to avoid foodborne illnesses. Be cautious about consuming unfamiliar plants or animals without confirming they are safe.

Physical Fitness

Staying Active: Maintaining physical fitness is important for handling daily survival tasks, such as gathering resources or maintaining shelter. Regular physical activity can also help maintain mental health.

Handling the Psychological Effects of Loneliness and Fear

Loneliness and fear can have debilitating effects, especially when prolonged isolation is combined with constant uncertainty. Managing these emotions is critical for maintaining resilience.

Coping with Loneliness

Self-Care: Find small ways to take care of yourself, whether it's through personal rituals, hobbies, or quiet reflection. These acts of self-care provide emotional relief in times of isolation.

Creative Outlets: Hobbies such as writing, drawing, or crafting can provide a way to express feelings and pass the time in isolation. These activities also serve as a mental distraction from fear and anxiety.

Facing Fear

Acknowledging Fear: It's normal to experience fear in a survival situation, but it must be managed. Acknowledging fear rather than ignoring it can help prevent panic and keep you grounded.

Focusing on the Present: When fear becomes overwhelming, focus on the immediate present and the tasks at hand. This approach helps prevent catastrophizing about the future and keeps anxiety in check.

Preparing for Year Two: Planning for the Future

As the first year comes to a close, survivors must begin thinking about the future. Planning for the next phase of survival and growth becomes the focus.

Expanding Resources

Agriculture and Livestock: By the end of the first year, survivors should begin thinking about sustainable food sources for the future. Expanding gardens, planting long-term crops, or introducing small livestock will ensure that resources are available for the coming years.

Building Community Networks: If possible, establish stronger ties with other survivor groups. Sharing knowledge, resources, and protection will strengthen your ability to survive as time goes on.

Long-Term Survival Plans

Learning from the First Year: Reflect on what worked and what didn't during the first year. Use these lessons to improve your survival strategies, focusing on areas where you need to build skills or gather more resources.

Hope and Resilience: Looking to the future, the focus should shift from mere survival to building a sustainable way of life. This might involve finding new ways to generate energy, expanding shelter, or teaching others the skills you've learned.

The first year of surviving a nuclear winter is an extreme test of endurance, adaptability, and resilience. Through the challenges of isolation, resource scarcity, and psychological strain, survivors learn invaluable lessons that prepare them for the future. By focusing on immediate needs, managing resources carefully, building emotional resilience, and planning for the long term, survivors can lay the groundwork for a more stable life in the years to come. While the first year is fraught with difficulties, it also provides the opportunity to learn, grow, and develop the skills necessary to rebuild in a new world.

Looking to the Future: How Humanity Can Thrive After a Nuclear Winter

As humanity emerges from the shadow of a nuclear winter, the question of how to move beyond mere survival and toward thriving becomes paramount. While the immediate aftermath of a nuclear event is marked by loss, scarcity, and the breakdown of societal structures, the long-term future holds the possibility for renewal, rebuilding, and growth. This chapter explores how humanity can not only survive but thrive in a post-nuclear world, by embracing new technologies, fostering cooperation, rebuilding ecosystems, and learning from the mistakes of the past. The goal is to create a sustainable, resilient future that honors both human needs and the environment.

Key Strategies for Thriving After a Nuclear Winter

Building a Resilient and Sustainable Society

Relearning the Lessons of Cooperation and Community

Embracing Technological Innovation for a Better Future

Rebuilding the Environment and Fostering Biodiversity

Creating a New Economy Based on Sustainability and Equity

Re-establishing Education and Knowledge Sharing

Redefining Governance and Leadership for a New World

Healing the Psychological Wounds of Nuclear Conflict

Fostering Global Cooperation to Prevent Future Catastrophes

Cultivating Hope: Envisioning Humanity's Role in the Future

Building a Resilient and Sustainable Society

The foundation for a thriving future lies in the creation of a resilient and sustainable society. Instead of returning to old systems that led to conflict and environmental destruction, survivors must build a new world that is grounded in sustainability, balance, and resilience.

Sustainability as a Core Principle

Resource Management: Post-nuclear societies must adopt sustainable practices in managing resources like water, food, and energy. Learning from the scarcity of the first year, survivors can embrace renewable energy, efficient water use, and sustainable agriculture to ensure long-term stability.

Circular Economies: Developing a circular economy, where waste is minimized and resources are continuously reused, is essential for reducing environmental impact and ensuring that future generations can thrive without exhausting the planet's resources.

Resilience in Infrastructure and Systems

Adaptable Infrastructure: New infrastructure should be designed to be resilient to future shocks, whether from environmental disasters, societal collapse, or other threats. This means creating modular, adaptable systems that can be repaired and repurposed easily.

Localizing Production: Reducing reliance on distant or centralized production systems by promoting local food production, energy generation, and manufacturing helps communities become more self-sufficient and resilient to external disruptions.

Relearning the Lessons of Cooperation and Community

If the nuclear winter taught one major lesson, it's the importance of cooperation and community. In isolation, survival is incredibly difficult, but by working together, sharing resources, and supporting one another, humanity can thrive.

Community Building

Collective Survival: Thriving in a post-nuclear world requires strong community bonds, where people work together to solve problems, share resources, and support one another emotionally and physically. Cooperative living is a key element of resilience.

Mutual Aid Networks: Communities should formalize mutual aid systems where everyone contributes according to their abilities and receives according to their needs. These networks help ensure that no one is left behind, and resources are used efficiently.

Inclusive Societies

Social Equity: A key to thriving is building societies where everyone is valued and included. This means ensuring that women, minorities, and marginalized groups have a voice and equal access to resources and opportunities in the rebuilding process.

Conflict Resolution: To prevent future internal conflict, communities should adopt conflict resolution mechanisms that promote dialogue and cooperation over violence or coercion.

Embracing Technological Innovation for a Better Future

Technology will play a crucial role in humanity's ability to thrive after a nuclear winter. However, the focus should be on developing and using technologies that promote sustainability, equity, and environmental health.

Harnessing Renewable Energy

Solar, Wind, and Hydro: Rebuilding energy systems based on renewable resources like solar, wind, and hydropower will help reduce reliance on limited fossil fuels and create sustainable, resilient energy networks.

Low-Tech Innovations: In the early stages of rebuilding, simple, low-tech innovations—such as water purification systems, hand-powered tools, and small-scale solar panels—can be used to meet basic needs while avoiding dependency on complex, vulnerable systems.

Technological Collaboration

Sharing Knowledge: Open-source technologies and shared knowledge platforms can help communities access and develop technological solutions that are locally relevant. By focusing on accessible and adaptable technologies, humanity can advance collectively.

Rebuilding the Environment and Fostering Biodiversity

Humanity's ability to thrive is intrinsically tied to the health of the environment. Rebuilding ecosystems and fostering biodiversity is essential for creating a livable world that supports both humans and wildlife.

Restoring Ecosystems

Reforestation and Land Rehabilitation: Rebuilding damaged environments through reforestation, soil restoration, and the clean-up of contaminated areas is crucial for creating fertile land and restoring ecosystems that support human life.

Encouraging Biodiversity: Fostering biodiversity by reintroducing native plants and animals helps rebuild ecosystems that are resilient to future changes. Healthy ecosystems are essential for clean air, water, and food production.

b. Sustainable Agriculture

Permaculture and Agroforestry: Adopting sustainable farming techniques like permaculture and agroforestry can provide food while improving soil health, preserving water, and promoting biodiversity. These methods can ensure that communities remain fed without depleting the land.

Creating a New Economy Based on Sustainability and Equity

The economic systems of the past were often built on exploitation and unsustainable resource extraction. Thriving after a nuclear winter requires rethinking economic systems to prioritize sustainability, fairness, and mutual benefit.

Resource-Based Economies

Value in Sustainability: Instead of economies driven by profit at all costs, post-nuclear societies should value sustainability and resilience. Resources should be managed in a way that benefits both current and future generations.

Barter and Trade Networks: Communities can develop local and regional trade networks based on bartering goods and services. This fosters cooperation and ensures that everyone's basic needs are met, reducing reliance on formal currency systems.

Equity and Access

Inclusive Economic Systems: Ensuring that everyone has access to resources and opportunities is critical. This might include implementing systems of wealth redistribution, shared ownership, or cooperatives that prevent the concentration of power and wealth.

Re-establishing Education and Knowledge Sharing

Education is the foundation of long-term growth and sustainability. Rebuilding education systems that prioritize practical skills, environmental stewardship, and community cooperation is key to ensuring future generations thrive.

Practical, Localized Education

Skills for Survival: Focus on teaching practical skills such as agriculture, engineering, and medicine. Knowledge-sharing should be a community effort, where individuals teach others what they know.

Learning from the Past: Education systems must also focus on history—teaching the causes and consequences of nuclear conflict to ensure future generations avoid the same mistakes.

Knowledge Sharing Networks

Open Access to Information: Create systems where information and knowledge are freely shared between communities, ensuring that innovation and learning are accessible to all. This can include digital libraries, local resource centers, and traveling teachers.

Redefining Governance and Leadership for a New World

Survivors will need to redefine governance to create fair, accountable, and transparent systems that avoid the power imbalances of the past. Governance should focus on cooperation, inclusivity, and sustainability.

Democratic and Inclusive Leadership

Community-Led Governance: Encourage governance systems that are community-led, with representatives chosen by and accountable to the people. Shared leadership models can prevent the rise of authoritarianism or unequal power dynamics.

Transparency and Accountability: Governance systems should prioritize transparency in decision-making, ensuring that leaders are held accountable for their actions and that all voices are heard.

Ethical Leadership

Compassionate Decision-Making: Leaders must prioritize the well-being of both their communities and the environment, balancing the immediate needs of survival with long-term sustainability and equity.

Healing the Psychological Wounds of Nuclear Conflict

The trauma of a nuclear winter will leave deep psychological scars on survivors. Healing from this trauma is essential for moving forward as individuals and communities.

Trauma Recovery

Mental Health Support: Providing mental health support through community dialogue, counseling, and social connection will be critical in helping individuals process the grief, fear, and loss they have experienced.

Cultural Healing: Rituals, storytelling, and shared community events can help survivors reconnect with their humanity and begin the healing process after the devastation of nuclear war.

Building Resilience

Emotional Resilience: Helping survivors build emotional resilience through supportive communities, regular routines, and mental health care will provide the psychological strength needed to face future challenges.

Fostering Global Cooperation to Prevent Future Catastrophes

One of the most important lessons of a nuclear winter is the need for global cooperation to prevent such catastrophes from ever happening again. The future must be marked by collaboration and peace-building.

International Peace Agreements

Preventing Conflict: Survivors must work toward creating global systems of cooperation that prioritize peace, disarmament, and conflict prevention. The lessons of nuclear war must lead to concrete steps to avoid future conflicts.

Sharing Resources and Knowledge: Cooperation between communities and nations will help ensure that resources, technology, and knowledge are shared to benefit everyone, preventing the inequalities that often lead to conflict.

Environmental Protection as a Global Priority

Global Environmental Governance: The world must come together to protect the environment, prioritizing climate action, biodiversity conservation, and sustainable development to avoid future ecological collapse.

Cultivating Hope: Envisioning Humanity's Role in the Future

Finally, thriving after a nuclear winter requires a vision of hope. Humanity's role in the future must be one of stewardship, compassion, and cooperation—working not just to survive but to flourish.

Creating a Vision of the Future

Hope as a Driving Force: Survivors must cultivate hope, not just for their own survival, but for the future of humanity and the planet. This hope drives innovation, collaboration, and the determination to rebuild stronger than before.

Imagining a Better World: Envisioning a future where humans live in harmony with each other and the environment, where technology serves the collective good, and where peace and cooperation are the norm can guide humanity toward a brighter future.

Humanity's Role as Stewards of the Earth

Environmental Stewardship: Humanity's role in the future must be one of environmental stewardship, ensuring that the lessons of the nuclear winter lead to a world where ecosystems are protected and nurtured for future generations.

Thriving after a nuclear winter is possible, but it requires a radical shift in how humanity approaches survival, cooperation, and sustainability. Humanity must learn from the past, embrace technological and social innovation, and work toward a future where both people and the planet flourish. With hope, resilience, and cooperation, a brighter future is within reach.

Checklist for Long-Term Survival (Up to 2 Years of Shelter Survival)

This comprehensive checklist is designed to help ensure that individuals or families can survive in a fallout shelter for up to two years. The items are organized into categories to cover the essential needs: food, water, clothing, protection, tools, power, and other vital supplies.

List of Food Supplies (2-Year Supply)

Non-perishable Food:

Canned goods (vegetables, fruits, meats, beans, fish)

Dried grains (rice, oats, quinoa)

Pasta and noodles

Dried beans, lentils, and legumes

Freeze-dried meals or dehydrated food packets

Powdered milk and eggs

Peanut butter and nut butters

Crackers and hardtack

Jerky (beef, turkey, etc.)

Cooking oils (olive oil, coconut oil)

Salt, sugar, honey, and spices for flavor and preservation

Grains and Baking Supplies:

Flour (wheat, corn, etc.)

Yeast (for bread-making)

Baking powder and soda

Dried fruits and nuts for energy

Meal Replacements:

Protein bars, meal replacement bars, or energy bars

Protein powder or meal replacement shakes

Supplements:

Multivitamins to cover potential nutrient gaps

Vitamin D and calcium supplements

Seeds (for long-term sustainability, if possible to grow inside):

Heirloom seeds for sprouting (lettuce, spinach, radish)

Fast-growing, low-light vegetables for indoor gardening

Clean Water Supply

Water Storage:

1 gallon of water per person per day (minimum 730 gallons per person for 2 years)

Water storage containers (large drums, sealed barrels)

Water bricks or stackable water containers for space-saving

Water Filtration and Purification:

Water filters (gravity-fed filters, portable filters)

Water purification tablets or iodine drops

UV water purifier (battery or solar-powered)

Boiling equipment (pot, portable stove for heating)

Water Collection Systems:

Rainwater collection barrels (if access to rainwater is possible)

Condensation traps (to collect moisture from the air)

Clothing

Basic Clothing:

Multiple sets of durable, long-lasting clothing (layers for changing weather)

Thermal underwear, socks, and hats for cold weather

Waterproof outerwear (jackets, pants, boots)

Lightweight, moisture-wicking fabrics for warmer conditions

Undergarments and Socks:

Sufficient supply of undergarments, wool socks, and t-shirts

Footwear:

Sturdy boots for protection

Comfortable shoes for inside the shelter

Extra laces, insoles, and waterproofing supplies

Protective Gloves:

Durable work gloves

Waterproof gloves for harsh conditions

Seasonal Clothing:

Thermal and insulated jackets

Wool blankets or thermal sleeping bags

Protective Equipment

Respiratory Protection:

N95 masks or other particulate respirators (for fallout protection)

Full-face respirators with extra filters (for extreme contamination)

Gas masks with replacement filters (for chemical or radiological protection)

Radiation Protection:

Potassium iodide tablets (to protect the thyroid from radiation)

Lead or radiation shields (for extreme exposure protection)

Eye and Face Protection:

Safety goggles or glasses

Face shields (to protect from airborne contaminants)

First Aid and Hygiene:

First aid kits (bandages, antiseptics, gauze, splints)

Wound care supplies (antibiotics, antiseptic wipes, burn cream)

Personal hygiene supplies (soap, toothpaste, toilet paper, sanitary products)

Waste disposal bags (for human waste, plastic liners for toilets)

Tools and Other Equipment

Shelter Maintenance Tools:

Basic hand tools (hammer, screwdrivers, pliers, wrenches)

Nails, screws, duct tape, and heavy-duty adhesives

Multi-tool or Swiss Army knife

Saw or hatchet for wood cutting (if needed)

Shovel for clearing debris or digging

Fire-starting Supplies:

Lighters and waterproof matches

Fire-starting kits (magnesium striker, tinder)

Cooking Supplies:

Portable stove with fuel (butane, propane, or kerosene)

Cooking pots and pans

Utensils, plates, and cups (durable, reusable)

Solar cooker (for outdoor cooking if possible)

Lighting:

Flashlights and lanterns (solar or battery-powered)

Extra batteries or rechargeable batteries

Solar-powered lamps or string lights for interior illumination

Emergency candles or oil lamps with fuel

Hygiene and Cleaning Tools:

Heavy-duty plastic trash bags (for waste management)

Cleaning supplies (disinfectants, bleach, soap)

Mops, brooms, and dustpans for shelter cleanliness

Entertainment and Mental Health:

Books, card games, or board games to pass the time

Writing supplies (pens, pencils, paper)

Power Generation

Solar Power:

Solar panels with battery banks for energy storage

Solar-powered battery chargers for small devices (radios, lights, etc.)

Backup Power:

Hand-crank generators (for charging radios or small devices)

Portable generators (gas or solar-powered) with extra fuel storage

Battery Storage:

Rechargeable battery packs and battery storage units

Power banks for charging devices (phone, radio, lamps)

Other Necessary Items for Long-Term Survival

Communication Devices:

Two-way radios or ham radios for emergency communication

Emergency crank radio (with NOAA weather alerts)

Whistles and signal mirrors for attracting attention

Personal Identification and Documents:

Waterproof storage for important documents (IDs, medical records, property deeds)

Copies of insurance and emergency contacts

Health and Medical Supplies:

Prescription medications (stored in a safe, dry place)

Over-the-counter medications (pain relievers, antacids, allergy medicine)

Antibiotics or anti-viral medications (if possible)

Eyeglasses or contact lenses with extra cleaning supplies

Security and Defense:

Non-lethal defense tools (pepper spray, batons)

Firearms and ammunition (if legally allowed and trained to use)

Home defense systems (locks, barriers, motion-sensor lights)

Mental Health and Comfort:

Comfort items (personal keepsakes, photographs)

Stress-relief tools (stress balls, meditation apps, or guides)

Journals or notebooks to track progress or express emotions

Nobody would hope that a nuclear survival situation will happen but if it does, I hope that this information helps you and your family to stay alive. Good luck.

About the Author

Andrew Parry is a dedicated researcher and author specializing in non-fiction works. His writing is driven by a deep passion for exploring the subjects that resonate with his core beliefs and concerns. Andrew's work reflects his profound commitment to addressing some of the most pressing issues facing humanity today, including the future of our species, the environment, and the looming threat of extinction.